S0-COL-941

To Bruce with best wishes
from the authors.

The
First 75 Years

Members of the 75th Anniversary Engineering Foundation Board of Directors

Frank F. Aplan
AIME Representative

Herman Bieber
AIChE Representative

Franklin H. Blecher
IEEE Representative

Richard P. Case
IEEE Representative

Edward J. Doyle
UET Representative

David Duquette
Member-at-Large

Richard S. Fein, V. Chmn.
ASME Representative

Gordon P. Fisher
ASCE Representative

H.S. Kemp
AIChE Representative

Fred Landis
ASME Representative

B.I. MacDonald
UET President

Forest Mintz
UET Representative

Satoshi Oishi
ASCE Representative

Robert J.E. Roberts
UET Representative

William M. Sangster
Member-at-Large

Henry Shaw, Chmn.
AIChE Representative

David Swan
AIME Representative

Stafford E. Thornton
Member-at-Large

L. Edward Wilson
ASCE Representative

Harold A. Comerer
Director

Jay H. Kelley
Chmn., Proj. Com.

Alexander D. Korwek
Executive Secretary

Charles B. Stott
Chmn., Conf. Com.

The First 75 Years

A HISTORY OF THE ENGINEERING FOUNDATION

Lance E. Metz

and

Ivan M. Viest

Engineering Foundation

New York · 1991

For information contact:
The Engineering Foundation
345 East 47th Street
New York, New York 10017

Copy editing, design, and production management by
Gannon Graphics, 10 Sun Valley Ct., Northport, N.Y.

Library of Congress Catalog Number: 90-085572
ISBN Number: 0-939204-44-4

10 9 8 7 6 5 4 3 2 1

First Edition

Preface

Since the start of the Industrial Revolution, the role of the engineer evolved throughout the 19th century to the point where the profession was regarded as almost heroic. Small wonder: engineers were responsible for many of the comforts and conveniences of modern life, and each year saw life-changing advances. Momentous changes took place in society during the 19th century, with engineers in the vanguard. Steam engines, electric light and power, farm machinery that allowed significantly greater productivity, industrial machinery used to manufacture a myriad of products and textiles, telephones, automobiles, electric street cars—these and other inventions radically changed society in just a few decades. There was little the engineer, possessed of imagination and skills that had so benefitted society, could not do.

Against this background, and with the knowledge that technological advances were producing ever more technological needs and problems, several engineering disciplines came together in the early part of the 20th century. In 1904 the United Engineering Society was chartered. Their first action was to build a large headquarters building at 29 West 39th Street in New York City where they could combine their library resources and operate their individual societies under the same roof. The founder societies were the American Institute of Electrical Engineers, the American Institute of Mining Engineers, and the American Society of Mechanical Engineers. In 1916, they were joined by the American Society of Civil Engineers. The American Institute of Chemical Engineers joined in 1958. The objectives of the society were to advance the engineering arts and sciences and to maintain a free engineering library. Representatives of the founder societies sat on its board.

The Engineering Foundation, celebrating its 75th anniversary in

1989, was created in November 1914 by the board of the United Engineering Society. Engineer-entrepreneur Ambrose Swasey provided the concept of the Foundation and gave it its first endowment funds. From its first days, the Foundation's aim has been the "furtherance of research in science and engineering, [and] the advancement in any other manner of the profession of engineering and the good of mankind." Led by men who were among the most highly regarded engineers in the United States, the Foundation has pursued education, the popularization of engineering, professional enhancement, multidisciplinary approaches to solving problems, and applications of engineering solutions to society's needs.

The history of the Engineering Foundation is a chronicle of technological problem-solving in the 20th century. The Foundation combined an august reputation with an earnest commitment to identifying those problems of society that could be helped by engineers; its programs changed over the years as its directors sought the most appropriate ways of furthering its aims, but it never abandoned those aims.

The numerous research projects sponsored by the Foundation throughout the years were frequently at the cutting edge of technology of the time, sometimes even prescient in their concerns. They were, however, not the Foundation's only significant achievements. The establishment of the National Research Council (NRC) in 1916, just prior to the United States' direct involvement in World War I; and the National Academy of Engineering (NAE) in 1964, at the height of the Cold War, must be considered as being of fundamental importance to the conduct of technological research in the United States. The NRC brought government into the business of sponsoring research, which before had been outside its province; the National Academy gave engineering and engineers the professional status they deserved. The Foundation prepared much of the groundwork for the formation of each and provided funding for much of the work done by the National Research Council during its early years.

For many years the Engineering Foundation maintained a close relationship with the engineering division of NRC, to the point of sharing office space so each could have a Washington D.C. and a New York City office. It enjoyed a similar symbiotic relationship with the National Academy of Engineering, which placed the profession of engineering on the same level as science.

The careers of many young American engineers had their start with research projects sponsored by the Engineering Foundation. In its first decades the Foundation believed in assisting as many worthwhile projects as possible. Even though projects did not need the vast sums of money that are today associated with research, many researchers had to supplement their small Foundation grants with additional grants from industry. Nonetheless, the prestige of Foundation sponsorship made even a small grant desirable. Repeatedly over the years the Foundation's board reassessed its sponsorship of research, often agonizing over where its money should be spent and whether it should itself initiate projects rather than confining itself to evaluating proposals presented by researchers.

Studies of metal fatigue, thermal properties of steam, concrete arch dams, bridge design, alloys of iron, impregnated paper insulation, enzyme engineering, and biomechanics were among the most significant of the hundreds of research projects done under the sponsorship of the Engineering Foundation. Most studies were less wide-ranging.

The face of America's cities changed as a result of studies by various research councils, again supported by the Engineering Foundation, that examined the merits of welding as opposed to riveting, and strengths of reinforced concrete and structural steel. Other councils studied paints to be used on steel, fasteners, waves and coastal erosion, corrosion, air pollution, metal properties, and building research. Engineers from the private sector and from universities participated in studies for the research councils; their findings were published and made widely available.

The Engineering Foundation is widely known for its sponsorship of conferences and symposia. Started during the 1960s, these proved to be a popular way to bring together engineers with similar interests from all parts of the world to exchange information. "Engineering Research for the Developing Countries" was the topic of the first conference, held in 1962 in New Hampshire. The conclusion of those in attendance was that the United States' engineering and scientific community had a responsibility to channel technical and educational assistance to those countries that needed it. Subsequent conferences demonstrated that the Foundation had a genuine concern with bringing together engineers to discuss their role in addressing society's problems. Among the many important topics were solid waste, air pollution, storm water management, mass transit, and groundwater protection.

Tables 2 and 3 In Appendix A list all the research projects and the conference topics sponsored by the Foundation.

Of all the prominent men associated with the Engineering Foundation, Ambrose Swasey stands in the forefront. But for his gifts and inspiration, the Foundation may never have been created. He was, as were so many of those involved with the United Engineering Society, a man with talent to equal his inquisitiveness. An inventive, brilliant engineer, Swasey was the co-founder, with his friend Worcester R. Warner, of Warner and Swasey, a firm that designed and manufactured machine tools for the production of interchangeable parts. Precision instruments of the most sophisticated kind of their day came to be produced by the firm. While they were building up this very successful business, they continued to pursue a second shared interest, astronomy and the building of astronomical instruments, which brought them contracts with some leading universities and observatories. Government contracts for precision military instruments began during the Spanish-American War; Warner and Swasey's innovations became an indispensable contribution to the Allies during World War I.

Swasey's wide-ranging endeavors made him a rich man. His love of engineering and his concern that engineers be accorded the respect due them manifested themselves in the promise of a gift of $200,000 in 1914 to the United Engineering Society, to be used to promote engineering as a profession and to provide seed money for research projects. The money was to be the initial endowment of the Engineering Foundation, but while Swasey conceived the Foundation and was ultimately to give it a total of $839,000, he did not control its direction.

For all but a few years, the Foundation was led by a director appointed by the board. In all, six men have served in this position.

Alfred D. Flinn became the Foundation's first director in 1922 after a distinguished career as a civil engineer. Many of the first research projects took their form under his guidance, and attempts to improve professional accreditation were begun during his term. As the director of the Foundation, he viewed serving as America's ambassador to the world's engineering profession as a major part of his responsibilities. He believed engineers must publicize their work, and during his tenure the Popular Research Narratives, introduced by the Foundation in 1921, became a monthly publication. These leaflets, containing concise descriptions of some research or discovery, or of a notable achievement in science or engineering, were widely distributed. Occasional reprints in the daily press, magazines and technical journals exposed discussions of engineering to a large audience.

A humanist trying to define the role of the engineer in society, Flinn was faced with some of the Foundation's most difficult times during the Great Depression, when numerous engineers were put out of work. With reduced funds available from its investments, the Foundation was forced to reduce the number of projects it sponsored, but was able to find money to help to retrain engineers who had lost their jobs.

Flinn died in 1937. He was replaced later that year by Otis Hovey, a prestigious bridge engineer who was less interested in the humanistic values espoused by Flinn and more in sponsoring fundamental research. During his few years as director he began a practice of visiting the research projects sponsored by the Foundation, but did so without the attendant publicity for the Foundation that Flinn would have arranged.

Edwin H. Colpitts became the third director of the Engineering Foundation following the death of Hovey in 1941. The war effort was paramount, and the Foundation once again put most of its resources into supporting the National Research Council, as it had during World War I. The Kilgore Bill, which would have taken scientific and engineering research away from the private sector and put it entirely under government control, was vigorously and successfully opposed by the Foundation during this period.

Following World War II, the Engineering Foundation floundered for some time, not knowing which direction to follow. In 1949 the board separated the functions of the directorship, and created the position of technical director to supervise the grants program. Frank Sisco, a strong believer in placing the Foundation's primary emphasis on research, was the first technical director. He was appointed full director in 1952.

Applications for research grants slowed down in the late 1940s and early 1950s, a matter of some distress to Sisco and the board. Sisco improved public relations, and was able to report an increase in both applications and financial support. Publicity for Foundation projects improved even further when Mary Jessup, on the staff of the American Society of Civil Engineers, undertook the writing of articles to explain Foundation activities to the founder societies.

In 1958 the board of the Engineering Foundation decided to change direction in giving grants for research projects. The amount the Foundation could give was so small relative to the vast funds available from other sources that small grants were to be discontinued while a new long-range plan was prepared. Late in 1959 the Foundation decided it would no longer serve as fiscal agent for any of the research councils, thus ending that longstanding relationship. The following month, Sisco resigned.

Harold Work, who became director in 1959, was faced with

implementing several policy changes. He was the first part-time director, expected to serve only six days a month, and so was less able to serve as the representative of the American engineering profession than his predecessors. Conferences became the focus of Foundation activities while he was director; these became so popular that a director of conferences, Sandford Cole, was appointed in 1966.

Work was given the responsibility of promoting the formation of the National Academy of Engineering. He was chosen as the first secretary of the National Academy after it was founded in 1964. After he assumed his new position in 1965, the Foundation managed without a director until 1984, when Harold Comerer was appointed.

Grants for research were not entirely discontinued after the 1958 decision to reevaluate them. Fewer were now awarded, but they were larger. Several grants were made for research into areas engineers were beginning to see as critically important for the future, including some in the early 1960s for studies of air pollution. Basic research grants are still awarded by the Foundation.

The Engineering Foundation is well known today for its Research Initiation Grants, which date from the middle 1970s and are awarded to new university faculty members in technical areas of interest to the founder societies. Table 4 in Appendix A lists all grants awarded through 1988.

The authors wish to acknowledge the assistance and cooperation received from numerous persons during the preparation of the book. Basic material was contributed by many but particularly by William H. Munse, Lynn S. Beedle and Janice F. Goldblum. The preface was written by Ann Bartholomew and the manuscript was typed by Suzanne Boyer. Three former Engineering Foundation chairmen, R. P. Generaux, A. B. Giordano and J. A. Haddad reviewed the manuscript; their comments were most helpful in preparing the final copy. The immediate past chairman, Henry Shaw, and the current chairman, Richard S. Fein, provided much needed encouragement and advice. And last but not least, the book could hardly have been written without the guidance and help from H. A. Comerer, A. D. Korwek, and R. J. Yacyshyn of the Foundation staff.

Contents

Chapter 5—SUMMARY AND CONCLUDING REMARKS

The First 75 Years

CHAPTER 1

The Beginnings of the Engineering Foundation 1914–30

1. AMBROSE SWASEY'S GIFT

On May 28, 1914, President Gano Dunn of the United Engineering Society (UES) reported to its Board of Trustees that it had been his privilege to attend a conference with an eminent engineer who had expressed his desire to present to UES a considerable sum of money for the advancement of the profession of engineering. It was the recommendation of the President that a special committee be appointed to consider and propose the best means for accepting the gift and for establishing and administering an engineering research foundation under the broad terms of the donor's expressed wishes. The appointment of such special committee, composed of two members from each Founder Society, was approved. At the donor's request, his name was withheld pending a satisfactory action on the offer.

The latter part of the 19th century saw a diversification of the engineering profession accompanied by the establishment of numerous engineering societies. At the turn of the century, the leaders of national societies in the United States perceived that cooperation among them could bring benefits to all. In response, they took steps toward securing a building with proper facilities for offices, a library, an auditorium and other meeting rooms for the various engineering societies making their headquarters in New York City. In 1895, a plan for such a joint home was submitted to Andrew Carnegie by William D. Weaver, a member of the Board of Managers of the American Institute of Electrical Engineers (AIEE). In response, Weaver's plan received a reply of warm approval and commendation. Three other prominent members of AIEE, T. Commerford Martin, Calvin W. Rice and Charles F. Scott are understood to have made suggestions of the same nature at a later date.

On February 9, 1903, Carnegie, as one of the contributors to the library fund of AIEE, attended its library dinner where he spoke of the need for cooperation among engineers. Scott, by this time president of the Institute, recounted the growth of the Institute and discussed the need for an engineering societies building. The next day Carnegie invited Scott and Rice, chairman of the AIEE building committee, to a general discussion of the idea of a union engineering building. He said that a scheme of that type should include the social as well as the technical interests of engineering. His attention was called to the fact that the Engineers' Club had just secured land on West 40th Street opposite the new public library. This locality appealed to Carnegie as being highly suitable and on February 14, after further conferences, he offered $1 million ". . . to erect a suitable union building . . ."

This generous offer was followed by a prolonged series of negotiations concerning the participation by different engineering bodies. Three societies—AIEE, the American Institute of Mining Engineers (AIME), and the American Society of Mechanical Engineers (ASME)—accepted the plan but the fourth, the American Society of Civil Engineers (ASCE), having built its headquarters on 57th Street only a few years earlier, declined. One year later, when plans and estimates indicated that a larger sum would be required, Carnegie wrote a letter on March 14, 1904 increasing the amount of his gift to $1.5 million. The only limitation was that the gift should be used for the erection of the building, while the cost of the land should be paid for by the Societies.

An informal joint conference committee of all four engineering societies and the Engineers' Club was formed early in 1903 to give preliminary consideration to building plans and to formulate a permanent organization. It was succeeded in June by a formally constituted joint committee that later adopted the name "The Engineering Building Committee." Charles F. Scott of AIEE was elected as its chairman. The committee was divided into two groups, one dealing with the engineering societies building and the other with the building for the Engineers' Club. Of the gift, $1,050,000 was allocated for the societies building and $450,000 for the club. The committee continued its activities until the dedication of the engineering societies building in April 1907.

The Engineering Societies Building as originally built at 29 West 39th Street in New York City

United Engineering Society

The AIME, which was as yet unincorporated, could not partake in the legal arrangements for the property on the same basis as the other engineering organizations. For this reason and others it was decided to create a joint holding or trustee corporation. Representatives of the Founder Societies—AIEE, AIME, and ASME—had introduced into the New York State Legislature a bill to create the United Engineering Society, which would be a corporate body that would provide a home for the engineering community and in other ways advance the engineering arts and sciences. The bill became law on May 11, 1904. The UES held its first meeting in the offices of ASME at 12 West 31st Street in New York on December 16, 1904, and established its temporary office as a guest of ASME. Albert R. Ledoux of AIME was elected the first President of UES.

By its charter, the United Engineering Society was empowered to take real and personal property by grant, devise or bequest, and to use, maintain, occupy, lease, mortgage, and convey the same. It was entitled to receive from any source money or other property to be used for its own purposes or for any other activity or endowment within the scope of its charter. The objectives of this Society were to advance the engineering arts and sciences in all their branches, and to maintain a free public engineering library. Its business was conducted by a board of trustees whose members were appointed by the Founder Societies. It had two departments, one of which was the Engineering Societies Library created by merging the separate libraries of the Founder Societies. The other department was organized in response to President Dunn's report of May 28, 1914 based on recommendations of the special committee referred to earlier. The special committee was chaired by Alex. C. Humphreys of ASME and was composed of Gano Dunn (AIEE), James F. Kemp (AIME), Charles F. Rand (AIME), C. E. Scribner (AIEE), and Jesse M. Smith (ASME) in addition to the chairman.

In 1915 negotiations were reopened with ASCE, which became the fourth Founder Society on August 10, 1916. Although the inclusion of ASCE was a welcome development, it also posed a problem since the engineering building was fully utilized by the three original Founder Societies and by several other technical organizations that were referred to in UES bylaws as Associates. The three original Founder Societies had grown from an aggregate membership of 8,500 in 1903 to 20,500 in 1916. The corresponding figures for ASCE were 2,700 and 7,900. Consequently, to provide office space for the civil engineers, three stories were added to the building. This additional space was first occupied in November of 1917. The cost of the addition was $300,000, of which ASCE paid $262,500 and the three original Founder Societies $12,500 each. This allocation was determined as being an equitable means of making these four societies equal contributors and partners in UES.

The ASCE was formally welcomed as the fourth Founder Society on December 17, 1917. Its president and board of direction were entertained at a luncheon, and a convocation was held in the auditorium in the evening. Charles F. Rand, president of UES and past president of AIME, presided. Addresses of welcome were made by Gano Dunn, past president of both AIEE and UES, Ira N. Hollis, past president of ASME, and Rossiter W. Raymond, secretary emeritus of AIME. William L. Saunders, past president of AIME and member also of ASCE and ASME, spoke for the profession at large. President George H. Pegram responded for ASCE and past president George F. Swain for distant members. Charles Warren Hunt, secretary of ASCE and vice president of UES, made the closing remarks. The new floors which were the headquarters of ASCE were opened for inspection during the evening.

By a vote on September 26, 1929, and with the approval of the four Founder Societies, the name of the United Engineering Society was changed to Engineering Foundation, Inc. Because of important objections, the whole matter was re-evaluated and, after securing the approval of the four Founder Societies, the name was changed once more: effective January 2, 1931, the corporation became known as United Engineering Trustees, Inc.

Engineering Foundation

Chairman Alex. C. Humphreys' special committee presented its findings to the United Engineering Society Board at its October 22, 1914 meeting. They proposed additions to the existing bylaws of UES to enable it to administer the fund and suggested that the fund be given some broad, inclusive name such as the Engineering Foundation. On November 19, 1914 the Board adopted a resolution including the following key clauses:

> First: That a fund to be known as The Engineering Foundation be and hereby is established;
> Second: That Ambrose Swasey, Engineer, of Cleveland, Ohio, be informed of the establishment of The Engineering Foundation and be furnished with a copy of these Preambles and Resolutions;

The Engineering Societies Building after the completion of the 1916–17 addition that was built to house ASCE.

> Third: That the Special Committee of six appointed by action of the Board of Trustees on May 28th, 1914, be instructed to prepare and submit a draft of additions to the By-Laws of the United Engineering Society, to carry into effect The Engineering Foundation.

It also stipulated that notice be given of the proposed additions to the bylaws as required.

All preambles and resolutions proposed by the special committee were adopted and a copy of the minutes of the November 19 UES Board of Trustees meeting was sent to Ambrose Swasey for his information and comments. Swasey expressed satisfaction with these plans in the following letter:

> November 30, 1914
>
> Professor F. R. Hutton, Secretary
> United Engineering Society
> New York City
>
> Dear Professor Hutton:
>
> I am pleased to acknowledge receipt of yours of the 20th instant, also copy of the minutes of the Board of Trustees of November 19th, and I appreciate the courtesy of the Board in submitting them to me before final approval.
>
> The name adopted, "The Engineering Foundation" is ideal and the plan of organization and administration, as given in the minutes, is along the broadest lines and most admirable in every respect. I have no suggestions or recommendations to offer.
>
> As soon as I am advised that the plan of organization of The Engineering Foundation submitted, has become the law of The United Engineering Society, I will be pleased to transmit to the officer designated by the Society, the two hundred thousand dollars ($200,000) which constitutes my gift to the Society for The Engineering Foundation; the income only of which is to be used for the purposes of the Foundation. . . .
>
> Very truly yours,
>
> (signed) AMBROSE SWASEY

The bylaws proposed by Humphreys' Committee concerning the establishment of the Engineering Foundation were adopted at the December 17, 1914 meeting of the UES Board of Trustees. The bylaws stipulated an Engineering Foundation Board of eleven members including:

The auditorium and library at the Engineering Societies Building.

President of the United Engineering Society, ex officio;
One member from each Founder Society who is a member of UES Board of Trustees;
One member nominated by each Founder Society;
Two members nominated by ASCE; and
Two members at large.

The gift—$200,000 worth of New York City bonds—was transferred to UES on February 24, 1915 in a private room of the Astor Trust Company in New York in the presence of Ambrose Swasey, five UES representatives—including the president—and their respective fiscal agents.

In 1929, when the United Engineering Society changed its name to the Engineering Foundation, Inc., the name of its research arm was changed to Engineering Societies Research Board. At the adoption of the corporate name United Engineering Trustees, Inc., the research arm reverted to the use of its original name: the Engineering Foundation.

2. CREATION OF THE FOUNDATION

The December 17, 1914 additions to the UES bylaws provided the framework for the creation of the Engineering Foundation. These new bylaws called for the establishment of a fund to be known as the Engineering Foundation and further stipulated that the Foundation was to be controlled by a board composed of eleven members who would be elected by the UES Board of Trustees. In addition, the new bylaws specified the regularity of board meetings, the election of the officers and the rights and obligations of the board. The new bylaws gave the Board rather broad powers and considerable latitude of action:

> The Engineering Foundation shall have the authority to receive, conserve, invest, hold in trust or administer such monies, or other property, as it may elect to receive from the United Engineering Society or any other society or organization, or from any person, or the estate of any deceased person, or from any other source.
>
> The Engineering Foundation Board shall have discretionary power under the by-laws in the disposition of the funds received by it.

Ambrose Swasey (1846–1937) Engineer, Entrepreneur, Philanthropist and the founder of the Engineering Foundation.

> The Board may use any part of its funds, and in any manner, which it deems proper for the furtherance of research in Science and Engineering, or for the advancement in any other manner of the Profession of Engineering and the good of Mankind.
>
> The Board may, by publications or public lectures or by other means, in its discretion, make known to the world the results of its undertakings.

Over the years, the phrase "for furtherance of research in science and engineering, or for the advancement in any other manner of the profession of engineering and the good of mankind" has become the motto of the Engineering Foundation. Apparently, its exact wording first appeared in the above quoted bylaws and was recorded in the minutes of UES Board of Trustees meeting held on November 19, 1914. Although it was not found in Ambrose Swasey's letters, the phrase actually reflected his ideas regarding the goals of the Foundation. The continuing effort by a succession of Foundation Boards to best interpret this phrase according to the changing conditions and values of the passing decades would do much to shape the Engineering Foundation's subsequent history.

A banquet in honor of Ambrose Swasey was held on January 26, 1915. Inaugural ceremonies were conducted the next Wednesday evening at the auditorium of the UES building as a public acknowledgement of Swasey's initial gift. This program was a notable one due to the presence of a large group of representatives of all branches of engineering and its allied sciences. UES President Gano Dunn opened the meeting by describing the organization and purposes of the Foundation. H. S. Pritchett, President of the Carnegie Foundation for the Advancement of Teaching made an illuminating address on research. Captain R. W. Hunt of ASME and AIME and Charles McDonald of ASCE presented appropriate remarks. A. C. Humphreys of the Stevens Institute of Technology accepted the trusteeship of the fund and particularly acknowledged the underlying donation of Ambrose Swasey. In answer to a special request from the Chair, Swasey made an informal address in which he thanked those who had cooperated with him in organizing the Foundation.

The eleven members of the first Board of the Engineering Foundation were elected at the March 25 meeting of UES Trustees. They are listed below with the initials after their names indicating the organization which they represented:

Benjamin B. Thayer, UES
Alex. C. Humphreys, UES
Charles E. Scribner, UES
A. R. Ledoux, AIME
Jesse M. Smith, ASME
M. I. Pupin, AIEE
Charles Warren Hunt, ASCE
J. Waldo Smith, ASCE
Edward Dean Adams, at large
Howard Elliott, at large
Gano Dunn, president of UES

The organizational meeting of this newly appointed board was held on April 15, 1915. Gano Dunn, a past president of AIEE and current UES president, was elected chairman and Frederick R. Hutton, a past president of ASME, was elected secretary of the Foundation (for the names and terms of office of all Board members and officers of the Engineering Foundation see Table 1 in Appendix A). The Board also sent a telegram to Swasey which expressed their appreciation of his contribution and pledged their devotion to his aims.

3. AMBROSE SWASEY

The death of Ambrose Swasey, on June 15, 1937, at age ninety-one, brought to a close the long career of an eminent scientist, engineer, manufacturer, and public benefactor. Of old New England ancestry, Swasey received an early training that instilled in him qualities of independence, responsibility, industry, and spirituality which were vital to the best development of his brilliant and curious mind.

Ambrose, the ninth of ten children of Nathaniel and Abigail Chesley (Peavey) Swasey, was born in Exeter, New Hampshire, on December 19, 1846. His forebear, John Swasey, arrived at Salem, Massachusetts from England in 1630. Swasey's childhood home was a New England country farm. Nathaniel Swasey was a progressive farmer with sufficient mechanical skills to make his own farm implements. In addition he was interested in the study of ancient history and astronomy. From his father, Ambrose doubtless inherited his trait of intellectual curiosity. From his mother he inherited his cheerful disposition and his vein of quiet humor with which for years he delighted his associates.

While all of his brothers and sisters attended academies or colleges, Swasey received his schooling from a little country grammar school which he attended only for a short time. For one with his keen mentality, industrious habits, and ability to find stimulating associates, this lack of formal education was no handicap. His work on the farm taught him the use of his hands, and his mechanical ability was soon evident as he progressed from making toys, to models, and proper tools such as a turning lathe and similar mechanisms. His parents had intended to send him to an institution of higher learning, but at the age of eighteen he chose

Gano Dunn—The first chairman of the Engineering Foundation (1915–17).

instead to learn the machinists' trade. In 1865 he entered the Exeter Machine Works as an apprentice and worked there for three years. During 1866 Worcester Reed Warner also came to Exeter as an apprentice. He soon became Swasey's closest friend. Warner and Swasey lived and studied together and planned for the future. At Exeter they soon found that they had astonishingly parallel characters which were not in any sense identical but rather complementary. Warner was big, bluff, and hearty while Swasey was small, frail, and shy. Thus began their life-long friendship and business association.

Business Activities

On the completion of his apprenticeship at Exeter, Swasey was employed for a time by the Grant Locomotive Works at Paterson, New Jersey, but he soon returned to Exeter and resumed his close association with Warner. In May 1869 both young men went to Hartford, Connecticut, and took positions with the Pratt and Whitney Company where they remained for eleven years. This company had already begun the manufacture of special machine tools for the production of various types of machinery. The change from the little shop in Exeter to the extensive establishment in Hartford was a great stimulus to the young mechanics. Swasey soon found himself in charge of the company's gear cutting department. In this work he distinguished himself by designing an epicycloidal milling machine for accurately producing true theoretical curves from which cutters for shaping the teeth of gears were made. He also invented a machine for simultaneously generating and cutting teeth on spur gears, thereby solving the difficult problem of making interchangeable systems of gearing. Undoubtedly, Swasey did many other unrecorded things to improve machines and the shop practice at Pratt and Whitney. His work as a skilled machinist thus transformed him into an engineer.

In 1879 Swasey was invited to succeed the late John E. Sweet, as professor of practical mechanics, at Cornell University. This proposal came as a surprise to him, but after much consideration he declined the offer. Thus Cornell failed to obtain the services of a professor of remarkable potential while industry and engineering gained a stimulating leader.

In the spring of 1880 Swasey left Pratt and Whitney and joined with Warner to establish the firm of Warner and Swasey. This company began in Chicago, but by 1881 it had moved to Cleveland where there was a better supply of mechanics who had the skills and workmanship required for the unusually fine quality of its designs. Under the partners' direction, the new company engaged in the design and manufacture of machine tools for production of interchangeable parts. At first such machine tools were regarded as special, but in time they became standard. The turret lathe was one of the principal and most innovative products of the firm, and in due course a line of standard turret lathes was established and maintained.

Astronomy had been the delight of Worcester Warner since early youth. His interest inspired his young friend Ambrose Swasey and soon this subject became an added bond between them. They had talked and thought and built small astronomical instruments for many years. It is not difficult, therefore, to understand that soon after they established themselves in business, they had designed a distinctly improved telescope mounting for the nine and one-half inch refracting telescope at Beloit College in Wisconsin. A year later, in 1882, they constructed their first big dome, which was forty-five feet in diameter, for the Leander McCormick Observatory at the University of Virginia. Their next achievement in this field was the design and construction of the mounting for the thirty-six-inch refracting telescope of the Lick Observatory at Mount Hamilton, California. Erected during 1887–88, Lick had the largest refracting telescope of its day and it was the first one to be adapted to the triple purpose of visual, photographic and spectrographic work. Its mounting proved so satisfactory that the U.S. Government commissioned them in 1890 to construct a similar one for the U.S. Naval Observatory at Washington, D.C. In 1894 Warner and Swasey were entrusted with the task of designing and constructing the mounting for the forty-inch telescope and ninety-foot dome of the Yerkes Observatory at Williams Bay, Wisconsin. The Yerkes instrument was the largest ever built.

At the beginning of the twentieth century new trends in astrophysics brought reflecting telescopes into great prominence, and Ambrose Swasey took an active part in their development. His design for the mounting of the reflecting telescope's secondary mirror enabled astronomers to change from the Newtonian to the Cassegrain arrangement with little danger of damage or loss of time. This device was first installed in the seventy-two-inch reflector of the Dominion Astrophysical Observatory at Victoria, British Columbia. In 1922, the Warner and Swasey Company built a sixty-inch reflector for the Argentine National Observatory. In 1923 it constructed the sixty-nine-inch reflector for the Ohio Wesleyan

University. The eighty-two-inch reflector for the McDonald Observatory at Mt. Locke, Texas was finished by their concern at the time of Swasey's eighty-ninth birthday in 1935. By this time, the Warner and Swasey Company had no competition in the construction of mounting mechanisms for large telescopes.

Meanwhile the making of transits, meridian circles, and other instruments of extreme accuracy was finding a place in the firm's work. The design and construction of these instruments involved the perfecting of methods and mechanisms possessing a degree of previously unknown refinement. For example, they developed a dividing engine for automatically graduating circles up to forty inches in diameter to be used in astronomical and other precision instruments. Tests showed that its greatest error was less than one second of arc or about one-third of an inch in a distance of one mile. In May 1937, Swasey was asked, "What, in your opinion, is the greatest thing you have done?" All at once his eyes sparkled, his face brightened more than ever, and it seemed that a war veteran was getting ready to describe a famous battle. "The highest type of construction and piece of work I ever did," he said, "was the dividing engine. When you take a spindle four inches in diameter and about twenty-five inches long with five-eighths inch taper to the foot, and make that spindle fit into a bearing easily, and when you drop it a thousandth of an inch it goes hard, you are getting down to a refinement about which we knew nothing in those times."

During the Spanish-American war, Warner and Swasey undertook the manufacture of precision military instruments for the U.S. government. This order resulted in a relationship which continued for several decades. As a result, Warner and Swasey solved for the government many of its problems arising in the use of optical instruments, including the development of instruments necessary for the fire control of guns, such as range finders, field telescopes, azimuth instruments, gun sight telescopes, telescopic rifle sights, and precision binoculars. During World War I, their activities in this field were extensive and their service to U.S. government and its allies accordingly great.

Philanthropy and Honors

Swasey was one of the founders of the American Society of Mechanical Engineers and served as its president in 1904. For several years he considered how best to make a gift to assist his fellow engineers in solving problems for which there were no ready solutions. After consulting with friends, he anonymously offered a gift in the spring of 1914 to the United Engineering Society. This gift was accepted and on December 17, 1914 the Engineering Foundation was established as a department of the United Engineering Society. As has been related earlier, the first meeting of its board was held on April 15, 1915. Thus the organization was in place for the administration of his gift.

Swasey's conception was unique. He committed his gift not to a self perpetuating board but to a group of national engineering societies. He avoided the inclusion of any personal name in the name of the Foundation so as not to jeopardize future gifts from other benefactors. He did not wait until the end of his life. He gave resources while he could enjoy the satisfaction of seeing his money at work. His first gift was $200,000. In 1918 he added $100,000 while in 1920 he gave $200,000 to which he added $250,000 in 1931 for a total of $750,000. After his death, it was found that in 1923 he had also begun to build up an additional trust fund for the benefit of the Engineering Foundation. At the time when this was delivered to the Foundation it amounted to over $89,000, which raised the total of his gifts to $839,000. He intended his gifts to be a nucleus of a large endowment for the Engineering Foundation, to which he hoped that many other persons would contribute as years passed.

Swasey's other benevolences were widespread. He made a generous donation toward the purchase of a building site for the headquarters of the National Academy of Sciences and the National Research Council in Washington, D.C. He was the donor, with W. R. Warner, of the Warner and Swasey Observatory of the Case School of Applied Science in Cleveland, Ohio. He also gave an observatory and a chapel to Ohio's Denison University and endowed the chair of physics at the Case School of Applied Science (now Case Western Reserve University), which was occupied by his friend Dr. Dayton C. Miller. He established a library for the Colgate-Rochester Divinity School at Rochester, New York. Overseas he funded a building for the Canton Christian College (later known as Lignan University) and a science building for the University of Nanking. For his native town of Exeter, he built the Swasey Parkway and in its square he erected a pavilion. To the little Baptist church at Exeter he donated two memorial windows and an organ which he unfortunately did not live to hear played. He once said: "Whatever success I have had, which enabled me to make these gifts, I owe not to myself but to the fact that I have had a good mother, a good wife and a good partner in the business."

Another factor which contributed to his success was his ability to select capable young men whom he systematically trained. He and his partner were among the pioneers in establishing an industrial apprentice school in 1880. "They call this the age of metals and machinery," Swasey used to say, "but the time will never come when the world will not need fine men more than it needs fine machinery." Toward this need he worked as assiduously as he did on building machines and telescopes.

It is only natural that many institutions should wish to do honor to a man of Swasey's character, attainments and admirable human qualities. He held the following honorary degrees: Dr. Eng., Case School of Applied Science, 1905; ScD, Denison University, 1910; ScD, University of Pennsylvania, 1924; ScD, Brown University, 1931; LLD, University of California, 1924; LLD, University of Rochester, 1925; and LLD, University of New Hampshire, 1930. He was an honorary member of the Society of the Cincinnati; the Cleveland Engineering Society (president, 1894); the American Society of Civil Engineers; the American Society of Mechanical Engineers (vice president, 1900–02; president, 1904; Medalist, 1933); the American Institute of Electrical Engineers; the Institute of Mechanical Engineers (Great Britain); the Institute of Mining Engineers (Great Britain); Société des Ingénieurs Civils de France; the New Hampshire Historical Society; officer, the Legion of Honor of France; and founder and honorary member of the Engineering Foundation.

In recognition of Swasey's high attainments in a diversity of interests, he received the John Fritz Medal, awarded for notable scientific or industrial achievement, in 1924; the Cleveland Medal for public service, in 1930; the Franklin Institute Medal, for an outstanding career in physical sciences, in 1932; the ASME Medal, for distinguished service in engineering and science, in 1933; the Washington Award, conferred upon an engineer whose work in some special instance or whose services in general have been noteworthy for their merit in promoting the public good, in 1935; the Hoover Medal, awarded to a fellow engineer for distinguished public service, in 1935; and the Verein Deutscher Ingenieure Medal, awarded to men who have excelled in scientific and practical work in the field of engineering, in 1936. Because of the high requirements demanded of the recipient of each of these awards, no more lasting tribute could be paid to Ambrose Swasey, no stronger endorsement of his life's work could be given.

Perhaps the most unique and pleasurable honor rendered to

Ambrose Swasey occurred on his eighty-first birthday when astronomer Dr. Otto Struve announced the naming of Asteroid 922 "Swaseya" which Dr. Struve had discovered on November 14, 1922.

Miscellaneous

The wide scope of activities in which Swasey was interested is shown by the list of societies in which he was an active member at the time of his death. They included: the American Astronomical Society; the British Astronomical Society; the American Philosophical Society; the National Research Council; the National Academy of Sciences; the Cleveland Chamber of Commerce (past president); the Royal Astronomical Society (fellow); the Western Reserve Historical Society (patron); and the New England Society of Cleveland and the Western Reserve (past president and life member). He also was a trustee of Denison University, Adelbert College, Western Reserve University, the Case School of Applied Science, the University of Nanking, the YMCA of Cleveland (past president), and the Baptist Education Society of the State of New York (president). In the business world, Swasey was president of the Caxton Building Company of Cleveland, director of the Cleveland Trust Company, and corporate member of the Society for Savings.

Friends of Ambrose Swasey who best sensed the fine simplicity, sincerity and humility of his character knew that he valued highly his relationship to the Baptist Church. Whatever success he may have achieved in worldly affairs and whatever satisfaction this may have meant to him were enriched by his spiritual tie to the best traditions of his family and early training. However spiritual and idealistic this tie may have been, it also bound him to the practical affairs of his church and creed. He served the Baptist Education Society of New York State as its president, the Northern Baptist Convention on its finance committee, and the First Baptist Church of Cleveland as its honorary president. Deeply religious, but with the calm sanity of one whose art served those who sought out the secrets of heavens, he is said to have remarked shortly before his death, "There is one thing no telescope will ever be able to do. It will never be big enough to see around the edge."

In 1871 Swasey married Lavinia Dearborn Marston who died in 1913. Although there were no children to survive him, he has left a heritage to his fellow men which will live for many years to come.

4. OTHER EARLY BENEFACTORS OF THE ENGINEERING FOUNDATION

Ambrose Swasey intended his gifts to form a nucleus of a large endowment to which he believed many other persons would contribute as the years passed. Financial contributions to the Engineering Foundation from persons other than Swasey appear to have started in 1917 with a gift from Edward Dean Adams, vice chairman of the Foundation. He joined Swasey in contributing $5,000 each to aid the Foundation efforts in financing the early operations of the National Research Council. When in the beginning of 1920 the Board was faced with a possible deficit of $16,000 incurred in carrying out its undertakings with the National Research Council, Adams generously offered to assume this contingent liability. However, the financial requirements for this work were eventually provided for by the income from the additional gift made by Swasey to the endowment in October 1920. As a result, the assumption of the contingent liability by Adams was never needed.

Various attempts were made during the Foundation's first decade to raise money for specific purposes or as additions to the endowment. These attempts generally met with little success until 1924, when the Henry R. Towne Engineering Fund was established by a bequest of $50,000 to the United Engineering Society in the will of Henry R. Towne. Towne, who had died on October 15, 1924, was an engineer and manufacturer who was also a past president of ASME and active in its affairs. A believer in the value of engineering research, he was a long time friend of Ambrose Swasey. His bequest was paid to UES on June 30, 1925.

On January 20, 1927 Edward Dean Adams, who had served for more than 10 years as vice chairman of the Engineering Foundation, provided UES with $100,000 for the purpose of establishing the Edward Dean Adams Fund. The income of this fund was to be divided into two equal parts, one part to be paid over to the Engineering Foundation Board and the other part to the Library Board, since Adams had also served as an officer of the Engineering Societies Library as well as of the Foundation.

The Engineering Foundation's endowment was further increased in 1927 by two additional gifts. A bequest of $9,525 from the late Seeley W. Mudd, member of AIME, was received in October. And a contribution of $10,000 par value of its own securities was received from the Central Hudson Gas and Electric Corporation at the conclusion of the Foundation's 1927 activities. The second

DEED OF GIFT

BY

EDWARD DEAN ADAMS

TO

UNITED ENGINEERING SOCIETY

for the use of

ENGINEERING FOUNDATION BOARD
and LIBRARY BOARD

Instrumentalities of the

FOUNDER SOCIETIES

AMERICAN SOCIETY OF CIVIL ENGINEERS

AMERICAN INSTITUTE OF MINING AND METALLURGICAL ENGINEERS

AMERICAN SOCIETY OF MECHANICAL ENGINEERS

AMERICAN INSTITUTE OF ELECTRICAL ENGINEERS

For the Furtherance of Research in

SCIENCE AND ENGINEERING

or for the advancement in any other manner of the

PROFESSION OF ENGINEERING

and the

GOOD OF MANKIND

UNITED ENGINEERING SOCIETY
ESTABLISHED IN 1904

NEW YORK CITY
JANUARY 20TH, 1927

Title page of the deed of gift by Edward Dean Adams to the Engineering Foundation contribution was accompanied by a letter from the president of the corporation, Thaddeus R. Beal, in which he stated:

> You may be interested to know that this amount was in fact earned by our employees by their more efficient operation of the physical properties under their care. It represents $10,000 of returned premiums paid during the past year on accident insurance, as the result of [the] Company's excellent record in preceding years, so that we take great satisfaction in making this gift as it is related, in some degree at least, to the higher standards of engineering and operating practice to which the work of the Foundation pertains.
>
> Also, it gives us pleasure to testify to our appreciation of the men who made the Foundation possible and of the work that is being accomplished for the benefit of the industry.

In a letter to Gano Dunn, who was a member of the Engineering Foundation Board, Beal indicated the intent of his company's action:

> The amount is too small to produce much income compared with the needs, but a multitude of independent gifts from small corporations would give the kind of support needed, and my own belief is that somehow business organizations in this Country, that have been built on the results of scientific research, will come to regard such support as an obligation as well as an opportunity. Certainly the Trustees, who give their valued time to the affairs of the Foundation, are entitled to the widest possible support, and with this in mind I should be glad to feel that our gift may be helpful to you.

5. INITIAL PROJECTS AND GRANTS

The second meeting of the board was held on May 20, 1915. Edward Dean Adams, an engineer-financier from New Jersey, was elected as first vice chairman, a position which he continued to occupy until February of 1926, at which time he was elected as an honorary member. The position of second vice chairman was created in February of 1916 when Michael I. Pupin, a past president of AIEE, was elected to fill it. At the same time, Calvin W. Rice was appointed acting secretary. The board's secretary, F. R. Hutton, resigned on September 14, 1916 and Cary T. Hutchinson was appointed the Foundation's first executive secretary at the salary of $5,000 per year. Free office space was granted by UES and Ambrose Swasey paid the first year of the executive secretary's salary. In December UES extended the granting of free office space for one year to aid the Engineering Foundation in carrying out its work in funding the NRC. In February 1917 M. I. Pupin was elected chairman and J. Waldo Smith became vice chairman.

To handle the research proposals which had already been re-

Presidents of the Founder Societies in 1926. Pupin served on the board of the Engineering Foundation and as its second chairman. Durand was the recipient of the first Engineering Foundation grant.

ceived by the Foundation, Chairman Dunn appointed a four-member Committee on Applications. A. R. Ledoux (AIME) was its chairman; and M. I. Pupin (AIEE), J. Waldo Smith (ASCE), and Alex. C. Humphreys (ASME) served as members. The committee was charged with reviewing applications and making recommendations to the board with regard to their disposition. They were also asked to recommend a standard procedure for handling these matters with respect to future applications. The first four applications were presented at the December 21, 1915 meeting and all were rejected. It was suggested that all future rejections should be kept confidential so as not to prejudice the applicant's ability to seek funds elsewhere. It was also agreed that rejections shall be made without prejudice to the value of the inquiry unless specifically directed otherwise. Action on applications 5 through 15 was deferred. The following procedure was then adopted for the evaluation of applications. The secretary first screens the incoming applications. He then sends the selected ones to the chairman of the Applications Committee who reviews and distributes them to committee members in accordance with their particular fields of expertise. Then the entire committee meets to discuss all applications received from the secretary and it formulates recommendations for board action. The committee recommended and the board concurred in ruling out applications aimed primarily at bringing monetary benefit to the investigator.

The first approval of an application was given on January 26, 1916. It was a proposal by C. C. Rosenberg of the Railway Signal Association to study the relays used in track signaling. The Foundation board recommended that the investigation should be carried out by a railroad company with suitable facilities and that the cost should be born by the industry. This project's fate is a mystery because no further references to it were found either in the minutes of the board meetings or in published literature, except as a line item in various lists of projects sponsored by the Foundation, such as Item 1 in Table 2, Appendix A. Table 2 lists all regular projects in chronological order.

Two applications to conduct research on the strength and wearing power of gears were accepted with the instruction to the secretary that their funding would be contingent on the avoidance of a duplication of effort. Upon the withdrawal of one of the proposals, a grant of $1,000 was awarded in February 1916 to Professors Guido H. Marx and William F. Durand of Leland Stanford Junior University (Item 2 in Table 2). A special machine was designed by Professor Lawrence E. Cutter and built by the University's machine shop. On its completion early in 1918, some preliminary tests of gears were carried out by Professors Marx and Cutter which yielded limited results. The grant made five years earlier having been expended, and the board being unwilling to devote further funds, the Foundation withdrew its support for the investigation. However, Professors Marx and Cutter continued their tests and reported the results at the annual meeting of ASME held in New York from November 30 to December 4, 1925. The report, entitled "Some Comparative Wear Experiments on Cast-Iron Gear Teeth" was co-authored by the two principal investigators Marx

A special apparatus for testing the wear resistance of gear teeth at Stanford University.

and Cutter, and by Boynton M. Green, all of Stanford University. They reported the following conclusions:

> The deductions indicated by the foregoing comparative tests are:
>
> (a) The standard depth, 20-deg involute tooth form appears to be a better one to resist wear than the standard depth, 14½-deg involute form.
> (b) The stub-tooth, 20-deg involute tooth form appears to be a better one to resist wear than the standard depth, 14 ½-deg involute form.
> (c) The standard depth, 20-deg involute tooth form appears to be a better one to resist wear than the stub-tooth, 20-deg involute form.
> (d) Series 3 indicates predominance of the effect of pressure angle over the effect of the ratio of arc of action to pitch arc.
> (e) Series 4 and 5 indicate the importance of the effect of ratio of arc of action to pitch arc in distributing the load over several teeth, when employing the same pressure angle; and indicate that this effect more than counterbalances disadvantages due to the longer paths of relative rubbing and higher frictional velocity.

6. NATIONAL RESEARCH COUNCIL

In April 1916, following the German submarine attack on the American steamer S. S. Sussex, President Wilson requested that the National Academy of Sciences, which had been chartered by Congress in 1863, bring into cooperation governmental, educational, industrial, and other research agencies in the dual interests of national defense, and of scientific and industrial research. This request resulted in the formation of the National Research Council (Item 3 in Table 2) as an organization of scientists, engineers, and educators.

On June 21, 1916 the Engineering Foundation Board conferred with several members of the National Academy of Sciences. At this meeting it was decided to aid the Academy in the task which had been assigned to it by the president. Since no funding had been provided for NRC, the Engineering Foundation agreed to devote its resources to support NRC for one year starting in September of 1916. In fulfillment of this agreement, the Foundation gave the use of its offices in the Engineering Societies Building, the services of its secretary, and its income, which was increased by special gifts contributed for this purpose by Ambrose Swasey and Edward Dean Adams. To strengthen the bond between the Engineering Foundation and the National Research Council, the secretary of each organization was made an ex officio assistant secretary, without

SOME COMPARATIVE WEAR EXPERIMENTS ON CAST-IRON GEAR TEETH

BY

GUIDO H. MARX

AND

LAWRENCE E. CUTTER

MEMBERS A.S.M.E.

AND

BOYNTON M. GREEN

ASSOC-MEM. A.S.M.E.

THE AMERICAN SOCIETY OF MECHANICAL ENGINEERS

29 WEST THIRTY-NINTH STREET, NEW YORK

To be presented at the Annual Meeting of the Society, New York, November 30 to December 4, 1925

Title page of ASME publication reporting the results of tests of gear teeth at Stanford.

salary, of the other organization. As assistant secretary of NRC, the secretary of the Engineering Foundation became a member of the Executive Board of that council.

Origin of NRC

The concept of the National Research Council was initiated by astronomer George Ellery Hale, director of California's Mt. Wilson Observatory, who had earlier organized the International Union for Cooperation in Solar Research. Hale also took a prominent part in the work of other international cooperative organizations. During his visit to Europe, he received extensive information on the steps taken in England and France in coordinating and mobilizing the activities of scientific men in their respective countries' war efforts.

The National Academy of Sciences afforded a natural means of starting a broad movement for coordinating the research activities in all fields of science and engineering in the United States. The Academy was the only scientific body that had a federal charter which provided for it to be the advisor to the government on scientific matters. During the more than fifty years since the Academy's founding, the government had called upon the Academy many times for reports in a great variety of cases. Some of them dealt with matters of great importance such as the report on the geology of the Panama Canal. Today, NRC continues to serve the U.S. government with studies such as those recently completed on the space shuttle disaster, and that on the effect of smoking on health.

Hale realized that to conduct and coordinate scientific research in all its ramifications, it would be necessary to be in close communication not only with this nation's educational institutions but also with the many government departments and industrial establishments that are involved in research activities. He accordingly proposed that the National Academy should initiate a movement in the interest of preparedness, using this term in its broadest sense; and that various other societies and scientific organizations, particularly the national engineering societies, should also be invited to cooperate with the Academy in the inauguration of this work. After preliminary discussions with several prominent members of the national engineering societies, an organizing committee was appointed by the National Academy of Sciences with G. E. Hale as chairman; and E. G. Conklin, professor of zoology at Princeton University; Simon Flexner, director of the Rockefeller Medical Institute; Robert A. Millikan, professor of physics at the University of Chicago; and Arthur A. Noyes, professor of chemistry at the Massachusetts Institute of Technology as members. The committee canvassed the subject thoroughly, holding conferences with many public bodies and, particularly, with the presidents of the engineering societies. Conferences were also held with the scientific bureaus of the government, with organizations of physicians and surgeons, with the Naval Consulting Board, and many other organizations which were devoted to science, both on the industrial and educational sides. It was clearly evident that promotion of research bearing upon military problems should not be considered alone, but that true preparedness would best result from the encouragement of every form of research, whether for military or industrial application, or for the advancement of knowledge without regard to its immediate practical bearing.

After considering many plans, the organizing committee reported to the National Academy of Sciences the following recommendations:

> That there be formed a National Research Council, whose purpose shall be to bring into cooperation existing governmental, educational, industrial and other research organizations of natural phenomena, the increased use of scientific research in the development of American industries, the employment of scientific methods in strengthening the national defense, and such other applications of science as will promote the national security and welfare.
>
> That the Council be composed of leading American investigators and engineers, representing the Army, Navy, Smithsonian Institution, and various scientific bureaus of the government; educational institutions and research endowments; and the research divisions of industrial and manufacturing establishments.
>
> That, in order to secure a thoroughly representative body, the members of the Council be chosen in consultations with the presidents of the American Association for the Advancement of Science, the American Philosophical Society, the American Academy of Arts and Sciences, the American Association of University Professors, and the Association of American Universities; that representatives of industrial research be selected with the advice of the Presidents of the Society of Civil Engineers, the American Institute of Mining Engineers, the American Society of Mechanical Engineers, the American Society of Electrical Engineers, and the American Chemical Society, and that members of the Cabinet be asked to name the representatives of the various departments of the government.
>
> That research committees of two classes be appointed, as follows: (a)

central committees, representing various departments of science, comprised of leading authorities in each field, selected in consultation with the president of the corresponding national society, (b) local committees in universities, colleges and other cooperating institutions engaged in scientific research.

This scheme of organization was reported to President Wilson who approved it with the following letter:

Washington, D.C., July 24th, 1916.

Dr. William H. Welch
President of the National Academy of Sciences
Baltimore, Maryland

My Dear Dr. Welch:

I want to tell you with what gratification I have received the preliminary report of the National Research Council, which was formed at my request under the National Academy of Sciences. The outline of work there set forth and the evidences of remarkable progress towards the accomplishment of the object of the Council are indeed gratifying. May I now take this occasion to say that the Departments of the Government are ready to cooperate in every way that may be required, and that the heads of the Departments most immediately concerned are now, at my request, actively engaged in considering the best methods of cooperation.

Representatives of Government Bureaus will be appointed as members of the Research Council as the Council desires.

Cordially and sincerely yours,

(signed) WOODROW WILSON.

The object of NRC, in its broadest terms, was the promotion and furtherance of scientific research. Therefore, the personnel of the research council had to be selected to coordinate the activities of technical, scientific and medical educational institutions, the government departments, and the industries.

In summary, NRC was formed at the request of the president of the United States by the president of the National Academy of Sciences in order to further scientific research in its broadest aspects. The organizational meeting of the Council was held on September 20, 1916 at the Engineering Societies Building in New York. Present were nearly all those individuals who have been requested by the president of the National Academy of Sciences to serve as members of the Council. G. E. Hale was made the permanent chairman, C. D. Walcott, secretary of the Smithsonian Institution, became first vice chairman, Gano Dunn was appointed second vice chairman and C. T. Hutchinson, who was already appointed secretary of the Foundation was, by agreement with the Academy, made secretary of NRC.

The NRC authorized Chairman Hale to appoint an executive committee of ten members. In addition, the president of the National Academy of Sciences, the chairman, and the two vice chairmen of the NRC were to be *ex officio* members of the executive committee. Hale appointed the following members: J. J. Carty, chief engineer of American Telephone & Telegraph Company; Russell H. Chittenden, director of Sheffield Scientific School at Yale; Edwin G. Conklin; Gano Dunn; Robert A. Millikan; Arthur A. Noyes; Raymond Pearl, biologist at Maine Agricultural Experiment Station; M. I. Pupin, S. W. Stratton, director of the Bureau of Standards; and Victor C. Vaughan, director of Medical Research Laboratory at the University of Michigan.

The operating policy of NRC was directed towards the encouragement of individual research efforts. At the first meeting of its executive committee, this policy was stated as follows:

> RESOLVED: That the efforts of the Research Council shall be uniformly directed to the encouragement of the individual initiative in research work, and that cooperation and organization, as understood by the Research Council, shall not be deemed to involve restrictions or limitations of any kind to be placed upon research workers.

The formation and work of NRC received the endorsement of many scientific and technical bodies, including the Woods Hole Biological Laboratory, the American Association for the Advancement of Science, the American Institute of Consulting Engineers, and the American Philosophical Society.

The principal duties of NRC were defined in the following abridged statements from the executive order issued by President Wilson on May 11, 1918 to perpetuate NRC:

> To stimulate research in the mathematical, physical and biological sciences and in the application of these sciences to engineering, agriculture, medicine and other useful arts;
>
> To survey the larger possibilities of science, to formulate comprehensive projects of research, and to develop means for dealing with these projects;
>
> To promote cooperation in research in order to secure concentration of effort, minimize duplication and stimulate progress;

To gather and collate scientific and technical information at home and abroad and to render such information available.

The membership of NRC was chosen with the view of making the Council an effective federation of the principal research agencies in the United States concerned with the fields of science and technology. The Council was organized in thirteen divisions of two classes: (1) six divisions dealing with the more general relations and activities of the NRC, including Government Division, divisions of Foreign, States, Educational and Industrial Relations, and Research Information Services Division; and (2) seven divisions of science and technology including Physical Sciences, Engineering, Chemistry and Chemical Technology, Geology and Geography, Medical Sciences, Biology and Agriculture, and Anthropology and Psychology.

NRC Division of Engineering

With the cooperation of the Engineering Foundation and a number of national engineering societies, NRC formed an Engineering Committee which was active until the cessation of hostilities. Soon after the armistice was signed in 1918, a Division of Engineering was substituted, to carry out the general purpose of coordinating the scientific resources of the entire country as regards to engineering. It would also secure the cooperation of those engineering agencies in which research facilities were available.

Under the leadership of Henry M. Howe, professor emeritus of metallurgy at Columbia University, the Division of Engineering was organized early in 1919 and a program of research initiated. In August, Howe relinquished the duties of the chairmanship, and Professor Comfort A. Adams, dean of the engineering school at Harvard University and a past president of AIEE, was chosen chairman, and Howe was elected as honorary chairman.

The original Division of Engineering included twenty-eight members: three from each Founder Society; one each from the American Society for Testing Materials, Illuminating Engineering Society, Society of Automotive Engineers, and Society of Western Engineers; and twelve members at large. By the fall of 1919 eighteen committees were established, each charged with a specific line of research, and together covering a wide variety of subjects of interest to engineers and the industries.

Cooperation Between NRC and the Engineering Foundation

At a special meeting of the Engineering Foundation Board held on September 5, 1917, its chairman read the following resolution adopted by the National Research Council at a recent meeting:

> As the year for which the cooperative agreement was made between the Engineering Foundation and the National Research Council will soon expire, the Research Council wishes again to record its high appreciation of the liberal financial and personal assistance which the Foundation and its members have afforded in the organization and support of the work of the Council in this first year of existence. It seems exceedingly desirable to maintain close relations between engineers and research workers which the Research Council has for the first time succeeded in establishing. Accordingly the Council wishes to express the hope that it may be practicable to continue some effective form of cooperation between the two bodies and it would be glad to appoint representatives to consider this matter with men designated by the Foundation.

At a meeting of the Foundation's board which was held on September 20, 1917, the following resolution was unanimously adopted:

> RESOLVED: That the Engineering Foundation receives with pleasure the resolution of the National Research Council expressing appreciation of the financial and personal assistance rendered to the organization and work of the Council during the past year. Reciprocating the Council's desire for the maintenance of close relations between workers in science and in engineering, the Foundation hereby declares that it will be its policy to continue the cooperation between the two bodies in all practicable ways that may be now, or may become, mutually available, such ways consisting of the interchange of helpful suggestions, advice, information, office representation and similar facilities, and in addition a recognition of community of purpose that shall promote in the field of engineering research increasingly intimate relations between engineering and science.
>
> The reciprocal designation of the Secretary of each of the bodies as an assistant Secretary of the other for the purpose of enabling both bodies to have offices in both New York and Washington is favorably regarded, and the Foundation welcomes and gladly accepts the offered assistance of the Research Council in the national corelation of the engineering research work of the Foundation to which the Foundation's resources will be devoted.

A special committee on relations with the National Research Council was appointed at the meeting of the Engineering Foundation Board held on February 13, 1919. W. F. M. Goss was its chairman and Charles Warren Hunt, Silas H. Woodard, and Frank B. Jewett were members. The committee report of April 10 stated:

> Your Committee on Relations with National Research Council, after several meetings and conferences, recommends the approval by the Foundation Board of the following proposals:
>
> 1. Engineering Foundation, recognizing the desirability of maintaining close affiliation with National Research Council, proposes to collaborate with the Council "for furtherance of research in science and engineering, or for the advancement in any other manner of the profession of engineering and the good of mankind."
> 2. To contribute to the above end, as part of the policy of Engineering Foundation, office space in Engineering Societies Building has been engaged at the expense of Engineering Foundation, in addition to its own requirements, to serve as the New York office of the Engineering Division of the National Research Council, beginning May 1, 1919, and the Foundation having brought its office to an adjacent room, in addition proffers to the Council and its Engineering Division, without charge, such secretarial services as the Foundation may from time to time determine [necessary].
> 3. National Research Council has proposed that its Engineering Division comprising in all not less than 23 nor more than 28 members (of whom at least 7 and not more than 12 shall be members at large), be so organized as to include at least 5 members of Engineering Foundation, and (including these 5) 17 members of the Founder Societies. Engineering Foundation accepts this proposal as well calculated to meet the mutual requirements of the Foundation and the National Research Council.
> 4. Engineering Foundation proposes to collaborate with National Research Council in the activities of its Engineering Division and to make such appropriations of funds as the Foundation may from time to time determine.
> 5. It is understood that all publications relating to research work in which the Foundation shall have participated, will be issued under the joint names of the Engineering Foundation and the National Research Council.

The report of the Special Committee was adopted by the Engineering Foundation Board. To implement its provisions, the Foundation secured two offices on the sixteenth floor of the Engineering Societies Building and the Division of Engineering moved into them in June of 1919.

By October 1919 sixty members of the Founder Societies were also members of NRC, and of those, eight were also members of the Foundation Board. The total membership of NRC was about 175.

One member of the Engineering Foundation Board, its Secretary, and thirty-one additional members of the Founder Societies were members of NRC divisions other than the Division of Engineering or served on its executive board. Engineers had, therefore, an influential position in the Council, and their activities had a large measure of responsibility for its success.

In a letter dated May 6, 1919, Chairman Goss of the Foundation inquired whether the Division of Engineering was prepared to recommend to the Foundation any specific research topics. Acting Chairman Clevenger replied on May 14 and suggested support of a research project into the fatigue phenomena of metals. On July 29, 1919 the division recommended that the Foundation undertake preliminary research projects into a number of subjects and underwrite their expenses to an amount not exceeding $1500. The subjects named were: New Hardness Testing Machine, Elimination of Somins from Steel, Uses of Cadmium, Uses of Alloy Steels, Pyrometers, Improvement of Metals at Blue Heat, Uses of Tellurium and Selenium, Neumann Bands in Iron and Steel, Heat Treatment of Carbon Steel, Pulverizing, Electrical Insulation, and Substitute Deoxidizers. On August 20 the executive committee took action to support these preliminary investigations.

Four NRC projects which were started during the formative years of the Council were of particular interest to the Engineering Foundation Board: the Advisory Board on Highway Research, Fatigue of Metals, Marine Piling Investigation and American Bureau of Welding. The first three are listed in Table 2 in Appendix A, thus indicating the Foundation's participation in their establishment in one form or another. While the Foundation did not participate in the establishment of the American Bureau of Welding, this organization was of basic importance to the subsequent development of Foundation activities in the field of welding. The Foundation appointed a representative to the Bureau in order to open a direct line of communication between the two organizations.

The American Bureau of Welding was established in 1920 to study subjects of practical importance to this industry. Professor Comfort A. Adams, director of the Bureau, organized eleven working committees which aided the rapid development of the art of welding. By 1923 the Bureau had published more than a dozen reports such as Specifications for Standard Tests for Welds, Resistance Welding Nomenclature, Present State of the Art of Butt Welding, and Welding of Oil Storage Tanks by Electric Arc. It had also started two major experimental investigations: (1) tests of welded steel tanks were conducted by the Committee on Pressure Vessels in cooperation with the U. S. Bureau of Standards, and (2) a study of welded rail joints was sponsored by the American Electric Railway Engineering Association (AEREA) and conducted at the Bureau of Standards. The first study supplied the Boiler Code Committee of ASME with authoritative data on which to base requirements for unfired pressure vessels. Twenty-two companies connected with AEREA cooperated in the second project. The Foundation's representative to the Bureau was D. S. Jacobus.

The Advisory Board on Highway Research (Item 11 in Table 2) was created as a cooperative effort of the NRC Division of Engineering, the U. S. Bureau of Public Roads, the state highway departments, the ASCE and other highway-oriented organizations and schools. It was founded on November 11, 1920 as the Advisory Board on Highway Research. It was renamed in 1925 as the Highway Research Board and in 1974 as the Transportation Research Board. Dean Anson Marston of Iowa State University was its first chairman and Professor W. K. Hatt of Purdue University its first director. Its first annual meeting was held in the Engineering Societies Building in New York on January 16, 1922. In an advisory capacity and as a clearing house, the research board has performed indispensable services to those engaged in highway transportation research and construction. It has also inspired broader studies of the whole subject of transport, and has made known its mission and activities to the faculties and students in America's engineering colleges. In cooperation with the American Association of State Highway and Transportation Officials it conducted occasional major tests of pavements and bridges, including the AASHO Road Test carried out during the years 1955–62 at a site near Ottawa, Illinois, which provided technical data for the design of the Interstate Highway System. Its annual meetings, held traditionally in January in Washington, D.C. became the focal point for research involved in the multitude of aspects of transportation design, construction and operations, and its publications a record of progress in the movement of both people and goods.

Fatigue of Metals

Failures of high speed machinery, particularly in airplane engines, during World War I were quickly brought to the attention of NRC. It was felt that a careful study ought to be made of the causes of these failures which were sometimes termed as fatigue of metals. After getting the advice of the most eminent people in the field, NRC organized the Fatigue Phenomena of Metals Committee (Item 9 in Table 2). Professor Herbert F. Moore of the University of Illinois, who had earlier investigated fatigue, was selected as its chairman, and a number of interested eminent metallurgists and scientists were appointed as members. The committee then prepared an investigative program and selected as the subject of their experiments a few commonly used steels which they thought were of the greatest importance to engineering work. Just about that time the Engineering Foundation, seeking to strengthen cooperation with NRC, came to the Division of Engineering and asked it to recommend a project which the Foundation could support. At its meeting on May 16, 1919, the Foundation board declared its willingness to appropriate a sum not to exceed $15,000 per year for two years to investigate the fatigue phenomena in metals, contingent upon submission by the NRC Division of Engineering of acceptable detailed plans for the conduct of this research. Such a plan was submitted and approved, and the work started at the laboratories of the Engineering Experiment Station of the University of Illinois under the direction of Chairman Moore. Two years later, this study was extended to cover nickel steels under the sponsorship of the General Electric Company. It became a quest for the fundamental reasons for the puzzling failures of machine parts, such as automotive car axles, springs, crankshafts, elevator and hoist ropes, and steel rails, which were subjected to frequently repeated loadings. With the increased use of high-speed equipment such failures became more numerous. The investigation pointed to the endurance limit as a reasonably firm basis for the design of structural elements which were frequently subjected to repeated loading.

In its fifth and following years, the investigation was extended to cover such areas as steels at elevated temperatures, the effect of stress intensification at a small hole, nonferrous metals, and a number of other important variables which were encountered in operational practice. The project's sponsorship was also broadened to include the Copper & Brass Association and several additional manufacturers. The results of the project were reported in detail in six research bulletins which were published by the University of Illinois Engineering Experiment Station. In 1927 a comprehensive treatise, *The Fatigue of Metals* by H. F. Moore and J. B. Kommers, was published by McGraw-Hill Book Company. The Engineering Foundation issued "The Manual of Endurance of Metals Under Repeated Stress" which was prepared by Professor Moore at the Foundation's request for use by research workers and by design,

Presidents of the Founder Societies in 1929. Sperry and Smith served on the board of the Engineering Foundation.

operating, and inspecting engineers. By this time, the investigation stimulated others to undertake work in the field of metal fatigue. To formulate a future course of action, a conference was held at the French Lick Springs, Indiana, conference center to which were invited all interested workers and contributors. A small committee was appointed to study the ideas expressed at the conference and to make recommendations to the Division of Engineering. The committee recommended to discharge the Fatigue Phenomena in Metals Committee and arrange with the American Society for Testing Materials (ASTM) or ASME to undertake further activities in this field. This recommendation was accepted in the spring of 1928 largely because it was felt that ASTM and ASME were perhaps better situated to undertake the coordination of the work of the large number of investigators and to encourage further studies in this field.

The investigation at Illinois was not unique: during the same period, work in fatigue of metals was in progress at the U.S. Naval Experiment Station in Washington, D.C. under the direction of D. J. McAdam, Jr., and at Wright Field in Dayton, Ohio under the direction of R. R. Moore and J. B. Johnson. Nevertheless, probably the greatest benefit of the University of Illinois work conducted under the auspices of the Engineering Foundation and the National Research Council was its encouragement of fatigue studies and testing all over the country. While it was first thought that fatigue of metals was not of primary interest to the structural engineer, the wide adoption of welding has made the structural engineer fatigue minded. Extensive fatigue test facilities may be found today not only in academic and governmental laboratories, but also in numerous industrial laboratories, and in research departments of major corporations. The latter are often equipped with specialized testing apparatus indispensable in the development of new and improved products.

Marine Piling Investigation

To cope with the marine borers which were constantly causing great damage to facilities in a number of ports in the United States, a national committee (Item 18 in Table 2) was organized by the NRC Divisions of Engineering, Biology and Agriculture, and Chemistry and Chemical Technology in cooperation with nine bureaus of the government, more than a score of railroads, a number of steamship and industrial corporations, several museums, munic-

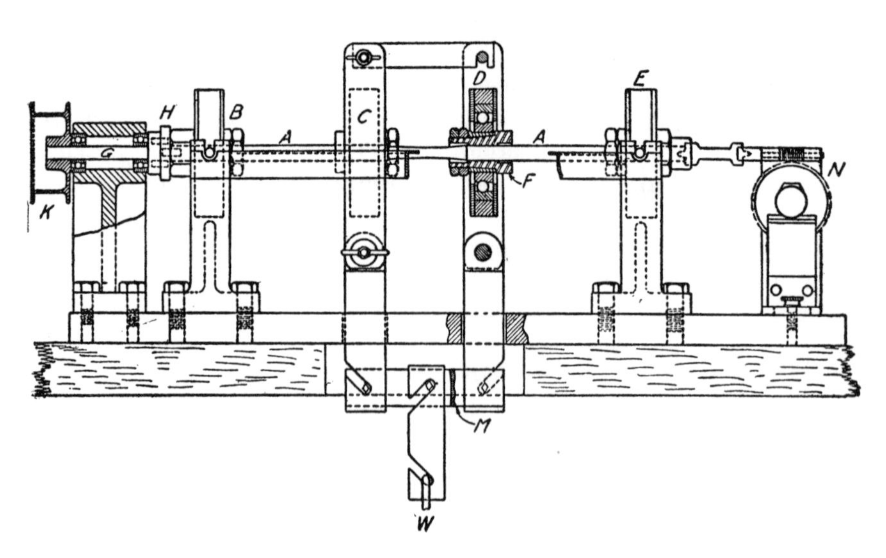

A rotating beam testing machine used at the University of Illinois to determine fatigue strength of metals.

ipal and state departments, technical societies, the Canadian government and Canadian railroads, and the Engineering Foundation. R. T. Betts became its chairman and W. G. Atwood was its director. The need for action became especially acute in San Francisco Bay because of damage by borers to piers, which amounted to fifteen million dollars during the 1919–20 season. The investigation combined basic research in marine biology and chemistry with studies of wood, concrete, and metals preservation to achieve protection of structures and reduction of losses from damage by marine borers. It also studied the deterioration of concrete and metals in or near sea water. More than 200 field stations for collection of biological, chemical, and engineering information were established at carefully selected points along the Pacific, Gulf, and Atlantic coasts, and on island possessions in the Atlantic, in the Pacific, and in the Caribbean. The field data were supplemented by laboratory and museum research.

The committee's work took two years in which the studies pertaining to wood piling were completed. Attention was given also to concrete and other substitutes for wood, but for lack of resources this part of the investigation could not be carried further than to collect information and plan investigations. The final report, "Marine Structures: Their Deterioration and Preservation," was issued in the spring of 1924. A number of railroads and several governmental departments were the chief contributors and collaborators. Their contributions totaled $53,000. Other funds, services, equipment, and supplies were conservatively estimated in excess of $100,000. The Engineering Foundation provided offices in the Engineering Societies Building in New York and contributed some other service support.

A summary of findings with respect to construction materials includes, among others, the following points. The engineer, in selecting materials for structures to be erected in sea water, must be governed in part by the life required of the structure and the traffic for which it is built, the materials available in the local markets, and the funds available for its construction. For temporary structures in harbors where borer attack is comparatively light, unprotected timber offers the most economical material. The proven durability of wrought iron structures is such that it should be given serious consideration whenever a long time service is required and wrought iron is available. Cast iron structures have even a better record . . . Impregnation of timber with creosote seems to give economical and efficient protection in cooler waters . . . The service by structures of plain and reinforced concrete shows great variation. When well built with good materials they are much more durable than timber structures. Also included in the report are studies of the effect of impregnation of timber piles with toxic materials and detail region-by-region results of investigations in thirty-one major harbors and coastal regions.

7. WORLD WAR I EFFORT

At the meeting of the board on December 21, 1915, the Committee on Applications reported that the U.S. Naval Consulting Board suggested to the Engineering Foundation to cooperate with it by giving preference to investigations which may result in benefit to the country, either in connection with its armed forces or otherwise. The board approved to follow this suggested guideline and, at the special meeting of January 26, 1916, requested M. I. Pupin to communicate this decision to the U.S. Naval Consulting Board and to inform them that, for projects sponsored by Engineering Foundation, the board decides on the topics to be investigated and on the suitable persons and places where such research might be conducted. However, the earlier offer of services by the Foundation board to the National Research Council had the effect of placing a one year moratorium on funding new research projects. This period ended by September, 1917 when the Foundation once again became free to devote its resources to other work. Two projects described below were carried out.

To aid in protecting ships from attacks by submarines, the Engineering Foundation joined with the New York Committee on Submarine Defense in making experiments on concealment of ships by means of water spray (Item 4 in Table 2). George B. Pegram of Columbia University chaired the New York Committee and board members M. I. Pupin, C. W. Hunt, and E. D. Adams joined Pegram in the supervision of the experiments. The idea of concealment by water spray was proposed to the Foundation by Howard P. Quick, a mechanical engineer from New York City. The tests were conducted under Quick's direction. A barge lent by U. S. Navy was equipped along one side and one end with spray lines and nozzles, and a number of tests of this apparatus were made in New York harbor. The results led to the conclusion that the method was not practical because it provided camouflage only under very limited weather conditions. Furthermore, while the spray obscured the view of the ship it also prevented observations by the ship personnel. An appropriation of two thousand dollars by the Foundation remained unused since all expenses were paid from other sources. The project was completed by the summer of 1918.

To investigate certain methods proposed for a secret directive control of wireless communication between ships, the sum of five hundred dollars was placed at the disposal of M. I. Pupin in May, 1918 (Item 5 in Table 2). The apparatus was constructed at Columbia University by Clarence G. Stone, Jr. and H. W. Farwell. The experiments gave fairly satisfactory results. Submarine detection research for the U.S. Navy, which was also in progress at the same institution, had to be speeded up greatly during the second half of 1918. This decision required the suspension of every other project and dedication of all laboratory facilities and manpower to the submarine problem. Apparently, the study of directive control was never resumed and no conclusions were reached from this investigation.

In summary, the Engineering Foundation made one overriding contribution to the war effort: its significant participation in the creation and nurturing of NRC during the first year of its existence. A. L. Barrows, executive secretary of NRC, acknowledged that help on September 8, 1939 as follows:

> The Research Council cherishes the relationships which it has had with the Engineering Foundation, and recognizes a constant debt of gratitude to the Foundation. We only hope that the Government and scientists and engineers generally may realize the value of the contribution which the Foundation made during the war to sustain the effort which this country was making to meet the emergency of those years. It was to this emergency that the resources of the Foundation were applied—the Research Council being but a tool or mechanism in the process.

And Frank B. Jewett, a long-time chairman of the Division of Engineering of NRC, commented in a similar vein in his letter dated March 10, 1949:

> To me the greatest thing which the Engineering Foundation ever did was to act at the time of World War I in allocating its entire income for a limited number of years to the National Academy of Sciences to enable it to establish the National Research Council. This to me was an act of statesmanship which has borne fruit in the past thirty years.

Changing its secretary and enlarging the Engineering Foundation board were the two principal organizational developments during World War I. The Foundation's Secretary C. T. Hutchinson resigned on December 13, 1917 and Alfred D. Flinn was elected to replace him. By an amendment to the bylaws of the United Engineering Society on April 15, 1918, the Foundation Board was enlarged to sixteen members and was composed of

UES president, *ex officio*,
four UES trustees, one representing each Founder Society,
two members nominated by each Founder Society, and three members at large.

Furthermore, the amendment provided for an executive committee of five members including the chairman, the two vice chairmen and two elected members. On April 27, Ambrose Swasey was appointed to represent the Engineering Foundation on NRC; from then on, he and M. I. Pupin were the principal links between the two organizations. The membership of the enlarged board was completed in June 1918 and that of the executive committee in September. The executive committee held its first meeting on October 2, 1918.

A few months before the cessation of hostilities, the U.S. government attempted to bring about a consolidation of national engineering societies. As the first step, they proposed for the Engineering Foundation to embark on a survey of engineering organizations. It was also proposed that the Foundation should become recognized as the representative of the engineering profession in matters concerned with research ". . . in all branches of engineering." These discussions continued almost to the end of 1918 when they were brought to a close by unfavorable responses from some Founder Societies.

8. DEVELOPMENT OF THE ENGINEERING PROFESSION

In February 1919 W. F. M. Goss was elected chairman of the Foundation. However, he resigned in March 1920 and was replaced by Charles F. Rand. A standard form of agreement was adopted at the same time to assure control of patents and publications resulting from projects sponsored by the Foundation. In September of 1922 the board created the position of director of the Foundation and appointed A. D. Flinn to it. In February 1925 L. B. Stillwell became chairman.

On December 10, 1925 the board adopted the following new

Presidents of the Founder Societies in 1923. Dwight and Jewett served on the Board of the Engineering Foundation.

policy platform with the approval of the Founder Societies and UES.

> Desiring to promote active and wisely directed research as a means to scientific and technical progress and believing that systematic co-operation by Engineering Foundation and the several Founder Societies is essential to any development of the research work of the Societies commensurate with the dignity, influence and resources of the Profession, Engineering Foundation, while reserving entire liberty of action under the authority conferred upon it by the Founder Societies, through United Engineering Society, adopts the following declaration of its present plan and policy:
>
> 1. Engineering Foundation regards engineering research as the preferred field for its activities.
> 2. It will select or approve specific researches, which it will assist by appropriation of funds or otherwise.
> 3. It will select for each project the agency, collective or individual, which it deems most effective.
> 4. It will assume no direct responsibility for the prosecution of any specific research.
> 5. It will co-operate with the National Engineering Societies and preferably support researches approved by it sponsored by one or more of them.
> 6. A member of Engineering Foundation, or of its staff, may be an advisory, but not an active member of any committee or other organization in immediate charge of a research assisted financially by the Foundation. This provision will not be retroactive.
> 7. Engineering Foundation reserves the right to require from committees or other organizations or individuals assisted satisfactory progress reports as a condition of continued support.
> 8. Engineering Foundation will co-operate with the several Founder or other national Engineering Societies in raising funds for the prosecution of approved researches.
> 9. It will endeavor to prevent conflict or overlap of research effort among the agencies which it supports or assists.
> 10. It will co-operate in securing information of the state of the art for use of committees of the Founder Societies or other agencies.

In February 1927 the board agreed to help underwrite the publication of Engineering Societies Monographs and, in October of the same year, began the Foundation's involvement in what was to become its largest single project, the "Alloys of Iron Research" (Item 32 in Table 2). In May 1929 H. H. Porter became the sixth chairman of the Foundation.

The bylaws governing the aims of the Engineering Foundation, as well as the early implementation of the program, placed emphasis on "furtherance of research in science and engineering," but they also called for activities leading to the "advancement of the profession of engineering." Five projects initiated during the period after World War I and prior to 1930 fell into the second category. Three of them dealt with matters related to personnel and two with questions of education and training.

At the December 12, 1918 meeting, the board considered a request for funding an investigation of mental abnormalities of engineering workers at Lynn, Massachusetts works of the General Electric Company. In February 1919 the Foundation authorized E. E. Southard, director of the Massachusetts State Psychiatric Institute, to make a preliminary investigation as to the part played by mental abnormalities in industry (Item 6 in Table 2). Upon the presentation of a report in May, showing satisfactory preliminary results, for which an expenditure of three hundred dollars was made, twenty-five hundred dollars was appropriated for research in mental hygiene in industry to be made under E. E. Southard's direction during the twelve months beginning June 1, 1919. The object of this research was to develop or discover methods for making psychopathic individuals useful in industry and to prevent them from becoming sources of disturbance, insofar as these ends may be attainable. The work was stopped in its preliminary stages by the sudden death of the principal investigator in February of 1920. However, work that had already been completed resulted in three reports which were published in the February, April, and June 1920 issues of *Industrial Management* and reissued by the Foundation as Reprints Nos. 1, 2 and 3: "The Mental Hygiene of Industry," "Trade-Unionism and Temperament," and "The Modern Specialist in Unrest."

The Foundation board realized that the research in mental hygiene of industry dealt with only one of the many elements of the industrial personnel problem. Therefore, in June 1919 the board addressed to NRC a letter proposing a coordinated broad research into problems of industrial personnel (Item 10 in Table 2). In response, the Council appointed a committee consisting of the chairman of NRC and of representatives of its divisions of Anthropology and Psychology, Educational Relations, Engineering, Industrial Relations, and Medicine to consider means of furthering the study of problems of industrial employment. Two successful conferences were held in Washington, D.C. which resulted in the proposal for the formation of the Personnel Research Federation.

Personnel Research Federation

In 1919 the Engineering Foundation made a modest contribution to the progress in the psychiatric field. Being aware that the mental hygiene of industry dealt only with one of the many elements of the industrial personnel problem, the Foundation proposed to NRC to embark on a coordinated broad research program. A joint committee was appointed which organized a conference held on November 12, 1920 in Washington, D.C. to which selected representatives of science, technology, industry, education and government were invited. The conference outlined a plan for a permanent personnel research federation (Item 12 in Table 2) which was referred back to the organizations represented at the meeting. A second conference took place in March 1921, at which time this Federation was established for the "correlation of research activities pertaining to personnel in industry, commerce, education, and government whenever such research is conducted in the spirit and with the methods of science." The first annual meeting was held in November 1921, and shortly thereafter the *Personnel Research Journal* was started as a monthly periodical of the Federation. Leonard Outhwaite of Columbia University became the director of the Federation in the Fall of 1921. He was also editor-in-chief of the journal. For reports and treatises too lengthy to be published in the journal, the Personnel Research Series of monographs and books was started. The first number, issued in January 1923, *Job Analysis and the Curriculum*, was authored by E. K. Strong, Jr. and R. S. Uhrbrock.

In 1924 Walter V. Bingham, director of cooperative research, Carnegie Institute of Technology of Pittsburgh, became the director of the Federation and the editor-in-chief of the journal, and in 1925 the Federation was incorporated in New York. Over the years the Federation continued the development of its functions as a coordinating agency and clearing house in the fields of research, studying problems affecting the effectiveness and well being of men and women at work in industry, commerce, transportation, public service, education, and other vocations. The staff conducted special studies including (1) interviewing as a means of collecting facts, (2) restriction of output by workers, and (3) susceptibility to accidents.

An outstanding investigation of accidents on street cars and bus lines was carried out in Boston. Its results produced a highly gratifying reduction in the number of accidents. The number of accidents in 1928 was 26.6 percent below the previous five year average. Serious collisions were reduced in the same proportion. These reductions were achieved in the face of increasing hazards, for the frequency of other highway accidents was mounting. A consistent effort was made to single out the accident-prone bus and car operators, and to help these men to become safe operators. Experience has shown that these men constituted twenty to twenty-five percent of the force and that they were involved in fifty to sixty percent of the accidents. No two men were exactly alike: one had to have a new pair of glasses, another was developing cataracts in both eyes, and still another, a widower with several children, was having accidents because he was undernourished and was worrying about the situation at home. Each one needed his own program for overcoming his difficulties and becoming a safe operator. Special medical examinations were made whenever indicated. About one in ten of these high-accident men was found to be accident-prone because of some physical handicap, such as changing eyesight or poor health. More than twice this number were accident-prone because of specific faults of operation of which they were unaware. About one in twenty needed general retraining. A most satisfactory feature of this safety work was the fact that very few men were discharged or disqualified from street car operation.

The study of accidents in Boston so improved the performance of the operators there that the company was awarded the Anthony N. Brady gold medal given by the American Museum of Safety for the best record in accident reduction for the year 1928. In addition, valuable scientific data was collected and a substantial cash savings was effected for the company. Claims for damages filed in 1928 were 1,040 fewer than in 1927. The investigation was conducted by C. S. Slocombe of the Foundation staff under the direction of Director Bingham. The results of this and many other studies were published in the *Personnel Journal* and thus had an effect on the American industry as a whole rather than being limited to the specific company involved in the study.

Early in 1934 the Federation undertook a reorganization in order to broaden its services to industrial and commercial companies. In addition to personnel changes, it restated its objectives as follows:

(a) To aid business organizations of all forms, technical societies, research and educational institutions, social agencies and governmental establishments in the cooperative solution of problems of personnel; and
(b) to utilize knowledge for the purpose of bettering the conditions and relations of men in their occupations.

A detailed review of the Federation's history revealed that the Foundation's Director A. D. Flinn was not only its creator but also a faithful supporter and a constant source of innovative ideas for its activities. Thus, it is not surprising that the Engineering Foundation's last report on the activities of the Personnel Research Federation appeared in the annual report for 1937–38, or less than two years after the death of Alfred D. Flinn.

Engineering Education Investigation

On the suggestion of the American Society of Mechanical Engineers, there was appointed at the December 1920 meeting of the Foundation a committee on Industrial Education and Training (Item 28 in Table 2) to examine the practicality and desirability of a thorough investigation of the education and training of men for and in the industries. W. F. M. Goss was its chairman and A. W. Berresford, I. N. Hollis, G. H. Pegram, G. B. Pegram and J. W. Richards were its members. Large funds were believed to be needed for an effective study on a suitable scale. The Engineering Foundation was prepared to offer its services in organizing and directing the proposed investigation, but could not provide sufficient funding. Other financial assistance was therefore sought. The committee submitted a report on May 9, 1921 outlining the project, and the final report on August 30 stating that it had been unsuccessful in seeking the funds. The committee was discharged but the discussions of the subject were continued outside the Foundation with the hope that means for making the desired comprehensive investigation might in time be obtained.

During the early 1920s, engineering education received considerable attention from engineers, industry, philanthropic foundations, and government, as well as from educators. Engineering education contributed greatly to the progress and prosperity of the country. However, engineering educators realized that methods and curricula must be constantly improved to keep in advance of the needs of the country and its industries. A project to improve engineering education was embarked on in the middle 1920s by the Society for Promotion of Engineering Education (SPEE) in cooperation with the Founder Societies. Funds for this investigation were contributed from many sources including the Carnegie Corporation of New York as the major contributor, industries, individuals, and several engineering societies. During 1927 the Engineering Foundation undertook to assist in financing the final stage, putting into practical use the knowledge gathered by this thorough three year investigation. Its contribution was $1,000 in 1927 and the same amount in 1928.

Charles F. Scott of Yale University, a past president of AIEE and SPEE, was chairman of the Board of Investigation and Coordination for the project. W. E. Wickenden and H. P. Hammond, both of SPEE, conducted the work in cooperation with more than one hundred engineering colleges. A thorough study was made of engineering education in this as well as other countries. Gathered facts were used to improve the training of teachers, and the preparation and selection of students before admission to engineering colleges, as well as for devising various forms of technical education to fit men for different vocational functions, and to improve numerous curricula and procedures. Provisions for more advanced professional training for some students and for greater emphasis upon economics and the humanistic aspects of engineering for other students also resulted, as well as stimulation of research, curtailment of the waste of funds and energies due to "mortality" in the colleges, better placement of graduates, and more effective and persistent liaison of educators with the profession and the industries.

A renewed interest in engineering education on the part of the Engineering Foundation lead in 1930 to the establishment of an Education Research Committee under the chairmanship of Harvey N. Davis, president of the Stevens Institute of Technology, charged with ascertaining the attitude of the engineering societies toward a project to comprise (a) fundamental research in engineering education, (b) vocational guidance especially for youth in preparatory schools, and (c) empirical investigations of objective tests and comprehensive examinations.

Interests as Guides to Engineering as a Vocation

Can engineers be differentiated from men of other vocations by their likes and dislikes? Furthermore, can men in major divisions of engineering be classified by these differences? These questions were considered worthy of study because they could develop means for guiding at least some youths at an early age into those vocations in which their natural traits would most probably lead to success. Edward K. Strong, Jr.'s study (Item 29 in Table 2) of this type at Stanford University with over 3,000 men in twenty quite varied

Presidents of the Founder Societies in 1921. Berresford served on the Engineering Foundation Committee on Industrial Education and Training.

occupations indicated that ". . . men engaged in any occupation have a characteristic set of likes and dislikes that differentiate them from men in other occupations." Subsequent research made it ". . . equally apparent that there are occupations which are quite distinct and yet nevertheless so related that men in one of these occupations cannot be differentiated very well from men in another of these related occupations." Tests covered by Strong's report indicated that the four main groups of engineers cannot be separated to any appreciable degree in terms of their interests. It may be more important to determine whether design engineers can be differentiated from those engaged in sales or administration regardless whether they are civil, electrical, mechanical or mining engineers.

In the study sponsored by the Engineering Foundation, a test group of 3,920 men in thirty different occupations, including 575 engineers, were asked to fill out vocational interest blanks containing 420 items which ". . . comprised 100 occupations, fifty-four amusements, thirty-nine school subjects, eighty-two activities, sixty-three peculiarities of people, forty-two miscellaneous items and forty estimates of present abilities and characteristics." Each man evaluated each item by indicating like, indifference, or dislike . . . It would appear that youths showing early interest in those activities preferred by the engineers tested are likely to become interested in engineering and, if so, should be encouraged. In contrast, youths indicating other interests should be discouraged from taking engineering courses. A few interest tests of college seniors were checked against their vocational choices. Of forty-four planning to enter engineering, forty-one percent rated A on engineering interests, and only four percent C, whereas of those not sure about engineering only half as many rated A and five times as many rated C . . . Of seventy-two students in a graduate school of business only five percent rated A in engineering interests, thirty-four percent B and sixty-one percent C; half of those rated A graduated from an engineering college and planned to combine engineering and business. Of 287 seniors, it was found nine months after the interest test that of those who were sure of their choice of vocation, seventy-one percent had selected the one on which their interest ratings were highest or second highest, and only twelve percent had selected a vocation on which they rated C.

9. SUCCESS OF EARLY ENGINEERING RESEARCH

Graphitic corrosion is the name which, for lack of a better designation, has been given to a form of deterioration of cast iron (Item 14 in Table 2). It had been known for many years that some iron castings, after long contact with sea water, were so changed that they could readily be cut with a knife. Similar changes had been observed in castings buried in soils impregnated with saline, aciduous, or alkaline water. Failures have occurred from graphitic corrosion in water, gas and oil pipe lines, parts of vessels, and other exposed objects. In extreme cases the deterioration has reached a dangerous stage in a few months; in other cases, not until many years have passed. Apparently, this change has often been attributed to the action of stray electric currents from railways. Possibly, the two kinds of corrosion have proceeded simultaneously in some places, only the more evident cause being suspected, and this fact has led to failure to separate the two. In August 1921, J. Vipond Davies, a member of the Engineering Foundation Board who had been studying the subject in the course of his practice for a number of years, sought the assistance of the Engineering Foundation. One or two members of the board and its secretary had experience with this form of failure. The Engineering Foundation conducted an extensive correspondence covering many parts of the country, and the Engineering Societies Library made a thorough search of the literature. These endeavors brought together much useful information, which, added to that previously in the hands of Davies, enabled him to meet his immediate problems. The results of Davies' studies of graphitic corrosion of cast iron were published as Appendix A of *Engineering Foundation Publication Number 8.*

The manufacture and assembly of the metal parts of machines and structures frequently result in internal or residual stresses which are additive to the stresses caused by loads. Thus, to determine the actual stress at any point, it is necessary to ascertain the magnitude of the residual stress at that point. Very little was known about such stresses since their determination required destructive testing of the parts. E. P. Polushkin, formerly chief inspector in the United States of the Russian artillery commission, submitted a research plan for determination of these residual stresses by nondestructive tests. On preliminary evaluation the plan appeared worthy of an exploratory study (Item 16 in Table 2). The work was carried out in the metallurgical laboratories of

Columbia University's School of Mines under the general supervision of A. L. Walker. These small scale experiments indicated that a successful attack upon this important problem would require much time and expense. As the results of these preliminary tests were mostly negative, the investigation was discontinued before it yielded any useful information.

Charles O. Gunther, professor of mathematics at Stevens Institute of Technology in Hoboken, New Jersey was granted $500 in 1929 to aid him in completing a six-year investigation of the principles underlying the identification of firearms and bullets, particularly in homicide cases (Item 36 in Table 2). Gunther evaluated scientific, engineering, and legal aspects of this subject and the results of his study were published by John Wiley & Sons, Inc. in 1935 under the title *Identification of Firearms.*

Wood Finishing Investigation

Another joint project of the Engineering Foundation and the NRC Division of Chemistry and Chemical Technology was the investigation first known as "Paint on Wood Research" and later renamed "Wood Finishing Investigation" (Item 15 in Table 2). The project's committee established its headquarters in Chicago at the offices of the National Association of Wood Using Industries. The secretary of the Association, William B. Baker, served also as secretary of the committee. To facilitate this project the Forest Products Laboratory of the Department of Agriculture and the Bureau of Standards of the Department of Commerce conducted a study of the literature, collection of data, and other preliminary work.

In September 1921, A. H. Sabin, a well-known paint chemist, and Cornelius T. Myers, an automotive engineer who had devoted much time to the study of wood as a material for vehicle wheels and bodies, suggested a thorough investigation of commonly used woods and the protective coatings which were applied to them. They also requested a study of the relations of various kinds of coatings to various kinds of wood and of the effects of various conditions of use on these coatings. They stated that although much was known about various kinds of woods, much less was known about paints and varnishes, and still less was known about the suitability of various kinds of paint and varnish for various kinds of wood under different conditions of use. Unfortunately, no comprehensive, systematic study had ever been made. The knowledge to be gained would effect substantial economies in the nation's $300 million dollar a year paint bill, as well as conserving valuable wood.

The study pertained to the painting and varnishing of the kinds of wood used in buildings in various parts of the United States. The principal work was done by the Forest Products Laboratory at Madison, Wisconsin. With the assistance of eight U.S. Forest Service Districts, the lumber associations, paint manufacturers and experts, and members of the Foundation's cooperating committee chaired by A. H. Sabin, much information was collected, and a study of the painting characteristics of different kinds of wood was planned. Panels of eighteen kinds of wood, about 16 x 36 inches, were prepared at the laboratory, and shipped to ten places selected so as to include widely different climatic conditions. There they were painted according to specifications and then exposed on test fences. There were eleven exposure sites:

Sayville, Long Island—National Lead Company
Palmerton, Pennsylvania—New Jersey Zinc Company
Washington, D.C.—U.S. Bureau of Standards
Gainesville, Florida—University of Florida, National Paint Manufacturers' Association
Milwaukee, Wisconsin—Pittsburgh Plate Glass Company
Madison, Wisconsin—Forest Products Laboratory
Fargo, North Dakota—University of North Dakota
Grand Junction, Colorado—Denver & Rio Grande Western Railroad
Tucson, Arizona—Southern Pacific Railway
Seattle, Washington—University of Washington
Fresno, California—W. P. Fuller Company.

In addition to the regular panels tested in the vertical position, the Madison fence contained a set of smaller painted panels at an angle of forty-five degrees to the horizontal for the purpose of comparing the rate of disintegration of the paint film in the two positions. A set of full-sized panels was also exposed with no film protection. Finally, several other panels were exposed to study the effects of miscellaneous variations in the painting practice.

The examination during 1925 of the painted test panels that had been erected in 1924 led to supplementary tests using seventeen kinds of priming coats on southern and western yellow pine, Douglas fir, eastern and western hemlock, western larch and cypress. The additional panels were made and painted at the U.S. Forest Products Laboratory in Madison, Wisconsin in the fall of 1925 and shipped to the same exposure sites as the 1924 panels. One half of each new panel was painted the same way as the 1924 panels; while the additional priming treatments under study were applied to the other halves of panels. In this way a direct comparison of the new method against the old was obtained. Further tests were started in 1927 and parts of some of the surfaces of the specimens first exposed in 1924 were repainted late in 1927 and in 1928. By the end of 1927, a total of nearly 1,500 specimens of about twenty-five species of wood were under observation.

The experiments made it possible to classify the softwoods in common construction use according to their painting characteristics, and to develop methods for selecting the proper lumber for painting, and improvements in milling lumber for uses in which exterior painting is important. The best kinds of wood for painting were known to be light in weight, with narrow annual growth rings, which were cut so that the principal painted surface was edged grain. Heavy wood, with wide growth rings and conspicuous summerwood, especially if the principal painted surface is flat grain and placed with the pith side out, holds paint relatively poorly. Characteristic chemicals in cypress and redwood seem to favorably influence the durability of paintings, but substances in

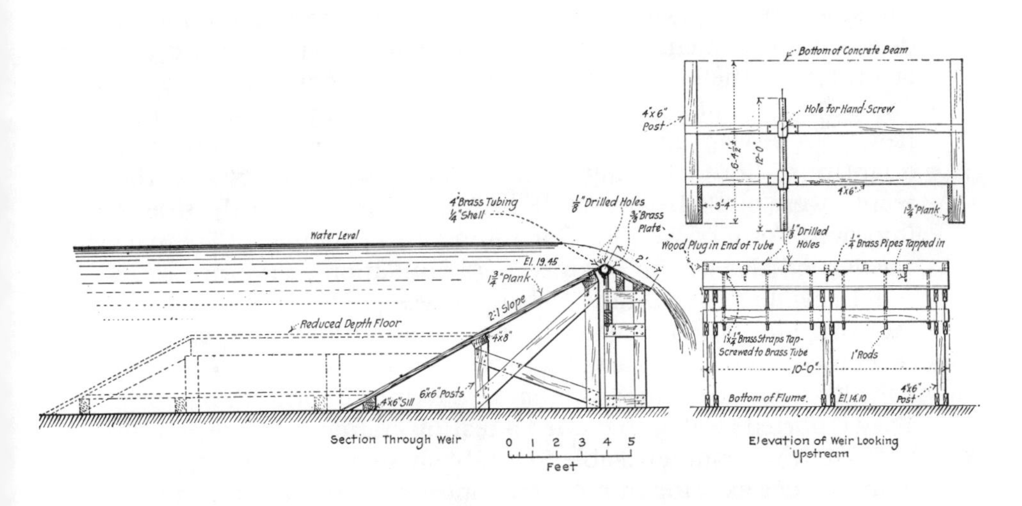

The new weir. The surface curve of water discharging over the weir is drawn from measurements.

the pine seem to shorten the life of some kinds of paint coatings. Woods that tend to warp, cup, or check badly when exposed to the weather required more frequent renewal of their paint coatings.

Variability in the behavior of coatings on different woods was not confined to the coating first applied, but persisted after the original coating had been worn out and the surface had been repainted. The tests showed that the durability of coatings on woods that are heavy and coarse in texture cannot be improved by adding more than the customary proportion of turpentine to the first or priming coat of paint or by replacing the turpentine with some other volatile liquid. Most special priming paints that were studied either failed to alter the durability of the coatings or shortened it. However, priming heavy, coarse-textured wood with a good aluminum paint before applying ordinary house paints materially increased both the durability and the protective power of the coatings.

Civil Engineering Projects

Many experiments have been made on weirs as means for measuring flowing water and other liquids. In December 1918 an appropriation not to exceed twenty-five hundred dollars was made for an investigation of this subject which was to be carried out under the direction of Clemens Herschel, in cooperation with the hydraulic laboratory of the Massachusetts Institute of Technology (Item 7 in Table 2). The tests, which were finished in two years, produced a new form of weir and a very simple formula for determining the quantity of liquid flowing over it. These experiments and their results were published by ASME in June of 1920 under the title "An Improved Form of Weir for Gaging in Open Channels." The paper was issued also by the Engineering Foundation as Reprint No. 4.

In December 1918 Julius Alsberg suggested the establishment of a testing station for large water wheels and other large hydraulic equipment (Item 8 in Table 2). S. H. Woodard, H. H. Porter, and C. Townley were appointed a committee to inquire into this subject. They reported in May that such a testing station was not practical because it was not advisable to establish a testing flume for small models since existing flumes could meet all current requirements. However, they found that testing of water wheels now in place would be useful and practicable.

A project for aerial surveying of mountainous regions was suggested in 1925 by F. B. Jewett and R. A. Millikan. It began with a staff exploration of the state of the art (Item 25 in Table 2) and concluded that aerial surveying was already a well developed, commercially available art.

At the board meeting of May 17, 1928, five-hundred dollars was appropriated to Professor George E. Beggs of Princeton University to study glass models of flat slabs as a part of the general investigation of the use of models for the prediction of stresses and strains in engineering structures (Item 34 in Table 2). The board confirmed the grant award at its October 18, 1928 meeting. The grant was refunded on September 29, 1930.

In 1923, the ASCE committee on steel columns, chaired by Professor F. E. Turneaure of the University of Wisconsin, undertook a review of the existing data on steel columns and the formulation of a supplementary test program (Item 20 in Table 2). Several progress reports were issued over the next ten years and

Apparatus and steel columns tested at the University of Wisconsin.

the final report was published in the *ASCE Transactions* in 1933. The object of the research was to digest all available useful information on the subject and to fill gaps in knowledge through experiments. The investigation covered various details of built-up columns, such as batten plates and lacing, and the response of a column as a part of a structure, including the end conditions, crookedness, and the effects of other members. The results were used in developing a new design formula for column strength for bridges, adopted by the American Railway Engineering Association. The investigation created a body of knowledge which increased the profession's confidence in the design of steel columns.

The files and annual reports of the Engineering Foundation contain no information on Item 31 in Table 2 aside from its listing as "Soils."

Concrete and Reinforced Concrete Arches

Bridges required at many locations span distances that are too wide for a single arch, and at some locations the height of the bridge requires tall piers to support multiple arches. For highway bridges arches of relatively slender dimensions are usually sufficient. Since the arches and piers of such multiple-arch bridges are tied together, the structure acts as a unit. The nature of the structure, therefore, introduces conditions that do not exist in single arch bridges, and gives rise to additional design problems.

The lack of data from experiments or observations on existing bridges with which to confirm or correct the theories used in arch bridge design lead A. C. Jani, an authority on the subject, to propose to the Engineering Foundation in 1922 a thorough investigation which should include the design, erection, and testing of an experimental structure. With the aid of the Advisory Board on Civil Engineering Research of NRC, this proposal was carefully examined and the recommendation was made to ASCE to establish a committee on concrete and reinforced concrete arches (Item 17 in Table 2) for the purpose of planning and supervising the necessary investigations.

Following Jani's suggestion, twelve years of study and of field and laboratory experimental investigations at an unprecedented scale were completed by the special committee of ASCE. Extensive experiments concentrated mainly on studies of stresses in hard rubber models of skewed arches and on tests of large-size model arch ribs which were supplemented with theoretical analyses.

Observations of pier movements were made on eleven bridges in service and effects of climatic changes were observed on two large bridges. Ohio State University, the University of Illinois, Princeton University, the University of California, the U.S. Bureau of Public Roads, state highway departments, and other bodies cooperated. Numerous papers on various features of the research were published. The final report "Laboratory Tests on Multiple-Span Reinforced Concrete Arch Bridges" by Wilbur M. Wilson was published in Volume 100 of the *ASCE Transactions* which was issued in 1935. This report contained data supporting fourteen conclusions related to the design and construction of straight and skew arches in single and multiple span bridges. Although the work of this committee was extensive, a large field for study remained for future investigators.

After authorization by the ASCE Board of Direction in May 1923, the committee, chaired by Professor C. T. Morris of Ohio State University, embarked on a well-planned program confined mainly to arch ribs and to arches with open spandrels. In spandrel-filled arches, the committee confined itself to the determination of

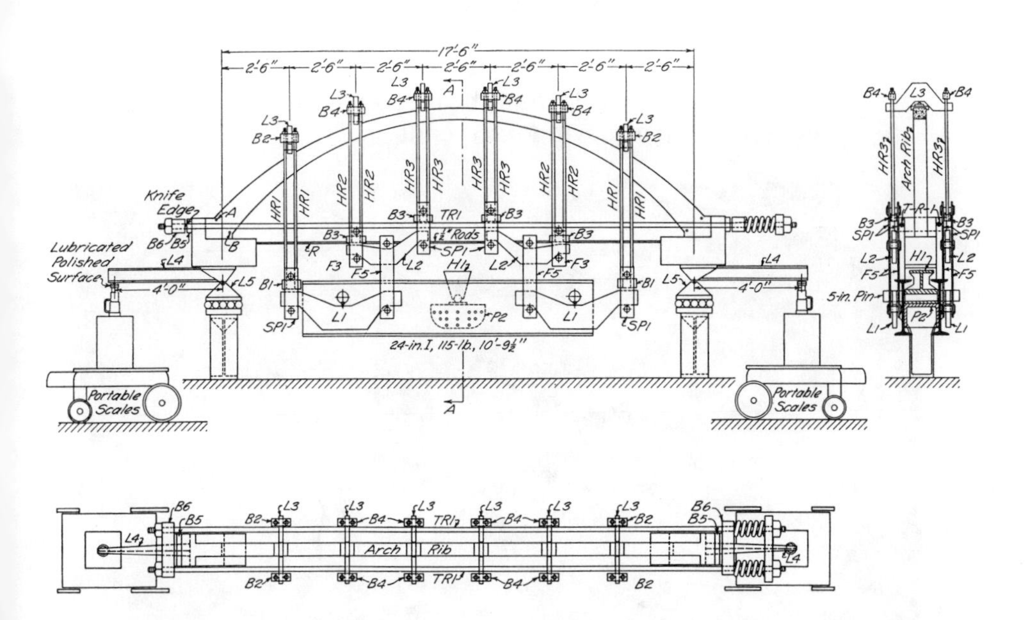

Loading apparatus for testing concrete arches at the University of Illinois.

reactions for skew arches. Much remained to be done on the distribution of stresses and the effect of spandrel walls. The data supported the following conclusions:

1. The elastic theory is adequate for determining the shear and moment in reinforced concrete arch systems without continuous superstructure. The results will be sufficiently accurate for purposes of design.
2. A considerable accidental variation in the moment of inertia and in the modulus of elasticity will not materially affect the position or magnitude of the thrust due to loads in a single span arch.
3. The failure of a reinforced concrete arch rib is improbable except as it may be subjected to a large moment.
4. The concrete in an arch rib will develop approximately the same ultimate strength as the same concrete in six-inch by twelve-inch cylinders.
5. If an arch rib is not subjected to lateral forces, its strength is not affected by its slenderness, so long as the ratio of the unsupported length in the arch axis to the width of the rib does not exceed twenty-five-to-thirty.
6. In multiple-span arch systems, the dead load unit stresses are sometimes greater than the corresponding unit stresses in a similar span having fixed ends.

Research on concrete and reinforced concrete arches at the University of Illinois.

7. A deck without intermediate expansion joints constructed so as to act with the rib in carrying loads reduces the magnitude of the live load moment at the springing where the decks have not increased the strength of the structure; and it increases the moment over the central portion of the span where the strengthening effect of the deck is greatest.
8. The moment at the springing due to shrinkage, time yield, temperature changes, and other causes equivalent to a change in span is greater for a structure having a continuous deck than for the same arch rib without a deck.
9. The height of the deck above the rib of an arch bridge does not greatly influence the stresses in the rib due to load, but increasing the height of the deck above the rib by reducing the rise of the arch greatly increases the stresses due to any cause equivalent to a change in span.
10. The elastic deformation of the pier in multiple-span structures increases the value of the live load and positive moment over the central portion of the span.
11. The basis of the pier and abutment of arch bridges should be designed so that the time yield of the soil due to dead load pressure will not cause piers to tilt.
12. The piers and abutments of arch bridges should be kept under observation with precision instruments during the construction of the superstructure, in order that any excessive motions may be detected.
13. For skew barrel arches, tests of three-dimensional models confirm the theory for computation of reaction components presented by J. Charles Rathbun.
14. The length of the rib of a reinforced concrete arch shortened gradually after the concrete set, due to many causes, including initial drop in temperature, shrinkage, rib shortening, and time yield.

All of these conclusions became extremely useful to bridge designers during the coming decade.

Arch Dam Investigation

The desire of engineers and owners of hydraulic power generation facilities for a basic investigation of arch dams was expressed in a March 3, 1922 letter from Fred A. Noetzli of San Francisco. The

Engineering Foundation's review of this proposal found a widespread interest and support for this type of research. The prospects of obtaining useful results were so assuring that the project was undertaken in early fall. An investigative committee composed of representatives from the U.S. Reclamation Service, the State of California, the City of San Francisco, ASCE, and the Engineering Foundation was established (Item 19 in Table 2). It was headquartered in San Francisco and its first task was the collection of information about existing arch dams. At the committee's first meeting, held on January 18, 1923, Charles Derleth, Jr., professor of civil engineering and dean of the College of Civil Engineering at the University of California, was elected as chairman, and Fred A. Noetzli, chief engineer in the firm of Bissel and Sinnicks of San Francisco, was elected secretary. Subcommittees were appointed on finance, on program and instruments, on publicity and publications, and on several existing dams to be examined. Dean

Test of a three-span reinforced concrete arch with a high deck at the University of Illinois.

Derleth resigned the chairmanship in 1925 and was succeeded by Charles D. Marx, professor emeritus of civil engineering at Stanford University of Palo Alto, California.

An experimental and theoretical investigation of arch dams was undertaken by the committee in order to obtain better bases for design and construction. Its program embraced extensive tests on a specially constructed sixty-feet high experimental dam, field observations on several dams in service or under construction, assembly and study of tests made by other persons, collection of information about important arch dams in many countries, and laboratory tests on models. The test dam was built on Stevenson Creek. Stevenson Creek was a tributary of the San Joaquin River and the dam site was located in the Sierra Nevadas at an elevation of about 1,800 feet above sea level, about 60 miles east of Fresno, California.

The natural conditions at the selected site were extremely favorable. The foundation conditions were excellent, with the bedrock consisting of solid granite having practically no overburden. The V-shaped canyon had fairly regular side slopes. The span of the dam was about twenty feet at its base and about 140 feet at its sixty-foot elevation. The flow of Stevenson Creek was normally diverted a few miles above the proposed dam site by the works of the Southern California Edison Company. However, a power tunnel of this company passed near the dam site, and an adit to this tunnel, equipped with a large valve and located just above the test dam, permitted the filling of the reservoir in a very short time. These exceedingly favorable conditions afforded the perfect control of the height of water in the reservoir without depending on the natural flow of the creek. The construction railroad and roads of the Edison Company also passed very near the dam site, and thus provided for convenient transportation to it.

Constructed of concrete, the Stevenson Creek Dam was sixty feet high and 140 feet long at the top. It was two feet thick in the upper thirty feet, increasing to seven and a half feet in thickness at the bottom. The dam was founded on the granite bottom and sides of the canyon, with the latter having been trimmed symmetrically to forty-five-degree slopes. The construction of the experimental arch dam was completed in June 1926, and many tests were conducted on it during the summer and early autumn by W. A. Slater, engineer-physicist at the U.S. Bureau of Standards. The reservoir back of the dam was alternately empty and filled with water at twenty, thirty, forty, fifty and sixty-foot depths. Thousands of measurements were made with independent instruments and the experimental methods were checked most satisfactorily. On November 27, 1926 the dam withstood, without apparent damage, a

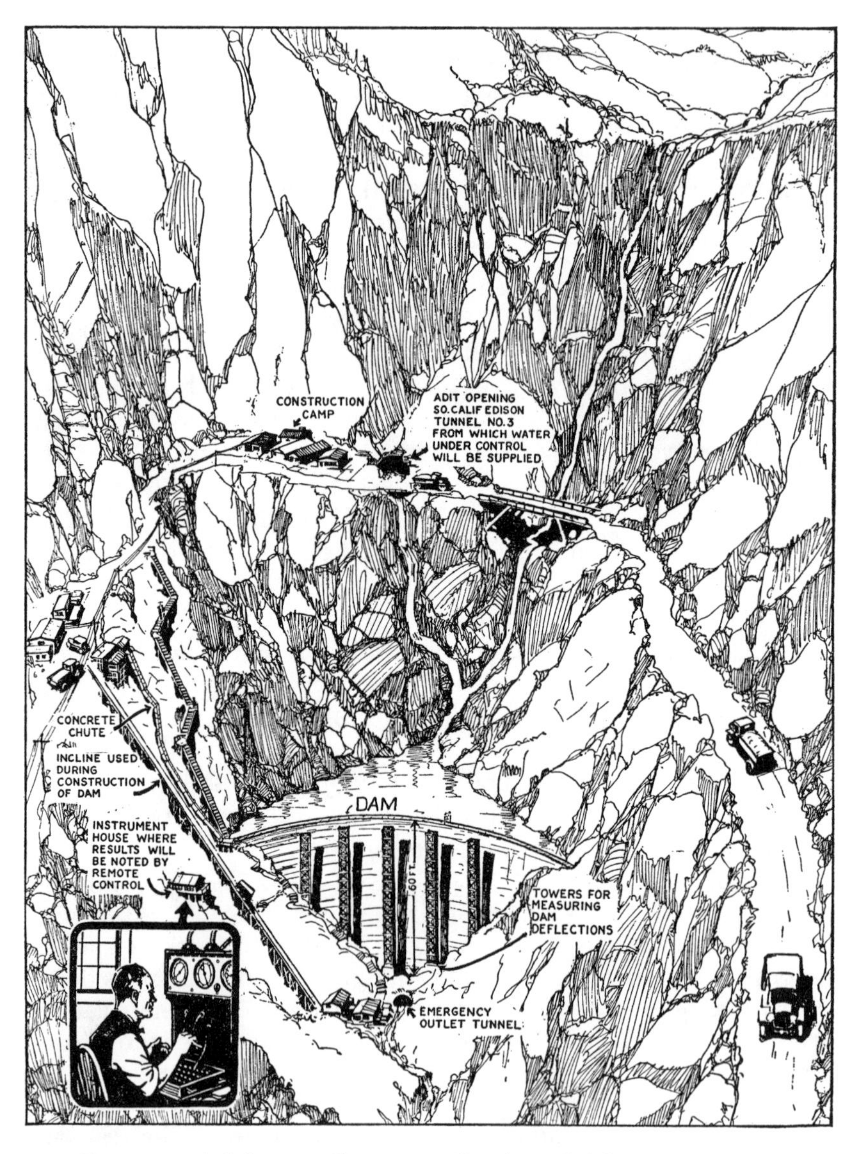

Experimental dam on Stevenson Creek in California.

flood which overtopped it to a maximum depth of at least three feet and deposited earth and rock debris against its back to a depth of thirty to forty feet. The results of the flood prevented the implementation of certain intended supplemental tests and added substantially to the costs of building the dam higher and refitting it for further testing. From the beginning of the investigation a number of engineers had expressed the opinion that the tests should be carried to the breaking of the dam. The flood made such further studies impractical because of the prohibitive cost escalation.

Besides tests of concrete that were made at the site, an extensive investigation of certain physical characteristics of concrete was made by Professor Raymond E. Davis in the Department of Civil Engineering of the University of California with the cooperation of the Portland Cement Association. This information was needed for interpreting the tests on the dam, but it also had a wide usefulness for other users of concrete.

Due to high cost only one design of a full-size experimental dam could be built, but there were numerous other forms of arch dams. To extend the scope of the investigation and to develop an inexpensive means by which engineers could test designs of existing or proposed dams, tests of arch dam models were undertaken at Princeton University and at the U.S. Bureau of Reclamation. The latter, in cooperation with the University of Colorado, tested a five-feet high concrete model of the Gibson Arch Dam. The Gibson Dam was a 200-foot high structure that had been built in 1928–29 by the Bureau for an irrigation project in Montana and during its construction was equipped with a large number of instruments for long-term measuring of temperatures, strains, and deflections. The tests were in the immediate charge of J. L. Savage, design engineer of the Bureau of Reclamation and Ivan E. Houk, research engineer with the Bureau. Professor Herbert J. Gilkey of the University of Colorado and Fredrik Vogt of the Norges Tekniske Hoiskole, Trondhjem, Norway, also collaborated in the experiments for a large part of the year.

At Princeton University, additional tests were made on two small-scale celluloid models of the Stevenson Creek Dam by Professor George E. Beggs, to determine at what height the dam would probably have been broken by pressure of water in the reservoir in back of it. One model represented the dam as eighty feet high and the other as 100 feet, the additional twenty and forty feet of height being uniformly two-feet thick, like the greater portion of the dam as built. These models were tested by the same methods as the ones which were used for the celluloid model of the sixty-foot dam and the results similarly recorded. It appeared doubtful whether the dam would have broken by water pressure alone, unless it had been built to a height somewhat in excess of 100 feet.

In view of difficulties involved in making periodic measurements on existing dams, most of which are located far up in the mountains, it was decided to start this part of the investigation on a limited scale, and to include during the first season of field observations only structures for which conditions for making measurements appeared to be particularly favorable. The investigation included the following dams:

Clear Creek Dam, Yakima Project, U.S. Bureau of Reclamation
Dam No. 6, Southern California Edison Company

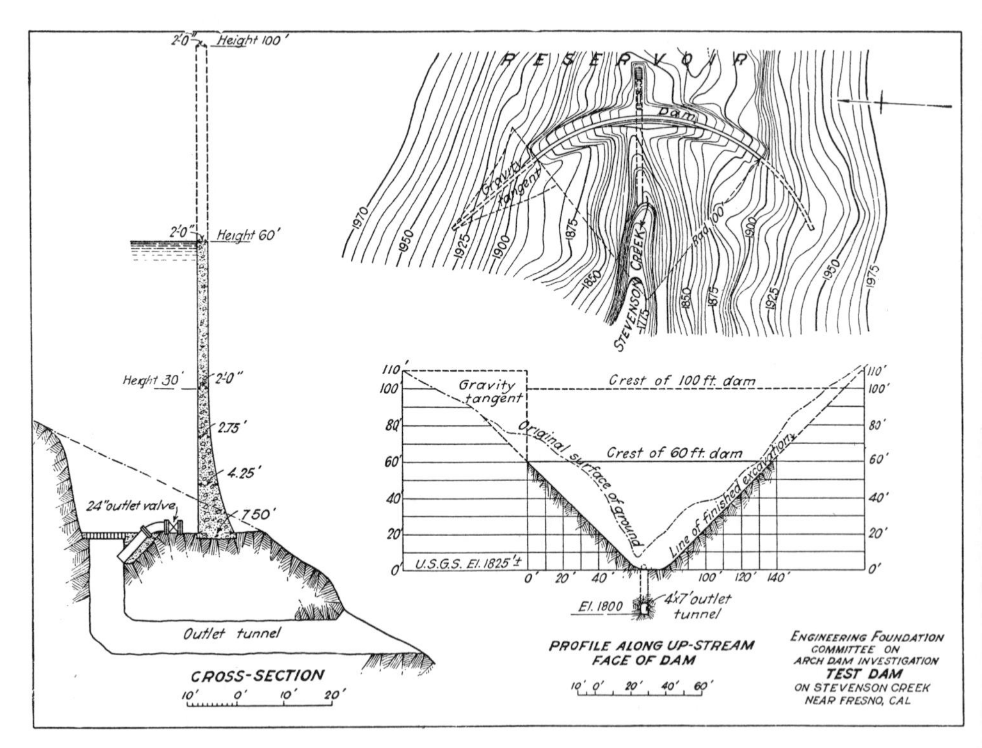

Cross-section, site plan, and profile along upstream face of the Stevenson Creek dam.

Lake Spaulding Dam, Pacific Gas & Electric Company
Lake Eleanor Multiple-Arch Dam, City of San Francisco.

The owners cooperated with the committee by furnishing all the necessary instruments and bearing all expenses incidental to a complete investigation of their dams in accordance with the program. The studies were later expanded to include measurements and observations on several large dams including the Emigrant Creek and Bull Run dams in Oregon, the Gibson Dam in Montana, and others.

The results of the Arch Dam Investigation were published in three volumes. Volume I appeared in May 1928 as Part III of the ASCE Proceedings. It contained an account of the design, construction, and tests of the Stevenson Creek Dam, and data on 154 arch and multiple-arch dams that were located in many parts of the world. Volumes II and III were published by the Engineering Foundation in 1934 and 1933, respectively.

The results of the Arch Dam Investigation had a far reaching influence on the design of concrete dams of both the arch and gravity types. The investigation experimentally established the validity of theory and focused the attention of many engineers upon problems of dam design. Out of the investigation came the development of instruments and methods of tests which have been employed in practically all subsequently built concrete dams. From the laboratory studies came results concerning the properties of concrete, which had usefulness in the whole broad field of concrete design and construction.

Earth and Foundations

Late in 1929 the Engineering Foundation made its first grant to this project. A special Committee on Earth and Foundations (Item 38 in Table 2) was organized by the ASCE in 1930 with Lazarus White as chairman. This committee continued and enlarged upon the work of an earlier committee on the bearing value of soils (Project 31 in Table 2). In 1936 the Society organized the Soils Mechanics and Foundations Division with Carlton S. Proctor as chairman, which assigned its activities to four committees assisted by the Engineering Foundation with grants of money:

Committee on Earth Dams and Embankments; L. F. Harza, chairman

Stevenson Creek dam: upstream face with reservoir nearly empty and downstream face with reservoir full.

Committee on Sampling and Testing; Joel D. Justin, chairman
Committee on Seepage and Erosion; Glennon Gilboy, chairman
Committee on Foundations; Lazarus White, chairman.

Much valuable work was done by these committees and their predecessors. Many articles and reports were published in the ASCE Proceedings and Transactions over the years. The results of these activities, which were nurtured by the rather modest Engineering Foundation grants, have been phenomenal. C. S. Proctor commented in 1939: "The infant science of 'soil mechanics' has made such rapid strides in growth as to have entirely changed the basis of engineering educational training and engineering design of foundations and earth structures. It has converted foundation engineering from an art to a science, has successfully challenged the old sacrosanct theories on earth movement and pressures, such as those of Coulomb and Rankine, and has provided an engineering tool for safer and more economical design."

Now, half a century later, Proctor's characterization still rings true. And what's more, the rapid growth of this field continued during the decades following the 30s: by 1974 when the ASCE division changed its name to Geotechnical Engineering, its enrollment was over 15,000, making it the third largest among the society's fifteen divisions.

The activities supported by Project 38 continued to 1948. The research findings of many of these activities were compiled in the book *Subsurface Exploration and Sampling for Engineering Purposes* authored by M. Juul Hvorslev and published in November 1949 by the Waterways Experiment Station in Vicksburg, Mississippi.

Electrical Engineering Projects

With the endorsement of the American Institute of Electrical Engineers, the Engineering Foundation began in 1926 to assist Professor John B. Whitehead of Johns Hopkins University in an experimental investigation of some of the fundamental physical problems related to the art of electrical insulation (Item 27 in Table 2). At that time considerable obscurity existed as to the internal behavior of dielectrics of various types when they were subjected to very high voltages. A particularly pressing problem was the improvement of the performance of the insulation of high voltage power cables. The goal was to produce cables of smaller dimensions, lower costs, and longer life. The best available material to reach these goals was found to be paper impregnated with various types of oils. Prior to the use of paper insulation, no satisfactory scientific studies had been made of the electrical characteristics of the available insulating oils and papers, nor as to how these respective characteristics influence the behavior of their combination as represented in the impregnated paper.

Some problems included in the research were:

1. Importance of residual moisture in cellulose paper, and of its elimination.
2. The nature and mechanism of dielectric loss in insulating oils, contributory factors and methods of control.
3. It was shown that these losses are conduction losses and subject to control through the paper and the oil.
4. Further studies showed that extremely small quantities of moisture in cellulose papers caused considerable variations in the dielectric constants of the papers.

These studies have been of much fundamental value to the electric power industry and their results have influenced, in substantial measure, the steady improvement in the design and manufacture of high voltage cable insulation, particularly as regards the control of the oil, control of the paper, and the knowledge of their joint behavior.

Mechanical Engineering Projects

The Bearing Metals Committee of the American Society of Mechanical Engineers, which was chaired by C. H. Bierbaum from Buffalo, planned further investigations of bearing metals and, at the same time, endeavored to raise funds necessary for the conduct of the study (Item 22 in Table 2). Although pledges received from the industry were fairly substantial, the total was judged insufficient for the intended research purposes and the committee was discharged.

The ASME organized in 1921 a committee which would investigate the strength of gear teeth. Wilfred Lewis was its chairman (Item 24 in Table 2). The committee was active until 1929 when it was reorganized and Ralph E. Flanders then became its chairman. The Engineering Foundation made its first grant to this project in 1924 and other grants followed. A special gear testing machine was designed and built, and many tests were made in connection with a variety of gear problems. Several reports and papers which summarized the results of these tests were published in *Mechanical Engineering*. In 1931 a particularly notable report was published by ASME under the title "Dynamic Loads on Gear Teeth" which gave valuable data in a form suitable for practical application. Another report, entitled "Qualitative Analysis of Wear," was published in the August 1937 issue of *Mechanical Engineering*. It dealt with surface fatigue. Professor Earle Buckingham, a noted authority on the subject, made the following statement regarding the significance of this work:

> "Largely as a result of published results, general practice in calculating gear-tooth loads and stresses has been changed materially. More consideration is given to accuracy and choice of materials for wear, as a result of the findings of this research, coupled with experience."

In 1927 the Engineering Foundation made a grant to Professors H. E. Hartig and H. B. Wilcox at the University of Minnesota in support of a research project on fluid flow (Item 30a in Table 2). In 1929 ASME organized a committee on the same subject under the chairmanship of W. F. Durand. This research was of such a basic and theoretical character that it was not of evident practical interest to industry. The small grant by the Engineering Foundation made it possible to complete the study and to report the results in *Mechanical Engineering*. According to Durand, "the purpose of the research was to explore the possibility of the use of acoustic methods for the measurement of fluid flow in closed conduits making use of the difference in the velocity of transmission of an acoustic wave in the fluid with and against the direction of the flow." The results indicated the possibility of accurately measuring fluid flow, especially when the latest type of equipment was used. This research thus clearly established the basis for a new means of the measurement of fluid flow in closed conduits. The adaptation of this method of measurement to special cases required further investigation, for which the research funded by the Engineering Foundation provided an admirable foundation.

Traditional research on wire ropes was directed toward methods of determining their strength while a project sponsored by ASME and partially funded by the Engineering Foundation (Item 35 in Table 2) was concerned with their replacement. The study was conducted by the ASME Special Committee on Wire Ropes chaired by Walter H. Fulweiler and financed by the Engineering Foundation. The committee received effective cooperation from the National Bureau of Standards as well as from the makers and users of wire rope. The committee collected samples of worn ropes from many users and supplied them to the Bureau for testing. The tests included 229 specimens taken from seventy-nine worn ropes. The data was published in 1936 in the National Bureau of Standards Research Paper RP 920 entitled "Inspection and Tensile Tests of Some Worn Wire Ropes" which was authored by W. H. Fulweiler, A. H. Stang, and L. R. Sweetman. The following is quoted from the abstract of the paper: "The condition and strength of each sample were determined. The strength was estimated using charts prepared by the Roebling Company. It was found that the estimated strength and the actual strength were nearly the same. These data indicate that the strength of worn ropes may be determined with sufficient accuracy for deciding when the rope should be replaced by measuring the length of wear on the outside wires and counting the number of broken wires."

The investigation "Woodworking Saws and Knives" (Item 37 in Table 2) was a project initiated by the ASME Special Committee on Saws and Knives during the period of 1926–29. The purpose of this committee, chaired by C. M. Bigelow, was to reduce waste and promote greater efficiency in the wood-working industry through improvement and standardization in the design and use of saws and knives. The Engineering Foundation was called on to contribute $2,000. No information could be found in the annual or the summary reports of the Foundation other than the project listing in Table 1 of the twenty-five-year report showing zero funding and a note that the funds were transferred to Project 50 in 1932. Thus the eventual outcome of Project 37 remains unknown.

Thermal Properties of Steam

The design of boilers, engines, and other steam appliances requires the knowledge of thermal properties of steam which may be found in steam tables. In 1921, when ASME research was inaugurated into the thermal properties of steam (Item 13 in Table 2), the steam tables relating pressure, volume, and temperature extended to pressures of 200 psi and temperatures to 380 degrees Fahrenheit, and the accuracy of the data was equal to about one part in 200. By the end of this investigation in 1934, the steam tables were extended to the maximum pressure of 3,200 psi and a maximum temperature of 750 degrees Fahrenheit, and were accurate to about one part in 2,500.

This cooperative research was inaugurated on the instigation of

George A. Orrok of New York and was sponsored by the American Society of Mechanical Engineers. It was inspired by the recommendation of a conference of forty-two engineers, which was held in

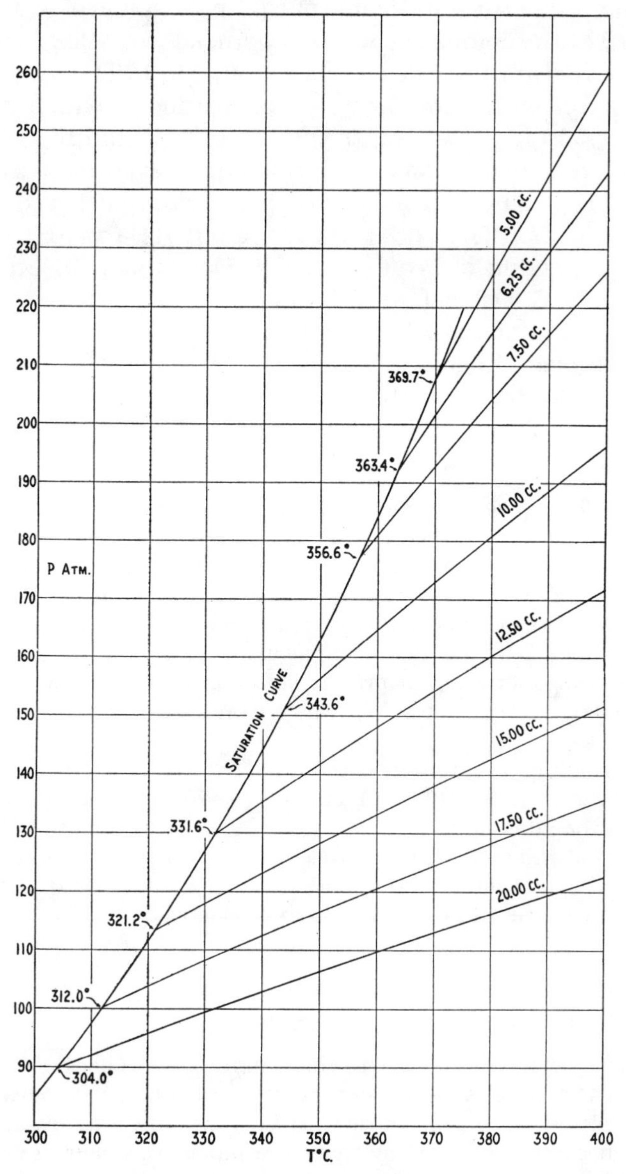

Steam research: isometrics and vapor pressure curve as determined at MIT.

June 1921 at Harvard University. An executive committee consisting of G. A. Orrok, chairman, A. M. Greene, Jr., and D. S. Jacobus was chosen to organize this research project and raise the necessary funds. Later, ASME established a Special Research Committee on Thermal Properties of Steam, of which G. A. Orrok was chairman. In 1929 the committee was reorganized with Alex Dow as its chairman. A Steam Table Fund was initiated in 1921 with a $1,000 contribution from the Engineering Foundation to which it added another $1,000 in 1934. Subscriptions to the fund by the interested industry, institutions, and individuals eventually exceeded $156,000. The value of contributions in kind are believed to have exceeded this sum. The work was carried out at Harvard University under the direction of Professor H. N. Davis, at NBS under the direction of N. S. Osborn, at MIT under the direction of Professor F. G. Keyes, and at the laboratories of General Electric Company where the project was assigned to J. H. Keenan.

As to the significance of the results, the following is quoted from "An Interpretation of the Research on Thermal Properties of Steam" published by ASME in 1930:

> It is difficult to place a monetary value on the importance of knowing accurately that results will come out as expected. The present demand of the power generating and using industries for steam at 1,200 pounds and the highest temperatures that materials will stand is evidence of their confidence in the accuracy of the information supplied by this investigation. If it had not been available and estimates had been over-sanguine or erratic, the present sound progress of these industries might have received setbacks which would have represented great economic losses.
>
> This investigation into the thermal properties of steam at high pressures and temperatures will not only produce far reaching improvements in the generation of power from fuels, but will materially affect the chemical and metallurgical processes dependent upon high temperatures and upon that property peculiar to vapors at the saturation point which insures the carrying on of such processes at a constant temperature.

Dean Arthur M. Greene, Jr., who served on Orrok's executive committee wrote in October 1939:

> . . . I have been connected with the committee on Thermal Properties of Steam . . . At that time there were great differences between the steam tables used in this country, in England, in Germany, and other European countries. In many cases the tables were merely extrapolations of formulae valid at lower pressures and temperatures and as a result the difference mentioned occurred.

. . . After formation of the committee and the inauguration of the work, an international conference on the properties of steam was formed and this conference fixed certain tolerances which should be allowed at different parts of the table. The work first obtained from Harvard University was used in the formation of a chart for the use of the General Electric Company to aid them in their work with high temperatures and pressures, and the information of this chart was freely given to others.

After a number of papers by Dr. Keyes and his associates, and Dr. Osborn and his associates, a preliminary table was made by Professor Keenan and upon the completion of further work, the final Keenan and Keyes tables of the properties of steam were issued. The great value of this table and the work of the investigators has been to give an

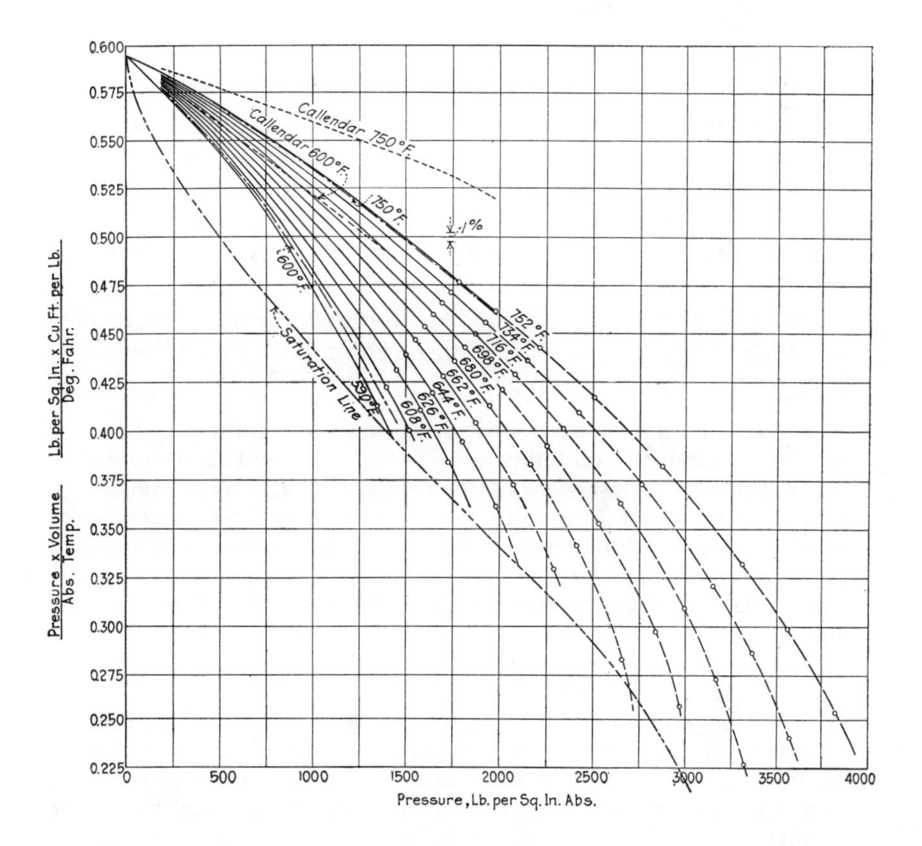

Steam research: pressure-volume-temperature relationship for superheated steam.

authoritative value to the various quantities used by the engineer at high temperatures and pressures, as well as properties of water below the point of vaporization.

The correct tables have aided us in determination of efficiencies of the proper design for heat-transfer apparatus, and for the computation of energy requirements in steam problems. Before the advent of the most recent table the values as mentioned before were those obtained by assumptions from data obtained under conditions different from those at which the tabular data would be used. This of course would lead to incorrect results, and as a result incorrect technical conclusions would be drawn from experimentation. The tables based on data at or near the point at which the table is applied give definiteness to our work and I feel that this as well as the properties of water below the point of vaporization have been the great contributions of the research in the properties of steam.

Lubrication

In 1915 ASME appointed a committee, chaired by Albert Kingsbury, to organize research on lubrication (Item 23 in Table 2). Beginning in 1924 the Engineering Foundation made grants in partial support of this research project which included many tests of a wide variety of lubrication problems. Many papers reporting the results of this research were published by ASME in *Transactions* and *Mechanical Engineering* and in other technical press.

The importance of lubrication research was discussed by Professor George B. Karelitz in 1939 while he chaired the ASME Research Committee on Lubrication:

> The Foundation has been generously supporting the work of the Research Committee on Lubrication of the ASME during a number of years. During the first years of this committee's activity it enjoyed the active support of Mr. Kingsbury. In the 1920s several grants were made to the committee for carrying out a number of research projects of general interest in the field of mechanics of lubrication, and the committee succeeded in collecting several thousand dollars from the industry to carry on the work. After the depression it was not possible to do this again. Three years ago the committee was reorganized, and during these years we have received grants from the Foundation. The money was spent mostly on the continuation of research on the effect of pressure on the viscosity of oil and on obtaining exact tables for pressure distribution in oil films in bearings. Both these projects are nearing the state where reports to the industry will be available on the results.
>
> However, the chief benefit from the Foundation's support is not in the immediate research data so obtained, but in the indirect stimulation of

general interest in the mechanics of lubrication, and in the existence of an active body free from any commercial interests. This group has been and is, no doubt, instrumental in drawing out from industrial concerns,

Presidents of the Founder Societies in 1928. DeGolyer and Gherardi served on the board of the Engineering Foundation.

> both mechanical and chemical, and publishing results which would otherwise have no adequate forum for presentation.
>
> The importance of this forum may be judged by the fact that in this country the theory of bearing design has been applied in industry to a degree without parallel in any of the European industrial countries.
>
> Experiments and investigations in the field of lubrication are expensive by their nature. Only large and powerful research laboratories can afford this kind of work. There is no doubt that the existence of the committee and its ability to carry on or encourage a little work through its own small funds is very useful in obtaining information from these laboratories, and in many instances instigating basic research in these laboratories.

The activities of the ASME Research Committee on Lubrication continued at least through the early 1960s. "Lubrication" was the Engineering Foundation's project of the longest duration.

Cutting Fluids for Working Metals

The ASME organized a Special Research Committee on the Cutting of Metals in 1924 and later set up a subcommittee on Cutting Fluids (Item 33 in Table 2). Professor O. W. Boston of the University of Michigan was the subcommittee's first chairman. During its first several years, the reports of the subcommittee were based on the personal experiences and research work carried on by its members. Starting in 1928 financial assistance was obtained from the Engineering Foundation which made a grant of over $6,000 and which was supplemented by industrial contributions totaling $4,000. A rather unique laboratory for testing cutting fluids in many different ways was developed. Many users of cutting fluids sponsored individual projects, and data that were not confidential were published. Two extensive bibliographies of the literature on cutting fluids were published, the first in 1930 by ASME and the second, larger list by Professor Boston in 1935.

The committee has cooperated with the American Society for Metals in developing a report listing machinability ratings of all SAE steels as well as a variety of other steels, irons, and nonferrous metals for turning, drilling, and threading. The subcommittee has indicated the recommended cutting fluid for each of these processes and materials.

An evaluation of the significance of the project was supplied by Professor Boston who wrote the following lines in 1939:

> My personal opinion is that the accomplishments have been commensurate with the cost of labor expended by the Foundation; the advance

of science and engineering has been effectively and definitely promoted; the data procured has been of real value to the public, as well as to science and engineering; the basic facts upon which engineering practice is founded have been developed and enlarged and standard practice promoted . . . and much of the information held secret only a few years ago is now being applied commercially.

Mining and Metallurgical Engineering Projects

To collect and study the methods used in all kinds of mining, AIME established a committee in the early 1920s. It was first chaired by J. E. Hodge and later by Professor R. M. Raymond (Item 21 in Table 2). The committee collected a great quantity of information and inspired a number of papers which were presented before AIME. From this mass of materials a volume of 925 pages was prepared by the Institute and published in 1925 as *Transactions of the AIME*.

Blast Furnace Slags Investigation

In 1926 Professor Richard S. McCaffery at the University of Wisconsin organized an investigation concerned with blast furnace slags (Item 26 in Table 2). The work was supervised by an AIME committee which was chaired by Charles H. Herty, Jr. The first grant by the Engineering Foundation to this project was made in 1926 and additional grants were made each succeeding year until the research was completed in 1931. The work resulted in five reports published by AIME: (1) "Research on Blast Furnace Slags," (2) "Determination of Viscosity of Iron Blast-Furnace Slags," (3) "Effect of Magnesia on Slag Viscosity," (4) "A Statistical Analysis of Blast-Furnace Data," and (5) "Air Discharge of Circular Tuyeres."

The following comments which summarized the impact of this project were written by Professor McCaffery in 1939:

The work on Blast Furnace Slags which we did at Wisconsin was selected as a project on account of the importance of determining those fundamental factors involved and their quantitative measurement. The most important of these factors is viscosity, which is the main factor beside the temperature involved in the reaction at the interface of the slag and metal in the furnace, and it is also of prime importance in the active operation of the blast furnace. These were the primary reasons that justified our selection of the particular problem.

The investigation we have carried out was a study and determination of the viscosity of lime, magnesia, alumina, and silica compounds and solutions, and this was done well beyond the ranges of chemical compositions which had previously been employed by blast furnace operators.

A blast furnace has been operated in the past to carry out primarily two functions: (1) to smelt a mixture of iron ore, flux (generally limestone or dolomite), and coke, in such a way as to produce the maximum economic production of pig iron, (2) along with the maximum desulphurization of the pig iron products.

It happens that the operation of a blast furnace to accomplish these two ends satisfactorily was not an easy matter due to the fact that desulphurization of the pig iron required the presence of a high lime slag and this could not always be done because the ore composition was such that the maximum economy of the operation was impossible.

When the reports on our investigations were published, they were not immediately utilized because most of the operators were not technically trained. However, a few years ago, at the blast furnace plant at Corby, England, they started in to operate their furnace, calculating the changes from the data presented in our reports and they operated the furnace to make the maximum production of pig iron without regard to its sulphur contents.

This pig iron which was high in sulphur, was tapped into ladles in which some limestone, fluorospar, and sodium salt (salt cake) had been placed. The hot incoming iron melted and mixed with this reagent, violently reacted and made a sodium base slag which took up the sulphur into solution, and when the iron was afterwards poured free from this sulphur bearing slag, the sulphur content of the metal had been reduced between sixty percent to eighty percent at a cost for the reagent of less than twenty-five cents U.S. currency per ton of pig iron produced.

This work at Corby was carried out by adequately trained technicians with the result that Corby produced pig iron at the lowest cost of any plant in the world, from an ore running about thirty-four percent iron content. The result of the disclosure of the Corby practice led others in this country to experiment with this method, and now it is being tried and experimented with in a number of American plants.

It may be of interest to know that our publications on blast furnace slags have been translated into French, German, Russian, and Japanese.

A past president of the Iron and Steel Institute of England has publicly stated that our research, which might have been looked upon purely as an investigation of 'academic' interest, had been proven to be of major importance to the great iron and steel industry.

Alloys of Iron Research

The most successful and second longest lasting project of the Engineering Foundation was undertaken to survey the important research work of the world, as reported in metallurgical literature on the properties and uses of carbon and alloy steels and cast irons. It also correlated this data and made it available to research men and engineers in the steel-producing and steel-consuming industries everywhere as a series of scientific monographs (Item 32 in Table 2). As a result of preliminary inquiries which began in 1927, a conference was held on May 24, 1928 in New York at the invitation of AIME. Following this conference, AIME appointed a Committee on Alloys of Iron Research to serve in an advisory capacity in the development of the plan and the conduct of the work. The committee, chaired by John Johnston, director of research and technology of the United States Steel Corporation, reported to the Institute that "the work to be done immediately is to assemble information of all kinds, from all sources, on iron and its alloys," and the work "should not be started unless a fund of at least $150,000 . . . can be secured." The Iron Alloys Committee, appointed by the Engineering Foundation and chaired by George B. Waterhouse, was organized as the governing body for the project and steps were taken to assure the necessary financing. Subscriptions soon exceeded the estimated needs and the committee made a start in building up its staff by engaging as editor Frank T. Sisco, formerly chief of the Metallurgical Laboratory of the Air Corps' Materials Division at Wright Field.

During the first twelve years of the project, the authors and consulting experts selected by the Iron Alloys Committee, and the New York editorial office wrote, critically reviewed, and edited ten regular monographs and one special volume, which were published at approximately one-year intervals by McGraw-Hill Book Company of New York. Early in 1942 the work on the project was discontinued on the grounds that it was not essential to winning the war, and it remained dormant for nearly five years. At the time the work was stopped the twelfth and thirteenth monographs of the original series were in process but remained unfinished. Late in 1946 the work was resumed on a small scale to finish the two uncompleted monographs of the old project and to prepare three new volumes which were made necessary by the advances in metallurgy during the war. The post-war volumes were published by John Wiley and Sons.

By the time the project was discontinued, sixteen regular monographs and one special volume were readily available to the researchers and engineers. They contained authoritative information on iron and its various alloys from all known reliable sources. The "Alloys of Iron" investigation was an excellent example of the value of the Engineering Foundation serving as a catalytic agent in establishing and initially financing a project that has proved to be of wide value to the industry and to the technical men. It also illustrated how readily cooperation can be attained on an industry-wide basis to carry on such a project once the need for it and its value have been demonstrated.

10. PROFESSIONAL STAFF, A. D. FLINN & SPECIAL PROJECTS

The first staff person employed by the Engineering Foundation was an acting secretary elected at the organizational meeting of the Foundation board, Professor Frederick R. Hutton. He continued in this part-time position until resignation in September 1915. Calvin W. Rice substituted for Hutton as needed. In September 1915 Cary T. Hutchinson was elected executive secretary. On his resignation on December 13, 1917, Alfred D. Flinn assumed the job.

Alfred Douglas Flinn

The death, on March 14, 1937, of Alfred D. Flinn, director, Engineering Foundation, brought to an end a career of exceptional engineering accomplishments, which encompassed, in the first half of his life, public works engineering and technical journalism, and, in its latter half, the administration of the properties and the research activities of the four national engineering societies.

Alfred Douglas Flinn was the oldest child of Matthew Bonner and Sarah Jane (Jones) Flinn. He was born in New Berlin in central Pennsylvania's Union County on August 4, 1868. He came of Irish-Scottish Presbyterian stock on his father's side, and among his mother's ancestors were the early settlers of Connecticut. He was one of six children of a busy country doctor who was often away from home on house calls; during these absences, as recalled by his brother, Professor Frederick B. Flinn, Alfred D. Flinn "was a great comfort to his mother" and helped her to take care of the younger children. This willingness to lighten the responsibilities of others was one of A. D. Flinn's outstanding traits.

Another outstanding trait was his seriousness of mind, colored, however, with a quick appreciation of humor. He only read serious books, and very rarely if ever attended the theater. He was reading Shakespeare assiduously before he was eleven years old. At about this time a younger brother, to whom he was very much attached, died from injuries inflicted by another boy. After the brother's death, Alfred assumed a very serious and grown-up attitude toward life. His brother believes that this incident had a great deal to do with this transformation, and his subsequent lack of appreciation of the frivolities of life. His guiding motive was to improve himself and those around him.

Flinn started his schooling in Reedsville, Pa., but was graduated from high school at Providence, R.I., High School, to which his family had moved. After graduating from high school, he worked for a short period in the clerical department of the Boston and Albany Railroad Company, changing later to employment in the

Alfred D. Flinn—First director of the Engineering Foundation, 1922–37.

largest hardware company in Worcester, Mass. He remained in this work until 1889 when, at twenty years of age, he was offered a junior partnership in the company. This brought him to a crisis in his life; he rejected a future career in the hardware business for one in engineering.

He entered the Worcester Polytechnic Institute in the fall of 1889, and was graduated in 1893 with the degree of Bachelor of Science in Civil Engineering. He also received the Salisbury Prize for the high quality of his work. His graduating thesis, "The Cippoletti Trapezoidal Weir," was based on large-scale experiments at the Holyoke Testing Flume. It was published as a paper by ASCE in 1894, and became immediately helpful in promoting the use of this means for measuring irrigation water in the United States.

Upon graduation from college, Flinn's first employment was with the firm of Rice and Evans, Civil Engineers of Boston, of which George S. Rice of New York Subway fame was the head. With this firm he worked for several months on early studies for rapid transit in Boston, embracing the so-called Alley Route; he also was engaged on surveys for a proposed large public park.

Engineer of Water Supply

The period of his life from the fall of 1893 to the end of 1917 was devoted mostly to water supply engineering, with the exception of a two-year interlude for editorial work.

During the winter and spring of 1893–94, Flinn was with the Associated Factory Mutual Fire Insurance Companies of Boston. In this job he was concerned with experimental work on sprinklers and the flow of water in pipes, and on factory surveys for fire prevention. He was then employed by the Boston City Engineer's Office in its Waterworks Division. There he conducted experimental work on all forms of fire-fighting apparatus used by the City of Boston; also on pumping engines and boilers, and the flow of water in large pipes. He also made studies for the design of cast-iron pipes, special castings, and valves; and routine surveys for field work on the construction of pipe lines, reservoirs, and pumping stations.

On the formation of the Massachusetts Metropolitan Water Works in 1895, Flinn became section engineer on surveys, construction, and studies within the metropolitan area, under Chief Engineer Frederic P. Stearns, past-president of ASCE. Later, he became principal assistant engineer in charge of the preparation of designs and contracts, the writing of reports, and various investigations, including experimental studies of dams, and the effect of ice on dams and other structures. These works comprised the Wachusett and Sudbury systems of additional supply for more than twenty communities, and included large dams, reservoirs, aqueducts, pipe lines, stand-pipes, and pumping stations. The most notable structures were the Wachusett dam and aqueduct, the Weston aqueduct, the Sudbury aqueduct, and the Chestnut Hill pumping station.

In October 1902 Flinn moved from Boston to New York City. He established his home in nearby Westchester. From October 1902 to August 1904, he was managing editor of *Engineering Record*, giving special attention to water supply and municipal engineering.

Until August 1905 Flinn was employed by the Croton Aqueduct commissioners as its general inspector. Under the general direction of Chief Engineer J. Waldo Smith, Flinn was engaged on the designs, specifications, and contracts for the completion of the New Croton dam and reservoir, the Muscoot dam and reservoir, the basin of the Jerome Park reservoir, and the Jerome Park pumping station.

He was also in charge of studies of rainfall and stream flow of the Croton watershed and of the selection of possible reservoir sites, to meet the city's future water needs. This work included the preparation of designs, contracts, and estimates of cost for the Cross River and the Croton Falls dams and reservoirs. These works cost several million dollars.

In August 1905, when Smith was called from the Aqueduct Commission to become chief engineer of the Board of Water Supply of New York City, he showed his appreciation of Flinn's peculiar talents for organization by taking him with him. During the next thirteen years Flinn served successively as division engineer, department engineer of the all-important headquarters department and deputy chief engineer. When Smith was absent from the city, Flinn acted as chief engineer. The construction work of the board during the years of Flinn's connection with it totaled $140,000,000, and included the building of the ninety-mile long Catskill aqueduct, the Ashokan and Kensico dams, two minor dams and four reservoirs, and studies for the Schoharie diversion. During this period, the combined forces of the board and all contractors engaged on the Catskill aqueduct reached a peak of 25,000.

Flinn's first responsibility was to recruit the Engineering Force of several hundred men. He personally interviewed a large proportion of the engineers after they had passed the required civil service tests. As department engineer, he was also instrumental in outlining surveys and field studies. He was responsible for the designs for practically all structures and equipment, and for the preparation of contracts and specifications. He was also responsible for development of cost estimates and for writing of special and annual reports. He supervised inspection of manufactured materials, and he selected consulting architects needed for various projects. The cooperation with consulting architects, general and special consulting engineers, geologists, landscape engineers, and sanitation specialists was also his responsibility. During World War I he served as the liaison with the military organizations guarding the waterworks. He also established and managed a special laboratory for research and testing of materials, and cooperated with other organizations on special research projects and tests relating to steels, brasses, cements, and other materials. He supervised preliminary studies, tests, and final design of the aerators for the entire flow in the Catskill aqueduct; water softening plants, filtration plants, and small sewage disposal works; and lastly, City Tunnel No. 1 and the connecting pipe lines.

Flinn was called on to recommend to Smith and John R. Freeman, past-president of ASCE, F. P. Stearns, also a past-president of ASCE, and W. H. Burr, his superiors, important changes in the materials of waterworks construction. Until 1905 metropolitan water systems had used massive masonry constructions, grade tunnels (the exception being the pressure tunnel of the New Croton aqueduct under the Harlem river), brick lining for tunnels, and cast-iron pipe siphons. As a result of Flinn's recommendations, the Board of Water Supply used rubble concrete and ashlar stone facing for its dams. It also utilized such innovations as pressure tunnels to a greater extent than had been used heretofore in New York, concrete lining for all tunnels, aqueducts of concrete without brick lining, steel pipe siphons with concrete linings and jackets, reinforced concrete siphons, removable reinforced concrete beams and slabs in gate chambers, reinforced concrete tile for chamber roofs, architecturally attractive superstructures, and landscaping and sodding for the tunnel spoil banks. It was due largely to his sense of aesthetics that the Kensico dam received a notable architectural treatment, and that the grounds below the dam were strikingly landscaped.

Flinn also undertook special investigations at several locations of the causes of defects in concrete and of "season cracking" of brasses. He supervised the studies of an improved flexible joint for large submerged pipe lines and developed special devices for closing the access and unwatering shafts of the pressure tunnels, including a novel scheme for unwatering; and of grouting of rock tunnels. He pioneered unusual methods of core drilling for exploration purposes, and methods of construction and materials for large earth dams. Finally, he supervised the studies of drainage galleries and other special features of masonry dams.

His duties included also a wide variety of administrative activities and contacts with public officials. He delivered the illustrated address descriptive of the Catskill aqueduct, the reservoirs, and structures, as part of the ceremonies by the New York City government at its celebration of the introduction of the Catskill water, in October 1917.

The value of his services to the Board of Water Supply is expressed in the resolution on his death, which reads as follows:

> Whereas, Alfred D. Flinn, formerly department engineer, Headquarters department, and deputy chief engineer of the Board of Water Supply, departed this life on March 14, 1937, now, therefore,
>
> Be It Resolved, that the Board of Water Supply expresses its deep sorrow at the death of Mr. Flinn, whose services to the City were vitally important in organizing the engineering bureau at the inception of the Catskill project and in designing the works and directing their construction.
>
> Be It Further Resolved, that this resolution be spread on the minutes of the Board of Water Supply and that a copy thereof be transmitted to Mrs. Flinn.

Near the end of his service with the Board of Water Supply, Flinn became a member of the board of direction of ASCE and served on the board of Engineering Societies Library. This was at the time of important changes in the library, and as chairman of a special committee, he investigated the various systems of classifications used in libraries and presented a plan for the reorganization of the library, and the unification of the book collections of the several societies into one well-arranged library.

Engineering Administrator

On January 1, 1918, Flinn resigned from the Board of Water Supply to take up the most congenial work of his life with the several joint organizations of the Engineering Societies. In this position he was eminently successful; and he worked in it happily and untiringly until his health failed. One of the members of the

1918 board of the United Engineering Societies who served two terms in this formative period, writes:

> Flinn came as a godsend to us. . . . In the early years before Flinn's advent, we were constantly in difficulties. . . . Flinn's coming in to those Boards made a very different picture of it. The business and efforts of those joint Boards went along much more smoothly than in the earlier days, and the "Foundation" and other organizations benefited very much.

The primary purpose in engaging the same person to be the executive officer of two or more organizations simultaneously was economy. None of the organizations participating in joint arrangements could then afford a separate administrative establishment. Operations were simplified and the work was reduced. For the Engineering Foundation and the Division of Engineering of National Research Council, the arrangement assured harmony of program and avoidance of duplication in closely related activities.

From January 1, 1918, to the time that his health began to fail in 1934, Flinn was secretary (executive officer) of the United Engineering Society and its successors, Engineering Foundation, Inc., and United Engineering Trustees, Inc. The position entailed supervision of the Engineering Societies Building; execution and records of actions of the Board of Trustees, including administration of trusts and other funds; and the provision of secretarial services to standing and special committees.

He was secretary of the Engineering Foundation from December 1917, and served as secretary and director from 1922, until his death. He initiated or aided in developing and financing such major projects as "Fatigue of Metals Research" (in co-operation with the National Research Council), the Arch Dam Investigation and the Alloys of Iron Research. He started the Education Research Committee and developed its pamphlet "Engineering: A Career—A Culture." He also provided assistance to special research committees of the Founder Societies and other organizations co-operating with them. His duties involved relationships with industries, universities, governmental bureaus, the National Research Council, the American Engineering Council, the Society for the Promotion of Engineering Education, and other bodies in all parts of the United States, and correspondence with other countries. The work included writing, compiling, or editing annual and special reports for the Foundation and its committees, the *Research Narratives* in leaflet and book forms, the *Directory of Hydraulic Laboratories in the United States* and other publications.

In co-operation with the National Research Council he aided in the establishment of the Personnel Research Federation, the Highway Research Board, and the American Bureau of Welding; and in conducting the Marine Piling Investigation (damage to structures by marine borers).

Flinn was secretary of the Engineering Council from January 1918 to the time of its disbanding on December 31, 1920, to make way for the Federated American Engineering Societies which later became known as the American Engineering Council. As such he dealt with "questions of general interest to engineers and to the public," some of which were: activities incident to the World War and reconstruction; classification and compensation of engineers; uniform state law for "Registration of Architects, Engineers and Land Surveyors"; improvement of the Patent Office, Engineering Societies Employment Bureau, and the National Public Works Department. An office was maintained at Washington, D.C. from January 1919 to the disbanding of the Council.

He was connected with the Division of Engineering, National Research Council, as its first vice-president from November 1, 1920 to December 8, 1921, and then as its chairman until February 15, 1923, when he resigned. His activities included facilitating the cooperation of the Engineering Foundation with other technical and scientific bodies, universities, industries, and governmental bureaus. He also monitored research related to testing of hardness of metals and other substances, uses of tellurium and selenium, highways, welding, molding sand for foundries, marine piling, pulverizing of material for technical processing, electrical insulation, high-speed tool steels, heat transmission, applications of psychology and psychiatry to industry, rock drill steels, magnetic analysis, and phosphorus and sulfur in steel.

Flinn was assistant secretary of the John Fritz Medal Fund Corporation from January 16, 1920 until he resigned in October 1935. He was elected secretary and treasurer of Daniel Guggenheim Medal Fund, Incorporated, on February 29, 1928. He resigned as treasurer on May 3, 1935, and as secretary on May 6, 1936. From time to time, he served temporarily other bodies whose routine work was committed to United Engineering Trustees, Inc., and only gave up these duties as ill health required.

Miscellaneous

Flinn received many honors for his work in engineering. The University of Louvain, Belgium, conferred the honorary degree of Doctor of Applied Sciences on him at its 500th Anniversary Celebration in 1927, and the Worcester (Mass.) Polytechnic Institute, from which he had received his degree of Bachelor of Science in 1893, made him an Honorary Doctor of Engineering in 1932. He was an honorary member of the Masaryk Academy of Prague, and held other Czechoslovak decorations.

He held membership in the following national societies:

American Society of Civil Engineers (director, 1917–1919)
American Institute of Mining and Metallurgical Engineers
American Society for Testing Materials
American Association for the Advancement of Science (fellow)
American Mathematical Society
American Iron and Steel Institute
Sigma Xi Scientific Fraternity

Flinn had always been interested in writing, and had a unique concise style which was easily recognized even though his name did not head the article. His hobby was to omit the article 'the' wherever possible. He never signed a letter 'Yours very truly' and often omitted the salutation. He considered middle initials superfluous.

In 1911 he was requested by McGraw Hill Book Company to undertake a reference book on waterworks. To aid him in this work, he recruited Robert Spurr Weston, Clinton L. Bogert, and John H. Gregory. Professor Gregory withdrew during the first year, but the other three authors worked together for five years producing the first edition of the *Waterworks Handbook* which was used extensively in the United States and other countries. With characteristic modesty, Flinn inscribed the preface as a compilation rather than an original contribution to literature.

He was justly proud of his originating and editing for twelve years the *Research Narratives* published by the Engineering Foundation, which were widely circulated and reprinted in this and other countries.

Flinn was always considerate of the working hours of his assistants, but not of himself; he was a most persistent worker. It was his habit, before ill health came, to be at his desk at home at 5:00 A.M., working for two hours at writing and editing before breakfast. In his early years, he kept Sunday free of his life work, but in later years, much of his time on Sundays was devoted to his absorbing writing of reports, and editorial work.

He was a pleasing and instructive public speaker. For two years in the 90s he gave courses on waterworks and water supply, and on sewerage to senior classes of Lawrence Scientific School, Harvard University. On invitation, he gave the Chester S. Lyman Memorial Lectures on Conservation of Water by Storage to the senior class at the Sheffield Scientific School, Yale University.

From his youth, he was interested in church work. From 1902 to 1908, he taught a Sunday school class in the 57th Street Baptist Church, in New York City. For many years, he was a member of the First Christian Science Church, of Yonkers, N.Y.

In 1900 Flinn was married to Mary Brownell Davis, of Boston, who survived him. During his last year he sold his spacious home in Park Hill, Yonkers, gave away most of his books and furniture, and retired to Heathcote Inn, Scarsdale, New York, where he died.

A memoir published in 1938 (Ref. 1.88) speaks of his character and his ideals:

> He was noted for his logical presentation of facts, accuracy of statement, and thoroughness of detail. These characteristics went to the extent of leaving a concise summary of his activities from which this memoir was largely compiled.
>
> Alfred D. Flinn was a Christian gentleman dominated by the highest ideals. His unfailing courtesy and thoughtfulness of others were outstanding characteristics. He worked unselfishly for the advancement of engineers and the profession of engineering. His influence toward these objectives has been great.

Special Projects

The bylaws of the United Engineering Society stated that the Engineering Foundation Board may make known to the world the results of its undertakings by publications, by public lectures or by other means at its discretion. The first step in publicizing the work of the Foundation was taken in February of 1917 with the issuance of the *Report on the Origin, Foundation and Scope of the National Research Council* which was the first publication of the Foundation. This report and the second publication entitled *Progress Report to the United Engineering Society* issued in October 1919 preceded regular publication of the annual report of the Foundation to the United Engineering Society which was started with

Publication Number 3 issued in March of 1921. Several of these annual reports included photographs of the presidents of the four Founder Societies; some of them are reproduced throughout this text. Other steps were taken in 1920 with the issuance of four reprints of papers reporting work sponsored by the Engineering Foundation, and the engagement of James T. Grady, director of the Department of Public Information at Columbia University, beginning November 1 as a public relations consultant. Through Grady's activities, news and other items appeared in the daily press in all parts of the country. Other statements about the Foundation appeared in many technical journals in America and in Europe. Some of the activities initiated by Grady, such as an aperiodic issuance of news releases about Foundation activities, have been continued to this day.

In January 1921 the semi-monthly publication was begun of a series of leaflets called *Research Narratives*, each of which contained a concise story of some research or discovery or notable achievement in science or engineering. Approximately 1500 copies of each issue were mailed to prominent men of means, leading executives and engineers in many industries, engineering societies, technical journals, libraries, colleges and others. A number of *Narratives* were reprinted by the daily press, magazines and technical journals, and thus had wide circulation. Starting in 1925 the frequency of publication was changed to monthly. A compilation of the first fifty *Narratives* was issued in a book form in 1924, of the second fifty in 1926, and of the third fifty in 1929. The last Narrative was published in December 1932. The publication of *Research Narratives* was discontinued in 1933.

In the early 1920s the Engineering Foundation's committee on hydraulic research compiled a directory of hydraulic laboratories. The Foundation made it generally available as Publication No. 5 entitled *Hydraulic Laboratories in the United States of America.* This directory included descriptions of forty-nine laboratories. The descriptions, which were supplied by the laboratories responding to a letter of inquiry, included general information such as the name and year of each laboratory's establishment, information about the type of work for which the laboratory was especially equipped and the possibility of work for or by outside parties, and past work of unusual importance. According to the directory, most of the forty-nine laboratories were associated with academic institutions and the remainder with the industry or the government. Among the listed laboratories eleven were established prior to 1900.

On December 4, 1924 Engineering Foundation joined with the Founder Societies, other learned associations, and several universities in celebrating the centenary of Nicolas Leonard Sadi Carnot's

THE
ENGINEERING
FOUNDATION

HYDRAULIC LABORATORIES
IN THE
UNITED STATES OF AMERICA

PUBLICATION
NUMBER 5

ENGINEERING SOCIETIES BUILDING
NEW YORK CITY
JUNE, 1922

Cover page of an early Engineering Foundation publication containing directory of American hydraulic laboratories.

publication of his great principle of thermodynamics. In January 1925 it issued Publication No. 9 *Carnot Centenary Commemoration*. The publication contained the proceedings of a commemorative meeting and other materials related to the subject of Carnot's contribution.

During the 1920s the Engineering Foundation established the custom of devoting its May meetings to promotion of friendliness among men who were interested in the application of sciences to the service of engineering and industry. On these occasions it told its friends in an intimate way of its achievements and aims, and presented through an able speaker a subject of timely interest in the domain of the applied sciences. The series included the following addresses:

Frank B. Jewett, president of Bell Telephone Laboratories and past vice chairman of the Foundation: Permalloy Cables, 1925

Edward Dean Adams, John Fritz Medalist and past vice chairman of the Foundation: America and Americans, 1926

John A. Mathews, vice president of Crucible Steel Company of America: Present Tendencies in Engineering Materials, 1926

Arthur D. Little, president and director of Petroleum Chemical Corporation and Arthur D. Little, Inc. of Cambridge, Massachusetts: Impending Change in Our Use of Fuels, 1927

The last listed lecture was printed as Engineering Foundation's Publication No. 14.

In response to an invitation from the University of Louvain in Belgium, the Engineering Foundation participated in the celebration of its 500th anniversary in June, 1927. Edward Dean Adams, honorary member, was the Foundation delegate. He also represented the Founder Societies and the Engineering Societies Library. When reporting on his participation, Adams proposed giving a carillon and clock for the tower of the new Louvain Library as a memorial to all engineers of the United States who gave their lives during World War I. On recommendation of the Foundation, the Founder Societies unanimously adopted this proposal and authorized United Engineering Society to appoint a joint committee to have full charge of the fund raising activities. Adams was chosen as its chairman. Besides the Engineering Foundation and the Engineering Societies Library, sixteen other national engineering societies cooperated. By strenuous exertion the engineers' memorial was installed in time to be dedicated along with the library building

The pioneering thermodynamic research of Nicolas Leonard Sadi Carnot was commemorated by a 1925 publication of the Engineering Foundation.

on July 4, 1928. The chairman and other members of the committee and many members of the American engineering societies attended the ceremonies. For three days the American engineers received many courtesies and delightful entertainment. In reciprocation of good will, the University conferred upon the chairman of the delegation an honorary degree of Doctor of Science. The engineers' memorial includes:

Carillon of forty-eight bells, about four octaves, the heaviest weighing nearly eight tons, known as The Library Bell of Louvain

Tower clock, with four dials of special design bearing forty-eight stars in memory of engineers from all states of the U.S.

Mechanism by which the clock automatically plays chime tunes upon selected bells of the carillon each quarter hour

Perpetual endowment of $10,000, of which the income is to provide for maintenance and operation of the carillon and the clock, the surplus being available for purchase of American engineering and scientific books

Memorial volume containing all names which could be collected of engineers of the United States who gave their lives in the war, descriptive statements about the memorial, and the names of the contributors of the fund which was collected for the memorial

Memorial inscriptions on the interior walls of the library tower and the Liberty Bell

Replica, weighing 377 pounds, of the Liberty Bell placed in the great hall of the library where it may easily be seen by all visitors

The total cost was approximately $90,000 all of which was donated by 3,000 engineers and friends in the United States.

References

1.1 "Minutes of the United Engineering Society Board of Trustees meetings."

1.2 "Minutes of the Engineering Foundation Board meetings."

1.3 "Engineering Societies Building, West Thirty-Ninth Street, New York City," Prepared for the Dedication Exercises, Apr. 16–19, 1907.

1.4 "Report on the Origin, Foundation and Scope of the National Research Council" (Engineering Foundation, Feb. 27, 1917).

1.5 Publication No. 2 (Engineering Foundation, 1919).
1.6 E. E. Southard, *The Mental Hygiene of Industry*, Reprint Series No. 1 (Engineering Foundation, 1920).
1.7 E. E. Southard, *Trade-Unionism and Temperament*, Reprint Series No. 2 (Engineering Foundation, 1920).
1.8 E. E. Southard, *The Modern Specialist in Unrest*, Reprint Series No. 3 (Engineering Foundation, 1920).
1.9 Clemens Herschel, *An Improved Form of Weir for Gaging in Open Channels*, Reprint Series No. 4 (Engineering Foundation, 1920).
1.10 Publication No. 3 (Engineering Foundation, 1921).
1.11 Publication No. 4 (Engineering Foundation, 1922).
1.12 *Hydraulic Laboratories in the United States of America*, Publication No. 5 (Engineering Foundation, 1922).
1.13 Publication No. 6 (Engineering Foundation, 1923).
1.14 *Terms of the Endowment and the Official Records of its Establishment*, Publication No. 7 (Engineering Foundation, 1924).
1.15 Publication No. 8 (Engineering Foundation, 1924).
1.16 *Carnot Centenary Commemoration*, Publication No. 9 (Engineering Foundation,1925).
1.17 Publication No. 10 (Engineering Foundation, 1925).
1.18 Publication No. 11 (Engineering Foundation, 1926).
1.19 Publication No. 12 (Engineering Foundation, 1927).
1.20 *The Manual of Endurance of Metals Under Repeated Stress*, Publication No. 13 (Engineering Foundation, 1927).
1.21 *Impending Changes in Our Use of Fuels*, Publication No. 14 (Engineering Foundation, 1927).
1.22 Publication No. 15 (Engineering Foundation, 1928).
1.23 *Cellulose. Progress in Czechoslovakia*, Publication No. 16 (Engineering Foundation, 1928).
1.24 Publication No. 17 (Engineering Foundation, 1929).
1.25 *Mr. Swasey's Reasons*, Pamphlet No. 1 (Engineering Foundation, 1926).
1.26 Pamphlet No. 2 (Engineering Foundation, 1925).
1.27 *Popular Research Narratives* (Baltimore, Maryland: Williams & Wilkins Company, 1924).
1.28 *Popular Research Narratives, Volume II* (Baltimore, Maryland: Williams & Wilkins Company, 1926).
1.29 *Popular Research Narratives, Volume III* (Baltimore, Maryland: Williams & Wilkins Company, 1929).

1.30 *History, Charter and By-Laws*, (Engineering Foundation, 1930).
1.31 Engineering Foundation, February, 1930.
1.32 United Engineering Trustees, Inc., The Engineering Foundation, Engineering Societies Library, 1931.
1.33 *Reports for Years 1929–1930* (United Engineering Trustees, Inc., 1931).
1.34 *Reports for the Year 1931* (Engineering Foundation, 1932).
1.35 *Reports for the Year 1932* (Engineering Foundation, 1933).
1.36 *Report for the Calendar Year 1933* (Engineering Foundation, 1935).
1.37 "1936–1937 Annual Report" (New York: Engineering Foundation).
1.38 "1937–1938 Annual Report" (New York: Engineering Foundation).
1.39 "1938–1939 Annual Report" (New York: Engineering Foundation).
1.40 "Twenty Five Years of Service," Engineering Foundation, Otis Ellis Hovey, Director.
1.41 "1939–1940 Annual Report" (New York: Engineering Foundation).
1.42 "1940–1941 Annual Report" (New York: Engineering Foundation).
1.43 "1941–1942 Annual Report" (New York: Engineering Foundation).
1.44 "1942–1943 Annual Report" (New York: Engineering Foundation).
1.45 "1943–1944 Annual Report" (New York: Engineering Foundation).
1.46 "1944–1945 Annual Report" (New York: Engineering Foundation).
1.47 "1945–1946 Annual Report" (New York: Engineering Foundation).
1.48 "1946–1947 Annual Report" (New York: Engineering Foundation).
1.49 "1947–1948, Annual Report" (New York: Engineering Foundation).
1.50 *Thirty-Five Year Summary and Annual Report 1948–49* (Engineering Foundation, 1949).
1.51 "1949–1950 Annual Report" (New York: Engineering Foundation).
1.52 "1950–1951 Annual Report" (New York: Engineering Foundation).

1.53 "1951–1952 Annual Report" (New York: Engineering Foundation).
1.54 "1952–1953 Annual Report" (New York: Engineering Foundation).
1.55 *Annual Report 1953–1954 and History of the Foundation 1914–1954*, (New York: Engineering Foundation, 1954).
1.56 "1954–1955 Annual Report" (New York: Engineering Foundation).
1.57 "1955–1956 Annual Report" (New York: Engineering Foundation).
1.58 "1956–1957 Annual Report" (New York: Engineering Foundation).
1.59 "1957–1958 Annual Report" (New York: Engineering Foundation).
1.60 "*October 1, 1958 to September 30, 1959 Annual Report*" (New York: Engineering Foundation).
1.61 William G. Atwood and A. A. Johnson, *Marine Structures: Their Deterioration and Preservation* (Washington, DC: National Research Council, 1924).
1.62 John Fritz Medal Presentation to Ambrose Swasey, 1924.
1.63 J. H. Keenan, "Progress Report on Development of Steam Charts and Tables from Harvard Throttling Experiments," *Mechanical Engineering* (February 1926), pp. 144–151.
1.64 "Progress in Steam Research," *Mechanical Engineering* (February 1926), pp. 151–60.
1.65 H. F. Moore and J. B. Kommers, *The Fatigue of Metals* (New York: McGraw-Hill Book Co., 1927).
1.66 J. B. Whitehead and R. H. Marvin, "Anomalous Conduction as a Cause of Dielectric Absorption," *Transaction of the AIEE*, Vol. 48, No. 2 (April 1929), pp. 299–316.
1.67 J. B. Whitehead and R. H. Marvin, "The Conductivity of Insulating Oils," *Journal of the AIEE*, Vol. XLIX (1930), pp. 182–186.
1.68 "Progress in Steam Research," *Mechanical Engineering* (February 1930), pp. 119–141.
1.69 "Dynamic Loads on Gear Teeth," *ASME Special Research Publication* (August 1931).
1.70 *Iron-Silicon Alloys Bibliography*, The Engineering Foundation (February 1931).
1.71 J. L. Gregg, *The Alloys of Iron and Molybdenum* (New York: McGraw-Hill Book Company, 1932).
1.72 F. E. Turneaure, A. F. Reichmenn, W. G. Grove, R. R. Leffler

and S. H. Widdicombe, "Steel Column Research," *ASCE Transactions* (1933), pp. 1376–1462.

1.73 E. S. Greiner, J. S. Marsh and B. Stoughton, *The Alloys of Iron and Silicon* (New York: McGraw-Hill Book Company, 1933).

1.74 J. L. Gregg, *The Alloys of Iron and Tungsten* (New York: McGraw-Hill Book Company, 1934).

1.75 J. L. Gregg and B. M. Daniloff, *The Alloys of Iron and Copper* (New York: McGraw-Hill Book Company, 1934).

1.76 J. S. Marsh, *Principles of Phase Diagrams* (New York: McGraw-Hill Book Company, 1935).

1.77 H. E. Cleaves and J. G. Thompson, *The Metal—Iron* (New York: McGraw-Hill Book Company, 1935).

1.78 W. M. Wilson, "Laboratory Tests of Multi-Span Reinforced Concrete Arch Bridges," *Transactions ASCE*, Vol. 100 (1935), pp. 424–470.

1.79 C. O. Gunther, *Identification of Fire Arms* (New York: John Wiley & Sons, 1935).

1.80 *Ambrose Swasey, Second Hoover Medalist* (New York: Hoover Medal Board of Award, 1936).

1.81 W. H. Fulweiler, A. H. Strong and L. R. Sweetman, "Inspection and Tensile Tests of Some Worn Wire Ropes," NBS Research Paper RP920 (1936).

1.82 S. Epstein, *The Alloys of Iron and Carbon, Volume I—Constitution* (New York: McGraw-Hill Book Company, 1936).

1.83 E. Buckingham, "Quantitative Analysis of Wear," *Mechanical Engineering* (ASME, August 1937), pp. 576–578.

1.84 F. T. Sisco, *The Alloys of Iron and Carbon, Volume II—Properties* (New York: McGraw-Hill Book Company, 1937).

1.85 A. B. Kinzel and W. Crafts, *The Alloys of Iron and Chromium, Volume I—Low-Chromium Alloys* (New York: McGraw-Hill Book Company, 1937).

1.86 "Ambrose Swasey Honored," *Mechanical Engineering Journal* (January 1937), pp. 38–44.

1.87 J. J. Nassau, "Ambrose Swasey, Builder of Machines, Telescopes and Men," *Journal of Applied Physics*, Vol. 8 (September 1937), pp. 595–601.

1.88 "Alfred Douglas Flinn, M. Am. Soc. C. E.," *Transactions ASCE*, Vol. 103 (1938), pp. 1787–95.

1.89 J. S. Marsh, *The Alloys of Iron and Nickel, Volume I—Special-Purpose Alloys* (New York: McGraw-Hill Book Company, 1938).

1.90 A. B. Kinzel and R. Franks, *The Alloys of Iron and Chromium, Volume II—High-Chromium Alloys* (New York: McGraw-Hill Book Company, 1940).
1.91 W. H. Wisely, *The American Civil Engineer 1852–1974* (American Society of Civil Engineers, 1974).
1.92 G. H. Marx, L. E. Cutter, and B. M. Green, "Some Comparative Wear Experiments on Cast-Iron Gear Teeth," ASME Preprint for the Annual Meeting of the Society in New York, November 30 to December 4, 1925.
1.93 Engineering Foundation, "Annual Report to United Engineering Society for Calendar Year 1920," a typed report dated January 6, 1921.
1.94 The Engineering Foundation, "Annual Report for Calendar Year 1933," a typed report dated January, 1934.
1.95 The Engineering Foundation, "Annual Report from January 1 to September 30, 1935," a typed report of October 10 and 24, 1935.
1.96 The Engineering Foundation, "Annual Report from October 1, 1935 to September 30, 1936," a typed report dated October 8, 1936.
1.97 Correspondence between the Railway Signal Association (C. C. Rosenberg, Secretary-Treasurer, and A. G. Shaver; C. H. Morrison and Howard Elliott, New York, New Haven & Hartford Railroad Company) and the Engineering Foundation (Secretary F. R. Hutton, Board Members Howard Elliott and A. R. Ledoux), August 1915 to April 1916.
1.98 Correspondence between Leland Stanford Junior University (W. F. Durand, Guido H. Marx and L. E. Cutter) and the Engineering Foundation (Secretaries C. T. Hutchinson and A. D. Flinn; ASME Secretary Calvin W. Rice), 1915 to 1925.
1.99 "Fatigue Phenomena of Metals: Minutes of the Meeting of the Committee on Reorganization," held on February 23, 1928 in New York. Minutes dated March 8, 1928.
1.100 Letter from W. Spraragen, Secretary of NRC Division of Engineering and Industrial Research, to the Committee on Fatigue Phenomena of Metals, April 11, 1928.
1.101 Letter from W. Spraragen to A. D. Flinn dated April 11, 1928.
1.102 A. D. Flinn, "Spray Concealment of Ships," a typed summary of the results of experiments, May 13, 1918.
1.103 Letter from M. I. Pupin to Alfred D. Flinn dated September 17, 1919.

1.104 *Arch Dam Investigations*, Vol. I, *ASCE Proceedings* (May 1928).
1.105 *Arch Dam Investigations*, Vol. II (Engineering Foundation, 1934).
1.106 *Arch Dam Investigations*, Vol. III (Engineering Foundation, 1933).
1.107 S. L. Case and K. R. Van Horn, *Aluminum in Iron and Steel* (New York: John Wiley and Sons, 1953).
1.108 A. M. Hall, *Nickel in Iron and Steel* (New York: John Wiley and Sons, 1954).
1.109 G. F. Comstock, *Titanium in Iron and Steel* (New York: John Wiley and Sons, 1955).
1.110 R. A. Grange, F. J. Shortsleeve, D. C. Hilty, W. O. Binder, G. T. Mottock and C. M. Offenhauer, *Boron, Calcium, Columbium, and Zirconium in Iron and Steel* (New York: John Wiley and Sons, 1957).
1.111 R. Moley and C. Jedel, "Engineer No. 1," *The Saturday Evening Post* (October 11, 1941), pp. 29, 110–116.
1.112 "Gano Dunn, Edison Medalist 1937," *AIEE* (New York, 1938).
1.113 "Gano Dunn," Obituary, *Electrical Engineering*, Vol. 72 (June 1953), p. 564.
1.114 "Gano Dunn is Dead; Noted Engineer, 82," *New York Times* (April 11, 1953).
1.115 "Michael Idvorsky Pupin," Obituary, *Electrical Engineering*, Vol. 54 (April 1935), p. 464.
1.116 *Transactions AIME*, Vol. LXXII 72 (New York: 1925).
1.117 *Ideas & Actions: A History of the Highway Research Board 1920–70*, (Washington, D.C.: National Academy of Sciences, 1971.)

CHAPTER 2

Support for the Engineer as a Professional Person 1930–59

The 1930–59 period of the Engineering Foundation history was best characterized as an era when support for the engineer as a professional person began to rival research projects as a focus of the Foundation activities. It was also an era that saw the Foundation overcome the financial difficulties that were caused by the Great Depression and make valuable contributions to national defense activities during World War II. During this period, the Foundation's first director, Alfred Flinn, passed from the scene and his able successors, Otis Hovey, Edwin Colpitts and Frank Sisco maintained and enhanced its activities and reputation. Finally, the 1930–59 period was notable for an ongoing process of self-evaluation by the Engineering Foundation board which ultimately resulted in a change of direction which brought about the creation of many of the Foundation's current areas of interest and funding.

1. ACCREDITATION BOARD FOR ENGINEERING AND TECHNOLOGY

The history of the 1930–59 era of the Engineering Foundation begins with the chronicle of its continuing involvement in efforts to improve engineering education, through its support of the Engineers' Council for Professional Development (ECPD).

The Engineers' Council for Professional Development was initially proposed on April 14, 1932. It was established on October 3, 1932 by the joint sponsorship of seven participating engineering societies: ASCE, AIME, ASME, AIEE, AIChE, the American Society for Engineering Education (ASEE), and the National Council of Engineering Examiners (NCEE). The ECPD was formed to help further the "enhancement of the professional status of the engineer." The ECPD also represented the engineering profession in matters related to the development, maintenance, and improvement of engineering education. Accreditation of the engineering curricula at participating colleges and universities became its principal tool to achieve these goals. Its formal accreditation activities began with the formation of the Committee on Engineering Schools under the leadership of K. T. Compton. The first listing of accredited undergraduate curricula was published in 1936. By 1940 all but eleven out of the 136 schools of engineering in the United States had submitted their curricula for evaluation by ECPD. The Committee on Engineering Schools was also presented with the task of implementing a comparable program for non-degree granting technical schools. To accommodate this new aspect of accreditation the Subcommittee on Technical Institutes was formed. On January 1, 1980, ECPD became the Accreditation Board for Engineering and Technology (ABET). By that date the number of sponsoring engineering societies had grown to seventeen.

The ECPD became the focal point of the Engineering Foundation's efforts to improve the training of engineers in the United States (Item 56 in Table 2). After February 21, 1935 ECPD assumed the duties of the Engineering Foundation's inactive Education Research Committee. In the same year, the Engineering Foundation granted $3,500 to ECPD to continue its accreditation activities with the condition that this grant should not constitute a precedent for additional grants. The grants disclaimer reflected an internal debate between the advocates of support for research projects and the advocates of grants to advance the humanistic aspects of the engineering profession that was to characterize board discussions for the next twenty years. This debate was also echoed in a letter written by founder Ambrose Swasey to the Engineering Foundation on December 6, 1933. In this letter he warned against allowing the Foundation to become too heavily involved in the support of other institutions including those activities of the Founder Societies which were not directly related to research.

The debate was brought about because a faction of the board felt that ECPD's activities did not conform to what they believed to be the purpose of the Engineering Foundation. Nevertheless, regular funding of ECPD by the Engineering Foundation was continued through 1958 for not only its regular accreditation activities but also for special projects. These special projects were particularly active after World War II when ECPD developed an intensive program to place returning soldiers as students in engineering schools. By the end of the 1958 fiscal year, when the Engineering Foundation's funding of ECPD was discontinued, the Engineering Foundation contributed a total of $118,000 which represented more than one third of ECPD's cash income for the entire period of Engineering Foundation support for its activities. However, by 1958 the Engineering Foundation's contribution represented less than ten percent of ECPD's income for that year.

The Engineering Foundation also made continuing efforts to insure that the visiting teams, which were sent by ECPD to evaluate engineering curricula, were composed of a mix of practicing engineers and engineering professors. The Engineering Foundation Board felt that without this mix the visiting teams could not determine the extent to which a school's curriculum reflected the needs of the real world and the challenges that confront practicing engineers on the job. This problem became particularly acute during the early 1950s when the employment demands of practicing engineers decreased their participation on evaluation teams.

The ECPD also received funding to develop a booklet for the Foundation which was entitled *Engineering—A Career, A Culture.* It was used to encourage high school students to choose engineering as their profession. On October 14, 1937 the board voted to further support ECPD by terminating its control of this booklet and by donating all remaining copies of it to the Council. By 1955 ECPD had become almost totally independent of the Foundation.

In 1934 a directory of organizations in the engineering profession was published jointly by the American Engineering Council and ECPD (Item 64 in Table 2). The Engineering Foundation sponsored the project but contributed no funds.

2. RESPONSES TO THE GREAT DEPRESSION

By 1933 the board of the Engineering Foundation had become aware of the deleterious effects of the Great Depression on the employment of engineers in industry. In response it attempted, through ECPD and other organizations, to aid unemployed engineers by providing guidance and retraining in order to help them to find jobs in new fields. At its November 16, 1933 meeting, the board voted to subsidize a program for retraining disengaged engineers (Item 54 in Table 2). This was organized in the closing weeks of 1932 and completed at the end of May 1933. With cooperation of five colleges of engineering and of business administration in the New York metropolitan district, Personnel Research Federation, Professional Engineers' Committee on Unemployment, technical societies, industries, individuals, and United Engineering Trustees, Inc., courses in ten subjects were given without charge under the auspices of the Foundation in the Engineering Societies Building, to more than 600 engineers. Attendance was high throughout the courses and the entire program proved to be of real benefit to those unemployed engineers who participated. Additional evidence of the concern that the Foundation expressed for unemployed or disengaged engineers is illustrated by the action taken at the board meeting of May 18, 1933, at which time it voted to subsidize the preparation of a report by Columbia University which would focus on the mental attitude of unemployed engineers (Item 57 in Table 2).

An extensive investigation of engineering education was initiated by the Society for the Promotion of Engineering Education in 1922 and concluded in 1934. A 'brief' summarizing the results of the investigation and related activities was prepared by Chairman Charles F. Scott. The brief appeared both in the *Journal of Engineering Education* for September 1934, and in a pamphlet reprint. Although the Carnegie Corporation of New York provided the major portion of the money required, the Engineering Foundation also contributed to it and obtained contributions from its Founder Societies and other sources. The work comprised a survey and a constructive critical investigation of the status of engineering education in the, approximately, 150 engineering colleges of the United States and Canada; an extensive study of engineering educational institutions in Europe, for purposes of comparison; a study of non-collegiate technical institutes for the training of supervisors, inspectors, and technicians in industry and construction; and a series of twelve sessions of summer schools for engineering teachers discussed below. William E. Wickenden was the director of the project, H. P. Hammond the associate director, and R. H. Spahr the leader of the special staff which studied technical institutes. Besides assembling a large body of useful information, the report contained numerous conclusions, recommendations, and suggestions. Two features were outstanding: it was an investigation of engineering education by engineering educators, and its immediate publication of data and reports in the Society's monthly journal led to immediate and general interest. The work exerted a strongly beneficial influence upon the trend and the detail development of engineering education and the related training of technicians.

The Engineering Foundation continued its earlier policy of making appropriations to the Personnel Research Federation. The Federation was dedicated to investigating the mental health of workers and relationships between employers and employees. At its June 11, 1937 meeting the Engineering Foundation's board granted $1,200 to the Personnel Research Foundation to fund a project entitled "Negotiations with Organized Employees" (Item 71 in Table 2). It should be noted, however, that the grant was made with several conditions such as the understanding that this grant would be the last one made by the Foundation to this agency and that the word 'organized' be removed from the project title. Eventually, the failure of the Personnel Research Foundation to accept these conditions caused the Foundation to withdraw its funding. On the other hand, the 1936 grant of $1,400 to the Personnel Research Foundation for the study of employer-employee cooperation resulted in a report which was published in the fall of the same year (Item 65 in Table 2). This report summarized the current status of the development of employer-employee cooperation. It was based on visits to industrial plants, on correspondence with governmental departments and labor organizations, and on conferences. It also discussed ways in which this method of industrial management could be developed further.

From 1927 to 1933 the Society for Promotion of Engineering Education conducted summer schools for engineering teachers with the financial aid of the Foundation and other contributors (Item 46 in Table 2). Twelve such summer seminars, held at several colleges, were attended by nearly one thousand teachers. The seminars covered the principal subjects of the engineering curriculum. For example, courses in economics were given at Stevens Institute of Technology and in the English language at Ohio State University. The course at Stevens was attended by teachers from twenty-three institutions in fourteen states.

Charles S. Slocombe of the Personnel Research Federation made a study of the problems that were incident to engineering advances (Item 70 in Table 2). The results of this study were published in the *Personnel Journal*. He concluded that the development of mass production industries, made possible through engineering research into the properties of metals, chemicals, plastics, and other industrial products, seemed to necessitate the growth of large organizations. On the human side, however, the growth of large organizations has in many cases resulted in diverting the energies and thoughts of professional engineers from the fields in which they were trained. More and more, engineers found themselves occupying positions of executive control and direction which required only a minimum use of their technical knowledge. This trend seemed to indicate that serious consideration should be given to providing some means by which engineers, in the course of their education could be trained not only in the technical aspects of their work but also in the administrative. Each time a new engineering method is developed and brought into use by industry, a series of human and social problems is also created. Engineers, therefore, if they are to continue to contribute to the development of the nation, should be trained to understand these problems and to aid in their avoidance or solution.

A study of the industrial system of the United States of America (Item 48 in Table 2) under twentieth century conditions of the arts, sciences, jurisprudence, finance, government, and international relations, was proposed by members of the Foundation Board in September 1931. These individuals felt that the unsatisfactory working of the industrial system obligated engineers, in cooperation with men of other vocations, to undertake endeavors for correction of its defects. The Foundation invited a group of able men of widely varied experience to advise whether a study should be undertaken. However no action came from this proposal.

Throughout the 1930s the Engineering Foundation was faced with the reality of a decline in the income that it received from its investments. These adverse financial conditions were reflected in an important general comment recorded in the minutes of the board meeting of May 22, 1936: "The reduction of the Foundation's income is especially unfortunate at a time when the research work is improving. Particularly encouraging are the attempts at more fundamental studies of phenomena underlying engineering problems in a search for principles which will provide surer bases for improving engineering practices than empirical experiments."

Items 39 in Table 2 involved an additional project that was a response to the Great Depression.

3. NEW PLATFORM ON GRANT POLICY

The continued development and implementation of the Engineering Foundation grant policy took a significant turn at the October 22, 1931 board meeting when the Research Procedure Committee was established. This new committee had the dual purpose of establishing the grant policy and evaluating grant proposals.

Another major trend that soon became evident in the development of the Foundation's grant policy was the increased emphasis that was placed on cooperative projects. The majority of the funding for these cooperative projects came from the Founder Societies or other engineering organizations that had originally sponsored these projects when they were first proposed to the Engineering Foundation. In line with this increased emphasis on cooperative projects was the use of contingency requirements for the granting of Foundation funding. In almost all cases, contingency funding requirements insured that a project would become both viable and independent of major Engineering Foundation funding at an early date. The use of contingency funding requirements also allowed the Foundation to spread its limited resources among the largest possible number of projects in an era when its income was sharply reduced due to the Great Depression inspired downturn in the value of its investments.

This decrease in available grant funding became a hallmark of Foundation activities during the 1930s. For example, it was voted at the February 15, 1934 board meeting to temporarily decline applications for research funds after March 1, 1934. This ban remained in effect throughout the remainder of the Engineering Foundation's fiscal year. However, the ban did not bring relief as a series of small financial crises continued to interfere with the Foundation's currently funded projects. At the executive committee meeting of September 12, 1933 it was noted that there was a temporary deficiency in the budget of the "Alloys of Iron Research" project. Since the Foundation did not have the funding that was needed to cover this deficit, it was forced to seek the financial aid of the American Iron and Steel Institute.

Another change in the Engineering Foundation's grants policy was the development of new wording in all grant agreements which was designed to protect the Foundation from any legal claims that might arise from a situation in which it had no funds available to pay grantees. This precaution was another indication of the financial problems that were caused by the Great Depression. A further evolution in the Engineering Foundation's grants policy occurred when Otis E. Hovey replaced the deceased Alfred D. Flinn as director. At his appointment, Hovey was directed by the board to institute a series of regular personal visits to all projects that were then being funded by the Engineering Foundation. This policy would remain in effect until 1959.

Like Flinn before him, Hovey was faced with the reality that the funds that were available to the Engineering Foundation to support projects were being seriously reduced. At the June 15, 1939 board meeting he stated, in response to a comment made regarding the lesser amount of grant recommendations of the Research Procedure Committee compared with the amounts that were requested in the grant applications, that these reductions had no bearing on the relative value or merit of these projects, but represented solely an attempt to spread the available funds as far as possible to conform with the somewhat reduced income of the Foundation. However, funding cutbacks continued to be necessary. In 1940 George B. Waterhouse, chairman of the "Alloys of Iron" project, agreed to continue in his position without a salary. This financial crunch was also responsible for the Engineering Foundation's decision to temporarily suspend all paid professional efforts to seek publicity for its activities.

Financial conditions also induced the Foundation to further modify its grants policy. At its June 20, 1940 meeting, the board expressed concern that it had no way of knowing the extent to which projects that it funded were receiving outside financial support. In response, it voted to add the following phrase to its grant contracts:

> In order that the Engineering Foundation might be better informed concerning outside assistance to its projects, a statement of income other than that received from the Engineering Foundation shall be submitted.

This new policy was designed to insure that the Engineering Foundation's limited funding would only be directed toward those projects where it was most needed.

The combined effects of changes in the Engineering Foundation grant policy, such as the emphasis on cooperative funding and the establishment of contingency levels of outside financial support, were summarized by Director Hovey at the board meeting of October 17, 1940:

> Engineering Foundation funded projects for 1940 totalled $91,373.66 which did not include the estimated value of in kind services and outside funding which equaled three times this amount.

The grant policy was once again changed at the March 19, 1942 board meeting when it was decided that all future patents, resulting from Engineering Foundation funded research should be dedicated to the public and that the Engineering Foundation could not accept any division of profits or royalties that may be derived from such patents.

Another problem that caused a change in the Engineering Foundation grant policy occurred in 1945 when the Welding Research Council claimed to be a formal part of the Engineering Foundation as a means of escaping a governmental regulation. This action caused the board on February 24 of that year to sharply limit the scope of its fiscal activities on behalf of the Welding Research Council and declare that this entity and all such future research councils would be considered to be completely independent of the Engineering Foundation.

A major factor in the continued evolution of the Engineering Foundation grant policy was a comprehensive re-evaluation of the objectives of the Foundation. In response to both external evaluation and internal self-examination, the board at its October 16, 1947 meeting decided that the Engineering Foundation's emphasis should be placed on funding broad fundamental projects rather than minor specific projects; and care should be taken in the future to avoid assuming long-term commitments such as had been done in previous years with such projects as the "Alloys of Iron Research" and "Welding Research Council." This policy was further refined at the March 3, 1945 board meeting when it was decided that

> the financing of promising projects until the work shows its value to attract industrial contributions has been considered to be the function of the Engineering Foundation. It acts as a 'catalyst' which starts needed research that would not otherwise be begun. Many subjects are of too broad and fundamental interest to attract one competitor. After the project is active and productive, competitors will and do pool their resources to keep it going.

4. CONTINUED SUCCESS OF RESEARCH GRANTS

The success of Engineering Foundation grant support for research projects continued to be evident during the 1930s and 1940s. Some of these projects would evolve into multi-year commitments and a few of them, such as "Lubrication" and the "Alloys of Iron Research" would span periods of more than a decade. In other cases, specific research grants would provide the basis for the Engineering Foundation's support for new fields of research. An example of this type of development is the $5,000 that was allocated during fiscal year 1930–31 to support AIEE sponsored research program on the properties of electric welding. This project was a precursor for the Engineering Foundation's later involvement in the "Welding Research Council."

An important aspect of the continuing success of the Engineering Foundation as a funding organization was provided by the growing awareness that it should be considered as the joint instrumentality of the Founder Societies for research coordination and funding. This concept was initially stated by Director Flinn at the board meeting of February 19, 1931.

Interdisciplinary Projects

A project "Study of Transportation" (Project 51 in Table 2) developed the preliminaries of a program that was to be undertaken by an impartial body. This project would deal with engineering and economic aspects of all forms of transport in the United States. The project was carried to the stages of planning, exploring and appraising sources of data and of tentative evaluation of the results in a few sample studies. During this preliminary examination it became evident that any such transport research must embrace consideration of the social aspects of industry, with which transportation is inextricably involved. Due to a lack of resources and various complications arising in this vast field, the project was cancelled.

Another interdisciplinary project was the "Engineering Index." The Engineering Index furnished to its subscribers a technical bibliographic service (Project 75 in Table 2). The staff, being located in the Engineering Societies Building, was able to use to the fullest extent the library there. It could also avail itself of the facilities of the other technical and scientific libraries in the city of New York. However, during World War II the European subscribers cancelled and limitations were placed upon the business of the Index in other countries outside of the United States. As a result, the management of the Index requested contributions from various sources to assure its continued and effective operation. The Engineering Foundation granted to it $500 for the year 1939–40 and $2,500 for the year 1942–43. In 1959, the Index requested a grant of $59,000 from the Foundation. The request was tabled pending formulation of a new Engineering Foundation policy on the use of its income.

Immediately prior to World War II, new and varied uses of plastics were proposed. As a result, it was essential to both the engineers and the public that the physical and mechanical behavior of plastics be subjected to thorough investigations. The Engineering Foundation contributed to this effort through its support of a project chaired by Professor B. J. Lazan of Pennsylvania State College (Project 81 in Table 2). Beginning in June 1942, the project was divided into three programs each of which had important bearing on war activities:

1. Development of a new, small capacity (+/- 3000 pounds) compact direct stress fatigue machine, especially suitable for testing structural plastics.
2. Static and dynamic creep tests of a canvas laminated phenolic plastic. The same material was used for wing tabs in the Martin B26 bomber.
3. Experimental work on the dynamic testing of rubber.

These programs were followed up by a comprehensive study of the structural behavior of a completely assembled small plastic-plywood trainer plane including non-destructive testing of the assembled plane and of its critical components, along with the destructive testing of several critical component parts such as the wings and the propeller. The Engineering Foundation sponsorship was limited to the original 1942 grant.

In October 1948 the Engineering Foundation, in response to a request from the Library Board, granted $5,000 to make a survey of the Engineering Societies Library (Project 100 in Table 2) to determine if possible the position of this library in comparison with other technical libraries of the United States. This study would also seek ways of increasing the value of the library and the services it rendered to the members of the founder societies and to engineers generally. At the same time it would explore possibilities of increasing the library's revenue. Richardson Wood, who had made a similar survey for the Crearar Library in Chicago, was employed to do this work. The survey was started on November 23, 1948, with a review of the services and facilities of the library. Wood then interviewed members and secretaries of the founder societies, editors, public librarians, special librarians, university librarians, heads of large corporations, representatives of foundations, of Engineering Index and other organizations. On the basis of these interviews, he prepared and mailed a sample questionnaire to one hundred engineers. From returns that he received, Wood prepared two other questionnaires, both of which proposed extensions and changes in the services of the library. One of the questionnaires was mailed to a sample of about 3,500 engineers taken from the memberships of the founder societies. The other was sent to the presidents of some 500 large industrial corporations. The percentage of returns was excellent. The results were tabulated and appraised and the final report was submitted to and accepted by the Library Board on the recommendation of the Library Study Committee on June 17, 1949.

An investigation of causes and methods of prevention of corrosion of water pipes (Item 101 in Table 2) originated in the Sanitary Engineering Division of ASCE. It was approved by the board of direction of that society in April 1949. The objectives were (1) the formulation of a comprehensive research program on the fundamental factors influencing the internal corrosion of water pipelines and (2) the application of the knowledge thus obtained to the development of remedial measures by the sanitary engineering profession. The end result would be a lowering of the large economic losses due to such corrosion. The organization of the project was delayed by the necessity of making a survey of other work that had been done in this field. During 1951 some funds were secured from industry, and these together with a small grant from the Foundation enabled the project to be started by December. The project was initially sponsored by ASCE from October 1951 to October 1953. After that, it was sponsored by the Foundation as an independent project at MIT under the direction of Rolf Eliassen. The project investigated corrosion protection through the addition of inhibiting chemicals to potable water supplies. On the assumption that an increased supply of oxygen increases the corrosion rate within the limits of oxygen concentrations found in natural waters, definite conclusions were drawn from theoretical fluid mechanics concerning the relation of corrosion rate to flow velocity (a) for flat plates in laminar flow, (b) for rotating disks in an infinite fluid, (c) for smooth pipes in turbulent flow, (d) for rough pipes in laminar flow, and (e) on the importance of Reynolds numbers as a criterion of similarity in corrosion processes. It was further concluded that the nature of the corrosive media may produce different effects when the velocity is changed, and that this behavior may be attributed to changes in the solubilities of metallic salts at different pH levels. Finally, velocity and size of the pipe affect the corrosion rate by changing the rate of supply of oxygen and other chemicals in the water.

Civil Engineering Projects

Structures

The investigation of the plastic flow of concrete (Project 41 in Table 2) was derived from the earlier "Arch Dam Investigation." During the construction of the experimental dam on Stevenson Creek in the California Sierras in 1925 and 1926, numerous concrete cylinders were made in order to determine the properties of the concrete that was used in the dam. The fifty-eight cylinders that were made of the dam's concrete were subjected to long-term tests at the University of California at Berkeley to determine their shortening with time under constant load. During the following years, many other cylinders were made and subjected to constant loading to investigate the effect of numerous variables such as the magnitude of sustained stress, the age of concrete at time of loading, and the moisture conditions of storage. By September 1941 when the Engineering Foundation's support of this project was terminated, a total of thirty-six separate test series were underway or completed. Each series was devoted to the studies of the effects of a few distinct variables on the characteristics of plastic flow of concrete. The results of these experiments were reported in nine papers that appeared in ASCE, ACI and ASTM publications. Perhaps the most interesting observation was reported in 1937 thusly: "After ten years of sustained compressive stress, plastic flow is still increasing at a measurable rate. In the extreme case the total flow over a ten-year period has amounted to more than 1.2 inches per 100 feet."

In 1955 a search was initiated to find alloy steels or structural quality steels which gave the promise of higher fatigue strength than the generally used low carbon steels when tested under conditions comparable to structural applications (Item 121 in Table 2). Considering the large number of alloy steels available, full-scale testing of even an intelligent selection would require an exorbitant investment of time and money. The solution of the problem was, therefore, to develop a small-scale type of specimen which could be used to evaluate a large number of steels within a reasonable time period and which could also be used to study any type of experimental steel. Tests were carried out on A-7, A-242, A-373, and T-1 steels with a notch configuration previously developed for a butt-welded joint. Where test results were available on full-scale joints of the same material, the results of the small scale samples were in agreement, with one notable exception. The results on T-1 steel indicated a much higher fatigue resistance than that obtained on actual welded joints. However, it was found that heat treatment of the small-scale specimen markedly lowered the fatigue resistance of T-1 steel.

At the recommendation of the Structural Division of ASCE a grant was awarded for a study of systemization of calculations that were utilized in structural analysis problems (Item 132 in Table 2). The major purpose of the research was to formulate equations that would describe the behavior of framed structures in such a way that high-speed electronic computers could be used to solve them. The main contributions of this work were (a) the development of the ability to analyze and design more complex and larger structures with greater efficiency and accuracy, and (b) to relieve the engineer from the tedium of routine calculations and thus free him for more creative work.

Projects 47 and 73 in Table 2 also deal with structures.

Fluids

The ASCE Committee on Hydraulic Research was organized in 1936. During its initial twelve years it was sponsored and partly financed by the Foundation (Item 67 in Table 2) as it carried out sixteen different projects of varying size and importance at several universities. When the Foundation's support ended on September 30, 1948, four of these projects were unfinished: "Studies of the Mechanics of Movement of Sediment Along Stream Beds" at the University of Minnesota, "Motion of Oscillatory Waves in Open Channels" at MIT, "Aerodynamic Properties of Structural Forms" at the Iowa State University and "Stability of Flow Stratified Due to Density Differences" at MIT. The large amount of data accumulated during the progress of these research investigations was presented to various hydraulic conferences and published by the American Society of Civil Engineers.

A study of the origin of storm surges (Item 118 in Table 2) and an attempt to forecast such phenomena was made at Columbia University's Lamont Geological Observatory under the auspices of AIME with Engineering Foundation support. If a warning of only a few hours could be given of an approaching freak wave or other abnormal tidal condition, the danger that these surges present to life and property could be greatly reduced. Two varieties of storm surges were investigated. An example of the first type was the disastrous June 26, 1954 Lake Michigan Wave in which there was a very rapid increase in water level for a short time. The other type was the slower rise of water level over a much larger area, particularly in coastal regions. The East Coast surge of November 6–7, 1953, was an example of this second type of surge. Research on a pilot program to determine whether damaging storm tides along the eastern seaboard could be forecast on an empirical basis was eventually completed. The results of this research indicated that such forecasts were possible, and forecast graphs were prepared which showed a useful relationship between wind velocity and storm tide. Initial studies of storm surges on this project were confined to Lake Michigan. The conclusions from these studies have been confirmed through a study of surges on Lakes Huron and Erie.

In 1947 the board of direction of ASCE requested Foundation sponsorship and financial support for the development of methods and instruments to determine characteristics of turbulent motion in water (Item 95 in Table 2). The object of this project was to develop (1) a suitable pressure transducer and associated instruments for the accurate and instantaneous determination of static and dynamic water pressures, with an accuracy maintained within plus or minus one percent and a dynamic response that is linear to pressure variations of at least 1,000 cycles per second; and (2) a practical technique for the measurement at a point of the average and of the instantaneous velocities, with an accuracy of plus or minus one percent and a response to pressure variations up to 1,000 cycles per second. An analysis of the four basic types of existing pressure transducers indicated that only the resistance wire strain gage and the variable capacitance types were suitable for the applications in question. Both the strain gage transducer and the capacitance gage were further developed and found to be suitable for the intended purposes.

Projects 72 and 120 in Table 2 also deal with fluids.

Soils

The study of mechanics of underground forces, which was begun in 1942, concerned the application of mechanics of soils to the solution of a problem in subsidence (Item 79 in Table 2). Before any substantial data had been obtained, it became evident that satisfactory work could not be carried on due to a wartime lack of competent research personnel. The project was cancelled.

The investigation "Highway Aggregates and Aerial Photographic Approach to the Determination of Soil Patterns," (Item 93 in Table 2) was carried out in 1948. This project was designed to examine and report on information that could be derived from an examination of aerial photographs and which could be expected to contribute materially to the intelligent planning of civil engineering projects. Unfortunately it is not know what benefits resulted from this project.

The Engineering Foundation, various industrial companies, and trade associations interested in the properties and uses of insulated cables organized a study of the thermal resistivity characteristics of soils (Item 116 in Table 2). The first stage of the program correlated all of the known information on (1) heat conductivity of soils with the granulometric composition, secondary structures and moisture content, (2) physical and chemical characteristics of water in soil under thermal potentials and (3) seasonal variations in soil moisture and soil structure and their influence on heat and moisture conduction in soils of various types. The second stage of the program involved laboratory work and the construction of field installations.

An investigation of explosive wave propagation and its effects on rock fragmentation was carried out at the Colorado School of Mines (Item 125 in Table 2) in 1956–58. It was found that the velocity of detonation and strength made little difference in the waves caused by confined drill-hole shots. However, with plaster shots a higher strength produced a larger compression-pulse amplitude, while the velocity of detonation produced no noticeable effect on the amplitude. With plaster shots, the magnitude of the compression pulse increased in almost direct proportion to the weight of the charge. It was also shown that the wave effect from charges detonated with extremely short delays can be calculated using the principle of superposition. On a free face, at equal distance from two charges, the largest seismic waves are obtained when the holes are blasted simultaneously.

Electrical Engineering Projects

Long experience has shown that the best materials for the insulation of high voltage cables were pure wood-pulp papers which had been impregnated with mineral oil. This combination met the extreme requirements imposed by the processes of manufacture, installation, and operation of such cables. This research on the stability of impregnated paper insulation (Item 66 in Table 2) started in 1936, was directed especially to the conditions bearing on the life of the insulation, particularly in the presence of the high stress which was imposed by the operating voltage. Of first importance in these studies was the development of a new method for detecting the first approaches to failure or breakdown. The application of this method to other studies in which the characteristics of the oil and of the paper separately and together have been varied, has led to important new indications by which the life and dielectric strength of the insulation may be increased. In particular, the analysis of the variation of the density of the paper has highlighted an important and hitherto unsuspected influence on the dielectric strength of the assembled insulation. Furthermore, new knowledge has been gained of the impregnating process as related to the properties of the oil and of the factors that caused complete failure. These findings and other results have all indicated directions in which the stability, life, and dielectric strength of the cable may be increased.

The factors which influenced the electrical stability of mineral oils under limited oxidation conditions were particularly important in the practical application of oil-paper insulation (Item 74 in Table 2). While a complete solution of the problem was not attained, a number of factors influencing the stability of mineral oils were determined and a complete system of tests, together with the necessary equipment, was developed. These studies led to the unexpected finding that the greatest amount of electrical deterioration is not produced by large or unlimited amounts of oxygen contamination but rather by comparatively small amounts of the order of eight percent of the oil volume. This finding was of practical importance as it proved that the greatest degree of electrical deterioration of oil was produced by amounts of oxygen of the order of magnitude which occur in sealed electrical equipment filled with undegassed oil such as transformers. Dielectric losses of the oil were of some importance for transformers, and they were a vital factor in cable operation both from an economic and from a technical standpoint. Two additional findings were important from a practical point of view: (1) the degree and type of refining greatly influence the electrical stability of the oils, and (2) certain addition compounds can very greatly influence the characteristics of an oil.

To obtain more precise information of vital interest to wooden pole producers, treaters, and users, a project was developed and partially funded by the Foundation in which approximately 600 wood poles and 18,000 small specimens were tested at the Forest Products Laboratory (Item 109 in Table 2) during the period 1954–59. The tests included five species of wood in both the untreated condition and pressure treated with preservatives. The tests reduced the ignorance factor in pole design and pole safety, provided more accurate means of rating poles for strength, and established a method of quickly and economically establishing the actual strength of new species of wood for poles. In addition to providing authoritative data for design, the tests resulted in a modification of the standard testing procedure for wood poles.

The Foundation also partially funded the study of the dielectric properties of gases, and of insulating liquids and solids, at high electrical gradients and at audio and radio frequencies which was carried out at Stanford University (Item 111 in Table 2) in 1954–56. First a voltage in the range of 1 to 111 kv, with a frequency span of 10 kc to 100 mc was produced and measured. Next, within these voltage and frequency ranges, the breakdown of air was investigated. Variables in this second phase included a gap configuration and spacing, and air temperature, pressure, and humidity. The study was then extended to other gases and to liquid and solid insulators.

Items 42, 49, and 69 in Table 2 are also electrical engineering projects.

Mechanical Engineering Projects

Fasteners and Bearings

A grant (Item 60 in Table 2) was made to the main research committee of ASME for the publication of a bibliography entitled "A Critical Review of the Literature Concerning Riveted Joints." This bibliography proved to be very useful to structural engineers.

A study of rolling friction was begun in 1945 because the available data on this subject were judged to be inadequate for future applications (Item 89 in Table 2). The study included (1) preparation of a comprehensive bibliography and (2) determination of rolling friction values from tests on mine car wheels that had been conducted earlier at the Bureau of Mines Experiment Station in Pittsburgh. At low speeds, the kinetic rolling friction was found to be less than static friction, but it increased with greater speed. Static rolling friction increased with the time elapsed while standing still. Kinetic rolling friction seemed to be slightly greater at the beginning of a run. Rolling friction decreased with increasing hardness, but increased with surface roughness.

The Engineering Foundation Board approved at its October 1954 meeting a grant of $2,000 to the ASME Research Committee on Mechanical Pressure Elements. The grant was to be used for organizational purposes in connection with an investigation of the design of Bourdon tubes (Project 107 in Table 2). A research program, including diaphragms and bellows in addition to Bourdon tubes, was also formulated, and five subcommittees were appointed—an Advisory Committee, a Finance Committee, two technical committees on Diaphragms and Bourdon tubes, respectively, and a Bibliography Committee. The overall objective of this project was to develop theoretical analysis and to compile and correlate empirical data relating to the design of mechanical pressure elements. The Subcommittee on Diaphragms also organized material for a technical session and arranged for presentation of a number of papers on diaphragm technology. Included in these technical sessions were a paper on the dynamic characteristics of diaphragms and a presentation of findings concerning the use of electronic computers to predict diaphragm characteristics, as well as a paper on the solution of diaphragm design problems by empirical methods. The Subcommittee on Bourdon Tubes supervised the conduct of a research project at Oklahoma State University on the characteristics of Bourdon tubes. This project was completed and a final report was issued in August of 1958. It provided an experimental check of the adequacy of certain theoretical treatments of Bourdon tubes and it also contained recommendations for future work.

Projects 44 and 143 in Table 2 also deal with fasteners and bearings.

Material Characteristics

In 1924 there was such a persistent demand for authoritative information on the high-temperature properties of metals that the American Society of Mechanical Engineers and the American Society for Testing Materials jointly sponsored a symposium on the effect of temperature on metals. The importance of elevated temperature properties was so evident that a research committee with members from both societies was quickly established. This joint committee was probably the pioneering group in arranging for concerted action in this field (Item 45 in Table 2). Since then this field of research has expanded greatly and it eventually also encompassed the field of low temperatures. From a relatively small group of a dozen members, the committee personnel grew into a group of more than one hundred of the country's leading authorities. Many research projects were underwritten by the committee with funds that it raised and, in addition, much work was carried out by committee members for which no appropriations were necessary. The Foundation sponsored the work of this committee and has supplied it with funds sporadically since 1930. The joint committee did a herculean job in devising reliable methods of testing metals at high and low temperatures; in determining the effect of these temperatures on the properties of a wide variety of metallic materials, especially steel; and particularly in making available, by preparing compilations, holding symposia, and in other ways, a vast amount of data of interest to engineers everywhere.

In 1934 a grant by the Engineering Foundation was used to assist in financing the preparation of a treatise entitled "Strength of Metals with Special Reference to Spring Materials and Stress Concentrations," by D. J. McAdam and R. W. Clyne (Item 59 in Table 2). Planned for publication in a single volume, the work included five parts: Part I—Elastic Strength and Modulus; Part II—Resistance of Metals to Fracture; Part III—Theoretical Stress Concentrations; Part IV—Fatigue of Metals, with Special Reference to Stress Concentrations; and Part V—Properties of Metals Available for Spring Materials. This project reappeared in 1944 (Item 85 in Table 2) when the Foundation contributed $500 toward the completion of the manuscript. It is not known whether the material was published or remained in the manuscript form.

The ASME Research Committee on Plastic Flow of Metals was appointed by the ASME Council in 1936 to study plasticity in the particular field of rolling steel (Item 68 in Table 2). This objective was later rephrased as follows: "Through study of the phenomena associated with the plastic forming of metals, to develop analytical or empirical relations which will provide a more complete understanding of the plastic forming processes, and which will assist both the design and the operation of equipment used for these processes." Since 1936 work on the rolling of metals has been going on almost continuously at the Massachusetts Institute of Technology. In addition, the committee sponsored and financed work on the flow of metal strip through circular dies at Case Institute of Technology, and on sliding friction, first at the research laboratories of Westinghouse Electric Company, East Pittsburgh, Pennsylvania, and later at the Illinois Institute of Technology. Much time and effort at MIT was spent in designing and constructing the special equipment that was required such as rolls, stress-measuring devices, torque meters, and other apparatus needed to determine contact stresses under carefully controlled rolling conditions. By April 1950 most of the operating difficulties had been eliminated, and a large number of tests followed. The data obtained from these tests were plotted to show the relation between normal roll pressure and the contact arc as dependent on the line of roll centers. The resulting curves showed the effect on pressure in the contact arc of the following variables: reduction in rolling, speed of rolling, lubrication, lubrication and speed, lubrication and transverse location, and material. The Engineering Foundation supported the project from 1936 to 1951, when the experimental work undertaken was practically completed. The final report summarizing the results was not finished until early in 1957. The Committee applied to the Foundation Board for a grant of $1,500 to publish this report and to distribute it to the contributing sponsors. This sum was granted, and the report "Rolling of Metals" was published during the first half of 1958.

To obtain new information on the relationship between internal friction and creep of metals, a research project was organized in 1940 at Washington State College at Pullman (Item 77 in Table 2). The work was greatly retarded due to the principal investigator's war effort work. The final disposition of this project is unknown.

Projects 61, 96, 129, and 142 in Table 2 also deal with material characteristics.

Steam

In 1931, following the installation of a steam-generating unit at Purdue University for operation at pressures up to 3500 psi, a special ASME Research Committee on Critical Pressure Steam Boilers was organized to guide and assist in research work related to steam generation at these high pressures (Item 50a in Table 2). It soon became apparent that the major problems associated with the use of steam at such pressures centered about the superheater and were concerned primarily with high steam temperature. The high-temperature properties of metals were then being studied by a joint ASME-ASTM committee. The new group undertook to investigate the general problem of the behavior of alloy steels in contact with high-temperature steam, especially from the standpoint of corrosion. During the following ten years, a number of papers dealing with the corrosion resistance of alloy steels in steam atmospheres at temperatures up to 1,800 deg F were published in the *Transactions of ASME*. This work was financed jointly by the Engineering Foundation, Purdue University, and the manufacturers and users of steam-generating units, high-temperature piping, and steam turbines.

ASME studies of steam were started at Purdue University in 1931, but they became dormant during the war years. These studies were reactivated in 1951 when the supervisory committee was reorganized and its title changed to the ASME Research Committee on High-Temperature Steam Generation. An extensive research program was outlined, and approximately $175,000 in cash was contributed by industry and by the Engineering Foundation (Item 50b in Table 2). There were also large contributions in services and materials from manufacturers and a very substantial contribution in design and fabrication services from the United States Navy through its Bureau of Ships. The principal objective of the committee was to determine performance characteristics of selected ferritic and austenitic alloys when exposed to the action of steam and to the products of combustion at temperatures higher than those encountered in modern power-plant operation. Four factors were investigated: (1) the nature, thickness, permanence, and thermal conductivity of the oxide film on surfaces swept by steam at temperatures between 1,100 and 1,500 deg F; (2) the

Equipment for high temperature steam generation tests, assembled for installation at Philip Sporn power plant.

resistance of selected alloys to the products of combustion of typical fuels; (3) the metallurgical stability of the alloys over long periods; and (4) the effects of repetitive temperature shock. The investigations were conducted at Purdue University, the U.S. Naval Experiment Station and the Philip Sporn power plant.

In 1944 the ASME Research Committee requested support for John R. Weske—at that time professor of Aerodynamics, Case School of Applied Science, but later occupying the same position at Purdue University—in the preparation of a book covering comprehensively fluid dynamics as applied to turbines and compressors (Item 87 in Table 2). The rapid technical development of turbines and compressors and their wide use made most timely the preparation of the proposed book presenting the basic dynamic theory. No information was found regarding the final disposition of the results of this project.

An extensive survey covering industrial companies active in the fields of power, refrigeration, chemical processing, and transportation had shown that there was definite dissatisfaction with knowledge regarding the thermodynamic properties of even the common gases such as oxygen, nitrogen, argon, hydrogen, helium, carbon monoxide, carbon dioxide, and water vapor. This survey also disclosed substantial interest in the thermodynamic properties of a large list of refrigerants, hydrocarbons, propellant gases, and dissociation products which have not been subject to systematic investigation. Attention was called to the almost complete lack of reliable information regarding the thermodynamic properties of gas mixtures. Finally, this survey revealed that a knowledge of the so-called non-thermodynamic properties, such as viscosity, thermal conductivity, diffusivity, and emissivity, was also urgently needed. The development of plans for an investigation of properties of gases and gas mixtures followed (Item 91 in Table 2). The investigation was carried out in three separate phases: (1) experimental determination of viscosity and thermal conductivity of the common gases in the high-temperature, high-pressure range; (2) experimental determination of thermal conductivity of gases at low pressures; and (3) the assembly and critical analysis of existing data, and development of skeleton tables with recommended tolerances and the like on an international scale. The experimental work was performed at Columbia University.

The art of steam generation and use has progressed well beyond the pressure and temperature limits of the investigations of the thermodynamic properties of water and water vapor, which constituted the basis of the international steam tables of 1934. The use of very high pressures and temperatures has since become common not only in steam power generation, including nuclear power, but also in the broad field of chemical processing. For some time the need for extended steam table data was so great that a number of industries found it necessary to prepare new tables by extrapolation. In response to this need a conference was held in September 1954, at which representatives of seven industrial nations in addition to the United States were present. The ASME, having reestablished its activity in this field in October 1953, was invited to serve as the secretariat of the international project, "Properties of Steam" (Item 113 in Table 2). The ultimate objective was to extend the 1934 tables to 1,500 deg F and 15,000 psia. How the scope of the program compared with the then available data is shown below. The basic experimental work was carried out at three institutions. Investigations were made at the California Institute of Technology of the Joule-Thompson coefficient and specific heat of steam; at Brown University of the viscosity of steam by the oscillating disk method, and of the speed of sound and relaxation rate in steam at high pressures and temperatures; and at Georgia Institute of Technology of the thermal diffusivity of steam by a new method of direct measurement of diffusivity. Another project at the last named school was concerned with measuring the viscosity of steam by the annular flow method.

Projects 53 and 108 in Table 2 also deal with steam.

Processes

During the 1930s, the University of Tennessee and the Tennessee Valley Authority coordinated joint experiments to increase the yield of cottonseed processing, to develop continuous processes in lieu of batch operations, and to adapt the cottonseed processing equipment for making peanut oil, linseed oil, and soybean oil (Item 52 in Table 2). Furthermore, in cooperation with the Mississippi Agricultural Experiment Station a simple and inexpensive method was developed for the recovery of protein prior to the oil extraction. Other experiments included the development of a plastic composition using cottonseed hulls and decortication of cottonseed by steam explosions. The results of this project proved to be of great commercial value.

The ASME Research Committee on Metal Cutting Data and Bibliography revised and compiled the "Bibliography on Cutting of Metals" (Item 58 in Table 2). The bibliography was published by ASME in the 1940s. To further supply up-to-date engineering information in the field of metal cutting to those who are employed in the mechanical manufacturing industries, shortly after the Second World War the ASME special research committee on metal cutting data and bibliography decided to revise the *Manual on Cutting of Metals* which was developed by the predecessor of this committee and published by the ASME in 1939 (Item 63 in Table 2). In view of the progress made in the ensuing years in the field of metal cutting the material contained in the first edition of the manual needed to be revised and enlarged. The revised manual was published by ASME in 1952.

The ASME Research Committee on Furnace Performance Factors

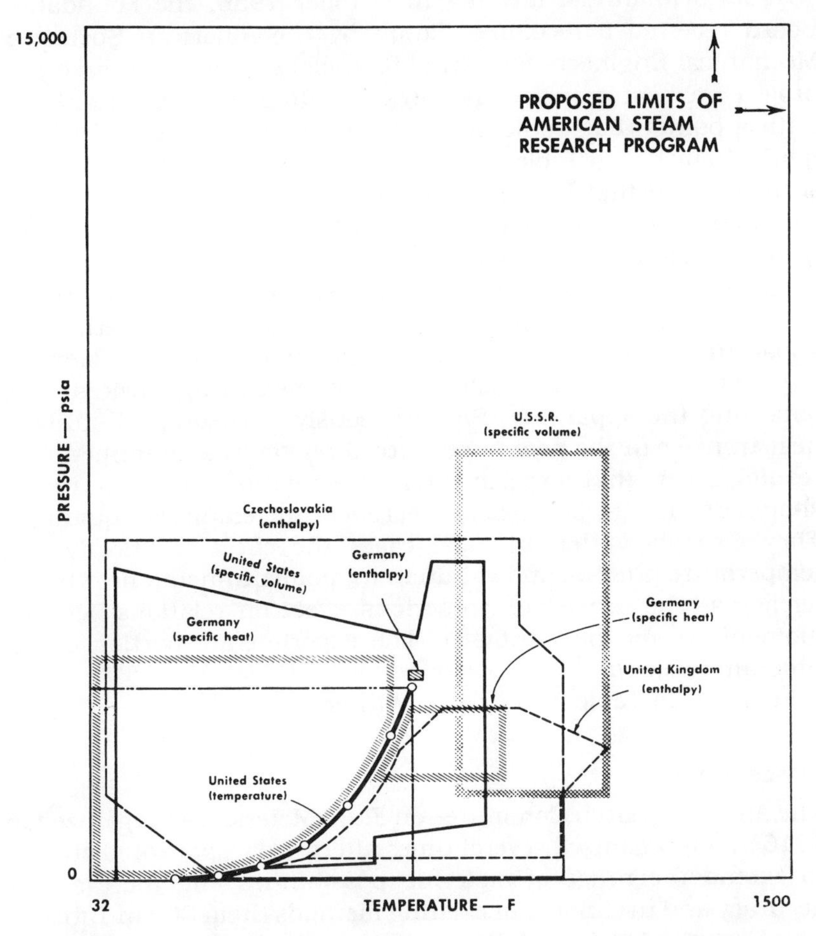

Pressure and temperature data for steam in 1954 and the proposed limits for Project 113.

was established in 1941 to develop and sponsor cooperative boiler-furnace investigations with the purpose of discovering heretofore unmeasured factors that influence furnace performance and design (Item 82 in Table 2). Its program was prompted and made possible by the mutual interest of boiler users and manufacturers in eliminating inefficient heating surfaces, and in pointing the way toward complete combustion within the least volume. The first phase dealt with pulverized-coal-fired furnaces of large capacity. Methods were established of measuring the amount and distribution of heat absorbed in such furnaces, and methods of measuring furnace-gas temperatures were improved.

At its semiannual meeting in October 1956, the Foundation Board received a recommendation by the American Society of Mechanical Engineers for partial financial support of a three-year project on heat transfer and pressure drop of fluids near their critical points that would be conducted at the University of Oklahoma (Item 124 in Table 2). It was acknowledged by those familiar with the field that little high-pressure heat-transfer work of any sort had been performed and that a fundamental study of heat transfer and pressure drop of fluids in the supercritical region near the critical point was urgently needed. The first step in the investigation was concerned with the design of the experimental apparatus for heat transfer, the design and construction of auxiliary apparatus for the calibration of measuring devices, and obtaining the apparatus. Simultaneously, a theoretical study of heat transfer in the near supercritical region was performed. The results show that experimental investigations in this region should provide the best test of theoretical and empirical equations. They also show that in most cases the effect of velocity and temperature profiles, and variation in momentum and heat transfer across the tube, had no serious effect on relations between momentum and heat transfer. The experimental portion of the program followed.

Item 128 in Table 2 also deals with processes.

Miscellaneous

The ASME Research Committee on Fluid Meters, first organized in 1916 and reorganized several times afterwards, sponsored studies to extend the range of metering possibilities and increase the accuracy and usefulness of existing methods (Item 30b in Table 2). The Engineering Foundation was first approached for financial support in 1927. This was granted and has been renewed repeatedly through 1968. Several subjects were covered by these studies over the years. Flow nozzles, volumeters, piping, discharge orifices, variable area meters, and hydrocarbon gas laws were among the principal areas of work.

The determination of thermal conditions during the quenching of steel was one of several projects that were planned for the Heat and Mass Flow Analyzer Laboratory at Columbia University under the broad title "Unsteady Heat Flow" (Item 80 in Table 2). The Heat and Mass Flow Analyzer Laboratory was established at Columbia University in recognition of the growing technical importance to mechanical engineers of heat flow in solids, particularly heat flow under unsteady conditions of temperature. This included heat flow in furnace walls and structures, heat flow through building walls, heat flow in heavy metal parts and structures, particularly while in the process of manufacturing (heat treating, for example), and heat flow conditions during welding. The procedure adopted in this particular research was based on an electrical analogy and differed from conventional thermal investigations in that electrical current and voltage take the place of heat flow and temperature, with electrical capacitance replacing heat storage values in unsteady-state heat flow studies. By this procedure it was possible to study the various phases of transient heat flow including time-temperature relations in solids. Events of hours in actual thermal duration could be compressed with graphical representation into the space of minutes so that interpretative analysis could readily be made. The experimental procedure thus operated as an electrical 'brain,' carrying out operations that hitherto have only been possible by tedious mathematical or graphical analysis. In 1950, upon recommendation by AIME, the Foundation made a grant to this project of $3,000 for the fiscal year 1951–52, contingent upon $15,000 being contributed by industry and others. This contingency was readily met, and the project got under way. The investigation solved some of the problems inherent in the well-known end-quench test and secured basic data on heat transfer during the quenching of steel.

The establishment of a monthly digest magazine of applied mechanics to take the place of a similar German periodical which suspended publication during World War II was authorized by the Board of Directors of ASME in 1947. An initial grant of $10,000 by the Foundation to aid in establishing the journal, was approved in October 1947 (Item 94 in Table 2). The first issue appeared in January 1948. In establishing *Applied Mechanics Reviews*, ASME had the cooperation of the Society for Experimental Stress Analysis, the Institute of Aeronautical Sciences, the American Institute of Physics, the American Mathematical Society, the American Society of Civil Engineers, The Engineering Institute of Canada, the Institution of Mechanical Engineers (Great Britain), the Illinois Institute of Technology, and the Office of Naval Research, which contributed funds to aid the journal during its first year. The Office of Air Research of the army and the Midwest Research Institute became sponsors during the third year. The Institute of Danish Engineers became sponsors in 1951. For two years the editorial office of Applied Mechanics Reviews was at the Illinois Institute of Technology in Chicago. On October 1, 1950 editorial responsibility was transferred to the Midwest Research Institute. By July 1, 1954 the accumulated deficit for the journal was $69,418, which was met by grants from an ASME development fund. Paid circulation reached approximately 3,000.

The need for more data on the corrosion, and deposits from gases of combustion had been apparent when, in 1954, ASME appointed a research committee to study the field and make recommendations for a research program (Item 114 in Table 2). The Foundation was first approached in 1954 for a grant of $1,000 for purposes of organization. This was approved by the board. In November 1956, the committee, after limiting its previous activities to cosponsoring papers on subjects of interest in its field, launched a campaign for solicitation of funds to finance a research project planned to be undertaken at Battelle Memorial Institute. The contract with Battelle was signed and the work started in 1957.

Additional miscellaneous mechanical engineering projects are included in Table 2 as items 55, 76, and 134.

Mining and Metallurgical Engineering Projects

Mining

Most problems involving underground and open pit mining and numerous civil engineering problems involving the behavior of earth and rock strata were solved on the basis of past experience or by rule of thumb methods. It was thought quite impractical to obtain even approximately accurate data for a scientific approach to their solution. At Columbia University, the idea was conceived of utilizing scale models of the same material and subjecting them to such forces in the laboratory that it would result in the model behaving in a manner similar to that of the full-scale structure in the field (Item 43 in Table 2). This was accomplished by substituting a centrifugal field of force for the gravitational one. Special centrifuges were designed and built so that models may be used to simulate conditions at depths up to 2000 feet when rotated at the proper speeds. Means were also added for observation of the model under test, for photographing model behavior and for using photoelastic equipment in conjunction with it, so that the centrifuge and accessory apparatus bear the same relationship to the solution of problems pertaining to heavy structures, as the test tube to chemical problems, and the wind tunnel to problems in aeronautics. The ground work had been laid for an approach to and a solution of many problems concerning the behavior of weighty structures, especially those composed of earth and rock strata. The following are some of the problems that were investigated: stress distribution in mine pillars and roofs; the development of artificial mine supports; effect of artificial support in longwall support; and time effects in the behavior of rock structures stressed beyond the elastic limits. The name of the project 'barodynamics' was coined to indicate that it dealt with the mechanics of heavy, weighty structures.

Comminution, or the mechanical size reduction through fine grinding of brittle crystalline materials, was one of the most important operations in industrial metallurgy. It was a large consumer of power and of many basic materials. Despite long experience with comminution, most mining and metallurgical engineers considered that fine grinding was only about five percent efficient. This inefficiency of fine grinding represented a large energy loss; increasing the efficiency was a fertile field for research. Much work was done on the mechanisms underlying fine grinding, but one field that had been largely neglected was that of the surface energies of minerals, millions of tons of which were ground annually. To make a start in filling this important gap, the AIME Committee on Research recommended to the AIME Board of Directors that a Foundation grant of $1,000 be made for improving and adding to equipment already installed at the metallurgical laboratories of the School of Mineral Sciences, Stanford University. The AIME Board gave the project its formal approval, and a grant was made by the Foundation to begin the work (Item 104 in Table 2). The grant was used for the construction of a calorimeter for the accurate determination of surface energies in brittle crystalline solids. The project was discontinued after only a year owing to a change in personnel at the School of Mineral Sciences at Stanford.

In 1951 the AIME Board recommended to the Foundation that a new research project on comminution be started at MIT to conduct fundamental investigations concerning mechanical size reduction in brittle crystalline materials. The objective was to obtain information that would lead to a clearer understanding of fracture mechanisms, size distribution, and the natural laws of size reduction.

Every year billions of tons of bulk solids are handled, stored, fed, blended, and rehandled many times on their way from mine or field to process plant and consumer. In most cases gravity is relied upon to produce flow of bulk solids from storage bin, hopper, or pile into feeder or chute. All too often materials would arch, hang up, and pipe, obstructing flow and reducing the live storage capacity. In 1953 the property which determines the flowability of bulk solids was discovered, and a means of measuring it was developed. This property, called the flow-factor, gave promise of placing the problem of bulk solids flow on a quantitative basis. Flow difficulties were so universal throughout industry that they were recognized by the American Institute of Mining and Metallurgical Engineers who supported a request for sponsorship of a research project on the flow of bulk solids. Recommendation to, and a grant from, the Engineering Foundation followed, and in September 1956 the Utah Engineering Experiment Station at the University of Utah, took up the execution of the project (Item 112 in Table 2). The American Iron and Steel Institute joined in sponsorship at a later date. A Bulk Solids Flow Laboratory was set up at the University of Utah campus and tests were carried out over several following years. The results of this project greatly improved the efficiency in handling bulk solids.

In 1954 the Council on Wave Research recommended that the Engineering Foundation support studies of forces exerted by oscillating fluid motion on cylinders (Item 117 in Table 2). It was reviewed and recommended by the Petroleum Division members of the AIME Research Committee, who emphasized its eventual importance to the petroleum industry in connection with off-shore drilling operations. AIME Board approval was obtained, and following these recommendations the Foundation Board approved a first-year grant of $1,500 to pay a graduate student working on the problem at the University of California. Materials, services, and supervision were contributed by the university. No outside support was sought for this project. The Foundation's Council on Wave Research was acting as an advisory committee. The objective of this project was to investigate the variation, with acceleration and

Reynolds number, of the drag coefficient for a circular cylinder.

The thickening process was one of two methods by which particulate solids suspended in liquids were separated from the liquid. In spite of the importance of the process, comparatively little had been published about it. Preliminary experiments at MIT indicated that the prevailing theory of thickening was largely incorrect and that plant design based on the then current experimental data with thickener columns twelve to eighteen inches in depth were far from being the most economical and effective. In the belief that thickening was a fertile field for engineering study, the Metallurgy Department at MIT submitted in October 1955, through AIME, a proposal for cooperative research in the field to be supported by the Foundation and by industry (Item 123 in Table 2). The investigation extended preliminary batch experimentation to hitherto unexplored variables, and designed and conducted experiments to evaluate pulp density and to study the operation of a continuous thickener. It was concluded from this study that thickening occurs according to two processes. In one, observed in dilute suspensions, a multitude of relatively small flocs settle by hindered settling. In the other, occurring in concentrated suspensions, there is one floc only, and it occupies the entire volume of the container. This floc becomes dewatered by squeezing out water like a sponge. If the settling column is tall and big enough, it will result in greater compaction at the bottom and sufficient dilution at the top to place the top of the column in the first category of thickening operations.

Projects 127 and 138 in Table 2 also deal with mining.

Metallurgy

In 1942 AIME established an advisory committee to organize and guide the writing of a text and reference book on the manufacture of steel by the basic open-hearth process. Despite the fact that more than 90 percent of the steel made in the United States was melted by the basic open-hearth process, there was no book in English that described the process or the underlying physical chemistry. Realizing that such a book would be of much use as a text in colleges, as an elementary discussion of the physical chemistry of steelmaking for metallurgists and executives in the industry, and as an accurate description of present-day operating practice for many skilled and semiskilled men in the industry and elsewhere, the Committee on Physical Chemistry of Steelmaking of the Iron and Steel Division in AIME undertook to sponsor the writing and publication of such a volume (Item 78 in Table 2). The book contained eighteen chapters written by about thirty experts who contributed their time through the cooperation of eight steel companies, three universities, and one research institute. Active work on the manuscript began late in 1941 and the finished book entitled *Basic Open-Hearth Steel Making* was published in 1944 by the American Institute of Mining and Metallurgical Engineers with the assistance of the Seeley W. Mudd Fund.

The project "True Tensile Properties of Metal in the Hot Working Range" was started in 1947 with the objective of securing data on the flow characteristics of metals at the usual forging and rolling temperatures (Item 96 in Table 2). Such data would be of value in working out a theory of plastic flow and could be used as a practical approach to a solution of the whole problem of forgeability. The project was approved early in 1947 by AIME, and Foundation sponsorship and financial support were requested and granted. Little outside support for this project was forthcoming and because of this lack of interest by industry and by trade organizations concerned with hot working, no Foundation grant was approved for the fiscal year 1949–50. The work done between the fall of 1947 and the spring of 1949, when the funds were exhausted, consisted of designing suitable equipment, which would facilitate securing test data more accurately and over a wider range of true strain rates than was previously possible. This task was accomplished, and test data was obtained on SAE 1020 steel over a range of temperatures between 1,900 and 2,300 deg F. The work by Professor Bartholomew and his associates on this project was excellent, and the results obtained with this equipment constituted a definite contribution to the knowledge of the flow of metals at elevated temperatures—a field in which there were few precise data available.

The availability of radioactive isotopes after the Second World War opened up new fields of research in metallurgy, one of which was diffusion in metals, long an important and difficult problem. At MIT, interest in diffusion was keen for many years. In 1947 a long-range program of research was worked out in this field, one phase of which involved using radioactive iron to study the self-diffusion of this element. The proposed project was approved for Foundation sponsorship by AIME, and the first Foundation grant—$2,500 to employ a research assistant—was made available on October 1, 1947 (Item 97 in Table 2). Some outside support was also available. A few years later, this project began to attract the attention of some of the government agencies interested in metallurgical reactions in the solid state; and two of these became sponsors of sizable projects on solid solutions and recovery and recrystallization, both of which can be considered as offshoots of the original diffusion program. Professor M. Cohen of MIT stated the objectives of the project as follows:

> We have started a long range Diffusion Program, with particular focus on the heat treatment of steel. The plan is to start with the determination of the self-diffusion rate of iron as a function of temperature, stress, impurities, grain size and fabrication direction. We have already constructed or purchased the essential equipment for this work, and have placed an order for radioactive iron with the Clinton Laboratories at Oak Ridge. Eventually, we want to determine the effect of carbon, nitrogen, and the usual alloying elements on the self-diffusion rate of iron, and then we shall proceed to diffusion rates of the various elements themselves in the presence of one another. Our over-all objective will be to secure the fundamental diffusion data underlying such heat-treating reactions as recrystallization, austenitizing, spheroidizing, tempering, and pearlite and bainite formation.

Engineering Foundation support of this project continued until 1959.

Most low-temperature data on steels available in the literature of the 1950s were for commercial materials and gave no fundamental information on the plastic behavior of high-purity iron and iron alloys. Furthermore it was then impossible from tests on commercial steels to determine the individual effect of such important variables as oxygen content and dispersion, grain size, form of the carbide, and the specific effect of alloying elements. The investigation "Mechanical Properties of Binary Alloyed Ferrites at Low Temperatures" (Item 98 in Table 2) was planned to aid in explaining the brittle behavior of iron at low temperatures and the beneficial or detrimental effects of other elements, in terms of the concepts of slip, twining, and cleavage stresses. The preliminary work on this project, including a comprehensive literature survey, the design of melting equipment and temperature controllers, and the development of methods of determining true stress-strain curves at low temperatures, was financed by the Bureau of Ships of the U.S. Navy. This preliminary work also included methods of making high-purity iron alloys and producing single crystals. Application for sponsorship was made to AIME early in 1948, and at its request the Foundation subsequently approved the application and granted $3,000. The major experimental difficulty in carrying on research on the properties of high-purity alloys of iron consisted of eliminating the impurities in the available commercially pure iron and at the same time avoiding contamination by the refractories that must be used for melting the iron and preparing the alloys. A large part of the early work on this project consisted of developing methods for producing alloys that could really be considered pure. This was accomplished principally by the use of special refractories and by melting and alloying in vacuum. The complete program for this project included the preparation and testing of high-purity alloys of iron with aluminum, silicon, carbon, oxygen, nitrogen, titanium, and zirconium.

At the March 17, 1954 meeting of the AIME Executive and Finance Committees, the Metals Branch of AIME was authorized to solicit funds to assist in the publication of a large number of high-grade papers reporting basic metallurgical research. The publication of these papers could not be sponsored due to the financial limitations of the society's regular publication fund. This need for additional funds to publish these papers was due primarily to the greatly expanded research activity in the field of physical metallurgy after the war. Between June 1, 1954, and February 1, 1955, approximately fifty industrial companies were invited to contribute to this fund, and the Engineering Foundation was requested to contribute $5,000. The total amount raised was $51,825 (Item 115 in Table 2). This fund was sufficient to publish approximately 600 pages by September 30, 1955, and to underwrite sixty-four additional pages for the November *AIME Journal*. The fund also left a balance of $16,404 for extra pages in 1956.

A study of surface diffusion of metals was approved by the research committee of AIME early in 1955 and was forwarded by this committee to the Foundation with the recommendation that it be supported because of its basic importance in metallic diffusion and in increasing our knowledge of the properties of metal surfaces (Item 119 in Table 2). The Foundation Board approved a grant of $3,000, provided the project secured $2,000 in other funds. This contingency was met by funds from the Office of Naval Research and the Research Corporation. Most of the equipment needed for the project, including radioactivity counters, X-ray diffraction units, furnaces, and vacuum pumps, was available at the Metals Research Laboratory of the Carnegie Institute of Technology. A second grant of $3,000 was made in 1956. The work was started early in 1956, but it was soon suspended due to personnel changes.

Project "Nucleation and Evaporation of Metals" (Item 131 in Table 2) was a continuation of the Foundation's earlier Project 119 of "Surface Diffusion of Metals," which had to be suspended because of personnel changes. The study endeavored to determine the behavior of condensed metal vapors on metallic substrates under ultra high vacuum (better than 10-10 mm mercury). The technique consisted of evaporating metal atoms from a wire filament onto a tungsten tip. With this technique, nucleation and evaporation characteristics as well as the mobility of the condensate were observed for various degrees of super- and undersaturation at various target temperatures. Although several papers were published in the field of condensation of metals on various substrates, information on the factors influencing heterogeneous nucleation of metal vapors on metallic substrates was lacking. The project was recommended to the Foundation by AIME for financial support. The Foundation Board granted $2,500 and the project also secured an industrial fellowship of $3,300.

5. DIRECTOR HOVEY AT THE HELM

The first director of the Engineering Foundation, Alfred D. Flinn was a visionary man, who was very much concerned with humanistic values. During his years at the helm, he did much to shape the course of the Foundation's development. He was responsible for many innovations, including the development of the *Popular Research Narratives*. At its March 22, 1937 meeting, the Executive Committee wrote the following tribute to Flinn:

> As a result of his wide influence in the engineering profession, Dr. Flinn accomplished a great deal making the engineer conscious of the broader implications of engineering and research and in bringing to the attention of the public the work of the engineering profession. He thus helped to integrate the activities of the engineer with many national interests.

At the same meeting, a search committee was formed to find a successor to Flinn who had died eight days before the meeting. During the course of its work, the financial problems of the Foundation led the members of this search committee to reevaluate the need for a full time director. In its report of September 17, 1937, the search committee recommended that Otis E. Hovey take over the directorship on a part-time basis at a salary of $5,000 per year. It was also decided that an essential part of the director's job would be personal visits, on a regularly scheduled basis, to all Engineering Foundation projects. The recommendation of the search committee was accepted, and Hovey became the second director of the Engineering Foundation. After having completed a distinguished professional engineering career, Hovey was less interested than Flinn in humanistic issues such as the place of the engineer in society and more interested in pursuing the sponsorship of fundamental research.

During his early years as director, Hovey was faced with the continuing problem that the income available to the Engineering Foundation was becoming less and less adequate to meet its needs. He attempted to alleviate this problem by reducing the amount of money that was given to each successful grant applicant in an attempt to spread the available funds as far as possible.

During 1939 Director Hovey began to implement the new policy of personal visits to Engineering Foundation funded projects. He reported to the Executive Committee on November 28, 1939 that he had made his initial visitation trip, the primary focus of which was the "Alloys of Iron" project based at MIT. Director Hovey did not share his predecessor's high value of publicity. Under his leadership, the Engineering Foundation cancelled its ongoing publicity contract with J. T. Grady of Columbia University and would in the near future rely on free newspaper articles. He also sold the remaining stock of the *Popular Research Narratives* to a discount sales outlet which disposed of them at a price of twenty-five cents per copy.

By the board meeting of October 17, 1940, Director Hovey could report that despite cutbacks in the amount of Engineering Foundation support, its funded projects received a total of $91,400 in aid from other sources. It is worthwhile to note that this total included the estimated value of in-kind services which amounted to three times the value of Engineering Foundation grants.

Otis E. Hovey

Otis Ellis Hovey was born in East Hardwick, Vermont, on April 9, 1864. His father, Jabez Wadsworth Hovey, was a farmer and schoolteacher who early imbued Otis with an interest in scientific pursuits and who was also determined that his son should enjoy all the educational advantages which he himself had been denied. Consequently, after finishing his public school education at East Hardwick, Otis entered the Chandler Scientific School at Dartmouth College, Hanover, New Hampshire, in 1882 and was graduated with the degree of Bachelor of Science in 1885. Throughout his undergraduate work, he attained grades of Phi Beta Kappa rank, but, because of a ruling of the fraternity, he was ineligible for membership. Appropriately enough, he was elected to honorary membership in this fraternity in 1937.

Otis was determined to become a civil engineer from his early boyhood. When in the early 1870s, a railroad location survey party ran its line through East Hardwick, young Otis was a constant and fascinated bystander, and it is quite probable that this childhood experience started him on the path to the high engineering achievement. After his graduation from Dartmouth, he decided to acquire some practical engineering experience before studying for his professional degree. For one year he worked as engineer of the Hoosac Tunnel and Wilmington Railroad Company in Massachusetts and Vermont. At this time he designed his first bridge, an eight-foot timber truss, relying on the fundamental background of mathematics and physics which he had acquired in his undergraduate work. Bridge design soon became his professional ambition and he spent the following year as draftsman and designer for the Edgemoor Iron Company at Wilmington, Delaware. The late Onward Bates, honorary member and past-president of ASCE, was then chief engineer of this company, and, when the time came for Hovey to enter engineering school, Bates offered him an attractive raise in salary and promotion "to test his determination." As it was throughout his life, Hovey's determination was firm, however; he declined the tempting offer and entered the Thayer School of Civil Engineering at Dartmouth College. Here he made a fine record, standing at the head of his class. He received the degree of Civil Engineer in 1889.

So outstanding was his record that the late Robert Fletcher, director of the Thayer School, recommended Hovey for a teaching position at Washington University in St. Louis, Missouri, while he was still seeking his engineering degree. Accordingly, Hovey was engaged there as an instructor in civil engineering, and he completed his final semester's studies by mail. In addition to his 'correspondence' course, he taught a heavy schedule of engineering subjects. He was also in charge of the university's testing laboratory, which was used both by the students and for commercial testing. This experience, although an interlude in his career, was indicative of his entire life, for he was a prodigious worker and possessed a remarkable capacity for both intensive and varied activities. He was an able instructor and an inspiring lecturer, both of which accomplishments meant much to him in later years when his contacts with younger engineers in his office gave him constant opportunity to teach. He was also widely requested as a visiting lecturer in engineering institutions throughout the East. After a year at Washington University, he resigned to enter his real lifework of bridge designing. In 1890 he took a position as design engineer for the late George S. Morison, past-president of ASCE, and he never again left this field through the following fifty-one years of his life.

During his six years' association with Morison, Hovey was responsible not only for the design of many small bridges and miscellaneous structures but also of several major projects. Considering the fact that he was only twenty-six years of age when he went to work for Morison, amazingly heavy responsibilities were placed on him. A few of the larger structures the design of which was entirely his responsibility were the approaches of the Memphis Bridge at Memphis, Tennessee; the superstructure of the Bellefontaine Bridge across the Mississippi River above St. Louis; the yards and structures for the entrance of the Chicago, Burlington and Quincy Railroad Company into St. Louis; the Alton Bridge across the Mississippi River at Alton, Illinois; and the Leavenworth Bridge across the Missouri River at Leavenworth, Kansas. During the

Otis E. Hovey (1864–1941), movable bridge pioneer and the second director of the Engineering Foundation (1937–41).

winter of 1895 he prepared a complete design and cost estimate for a 3,200 ft. suspension bridge across the Hudson River at New York, N.Y. The Brooklyn Bridge was then the longest span ever built, and here was a thirty-year-old engineer actually designing a span more than double the length of that great bridge. It is interesting to note that neither the span length nor the project of bridging the Hudson River, visualized by Hovey, was accomplished for nearly forty years.

In 1896 he became engineer for the Union Bridge Company at Athens, Pennsylvania, which was later to merge with other firms to form the American Bridge Company. Here he was in complete charge of designs and shop and erection drawings for a wide variety of bridges and buildings. Hovey was also engaged in estimating and contracting for new work for the company. On one occasion, he traveled to London, England, where he contracted for a number of bridges which were erected in South Africa. During this decade Hovey gained much experience, and no better training for his long duties with the American Bridge Company could have been obtained than his years as design engineer with Morison and the Union Bridge Company.

Bridge Engineer

His thirty-four years with the American Bridge Company constituted by far the largest part of Hovey's professional career. He entered the employ of this company when it was formed in 1900. From 1900 to 1904 he was engineer of design for the American Bridge Company at Pencoyd, Pennsylvania.

The United States Steel Corporation, of which the American Bridge Company became a subsidiary, was formed in 1901. During this difficult period Hovey's organizing and engineering ability was of inestimable value in shaping the course of the engineering department of the latter company. In 1904 his office was transferred to New York, and in 1907 he was appointed assistant chief engineer of the American Bridge Company, which position he held until 1931 when he was appointed consulting engineer and was relieved of many of his administrative duties so that he could devote his entire time to the important engineering problems which arose during the course of the company's projects. In 1934, at the age of seventy, he retired from service with the United States Steel Corporation.

During this thirty-four years of service with the American Bridge Company, Hovey's duties were confined principally to bridges and especially to movable bridges. The practices in the design and construction of movable bridge machinery in the various plants which formed the company were widely divergent, and it was his no small task to unify them. In so doing Hovey issued a series of notes over a period of several years on the design of heavy load moving machinery. These notes became the text from which the several offices in the company designed movable bridges and their operating machinery. Many of these notes were later incorporated, together with other data, in Hovey's two volume book, entitled *Movable Bridges*, which is considered the most authoritative work on the subject ever published. In 1935 he also published another authoritative book *Steel Dams.*

Another one of Hovey's specialties was locomotive turntables. All the engineering problems on this subject in the company found their way to his desk. He designed and patented a series of turntable centers or pivots which were built in considerable number by American railroads. The different types were named with his initials, the 'O-Center,' 'E-Center,' and 'H-Center.'

In 1904 the company commissioned Hovey to visit London and Constantinople, Turkey, to investigate structural steel trade conditions, particularly in connection with a proposed bridge over the Golden Horn at Constantinople.

The American Bridge Company had several difficult and interesting contracts, the engineering for which fell in Hovey's province: the construction of the Panama Canal emergency gates in 1912, the strengthening of the Williamsburg Suspension Bridge in New York City in 1914, and the renewal of the four large counterweight sheaves on the vertical lift span over the Willamette River in Oregon. All of these projects necessitated the intricate designing of special tools and equipment to enable the work to be done without taking the bridges out of commission. Hovey was justly proud of these jobs, not because they brought him any personal glory, but because they were difficult engineering problems to be solved.

As Hovey once stated, an engineer for a fabricating company cannot take public credit for his own individual ability. Any glory resulting from his efforts, no matter how arduous, must be subordinated to the welfare of his company. Thus, credit must rest with the engineers employed by the owner of the project. This troubled Hovey not one whit, as he loved engineering and the effort entailed in the solution of engineering problems. Those who were associated with him remember the many times when, after weeks of hard work, arduous and difficult computations had been solved successfully, and joy, pride, and elation shone from him and infected everyone near, not because of any praise that might come to him, but because of the pure pleasure of having done a difficult job to his own satisfaction.

Throughout his thirty-four years with the American Bridge Company, one of his greatest interests lay in helping his younger colleagues along in their chosen profession. His associates had their lives greatly enriched by his character and their professional enthusiasm fired by his example. In the early days of the company and throughout his incumbency as assistant chief engineer he knew practically every engineer and draftsman in the company, and lost no opportunity to seek advancement for the deserving. When he visited the various offices and drafting rooms, his stalwart figure and genial and kindly manner drew all to him like a magnet. He was never known to exhibit impatience when he was asked for either professional or personal advice.

His retirement from service with the American Bridge Company was accepted with a keen sense of loss by his associates. They missed not only his technical advice, but also his example as a gentleman who always lived up to a very high code of ethics and sense of moral and religious responsibility.

Hovey's Other Pursuits

With his characteristic energy and enthusiasm, he opened an office of his own as a consulting engineer, also maintaining all his contacts and taking part in committee work for the various engineering societies of which he was a member. In 1930 he was appointed to the board of the Engineering Foundation, and in 1937 was elected director, following the late Alfred D. Flinn. Hovey's duties necessitated considerable traveling for the purpose of visiting laboratories and shops where experiments were being conducted. His great knowledge of the properties of engineering materials and his experience in testing made his advice invaluable in this work. His long term as treasurer of the American Society of Civil Engineers gave him valuable training in the handling of the funds of the Foundation, and his engineering ability was helpful in the directing of its activities. Because he was a master of the English language far beyond the abilities of most engineers, the annual reports which he prepared are examples of conciseness of arrangement and clear detail that are difficult to equal. His dictation was truly remarkable, being very fast, yet accurate. His choice of words and arrangement of phrases, sentences, and paragraphs were almost perfect, and he had the peculiar ability to lay emphasis when and where needed. He had a distinct horror of split infinitives.

In addition to his regular duties, Hovey was able, because of his robust health, to engage in many other engineering activities (for which he is perhaps more widely known) such as writing books; lecturing at Yale and Princeton Universities, and at Dartmouth College; and serving as a member of the board of overseers for the Thayer School of Civil Engineering connected with his alma mater, Dartmouth College. The broad scope of his engineering interests was partly indicated by the memberships he maintained in several professional societies. In addition to his membership in the American Society of Civil Engineers, he was also a member of the American Society for Testing Materials, the American Railway Engineering Association, the American Welding Society, the American Institute of Consulting Engineers, the Thayer Society of Engineers, and the Engineers' Club of New York.

Hovey was also active in many pursuits not associated with engineering. He was an enthusiastic member of the New England Society of New York City. He was also interested in amateur photography. Always loyal and devoted to Dartmouth College, he kept in close touch with all his classmates by personal correspondence and regular attendance at class reunions. A deeply religious man in the highest sense of the word, for a long period he was a deacon in the Presbyterian Church. His participation in these varied activities always took the form of service and contribution and might be characterized by his own words when he said, "The secret of contentment and happiness is hard work minus worry and jealousy."

His most cherished avocation was music. Hovey was both an enthusiastic and a talented flutist and he was a longtime member of the New York Flute Club, at one time serving as vice-president of this organization. Many members of the Society remembered the performance at the convention in Denver, Colorado, during the summer of 1940, in which he participated, accompanied by his daughter, Ellen Hovey Davis. A previous occasion was reported as follows:

> . . . he assisted with J.E. Greiner [M. Am. Soc. C.E.] playing his Stradivarius violin and with Ralph Modjeski [M. Am. Soc. C.E.] at the piano, in an evening of music which is still vividly remembered not only for its artistry but for the celebrity of the performers.

Hovey's eminence in the engineering profession and his high scholarly ability were recognized in various official citations. As previously mentioned, he was elected an honorary member of Phi Beta Kappa at the 150th anniversary of that organization in 1937. Ten years previous to that, he was awarded the degree of Doctor of Engineering by his alma mater, Dartmouth College. In 1933 he was honored with the degree of Doctor of Science by the Clarkson College of Technology at Potsdam, N.Y. Finally, in 1937, he was elected to honorary membership in the American Society of Civil Engineers. The words of John P. Brooks in presenting Hovey for honorary membership in the Society, give a picture without which no description of Hovey could be adequate. He said in part:

> He is a man born to wealth—to a wealth of character as sturdy as his native hills; to a wealth of ancestry that with others made our early history what it is; to wealth of family and local traditions honoring industry and honesty; to a wealth of appreciation of what is fine and beautiful in art and nature; and to a wealth of mentality and ingenuity worthy of his pioneer forebears.

In his home life, Hovey found the peace and happiness which were the natural expression of his serene and loving character. Indicative of the part played in his life by his wife, Martha Owen Hovey, was the dedication of his book on movable bridges which read: "To my wife, who inspired the beginning of its preparation and whose encouragement has made its completion possible, this book is affectionately dedicated." He was survived by his wife and their two children, Ellen Hovey Davis and Otis Wadsworth Hovey.

Hovey himself can hardly be described more clearly than in his own words expressing the ideals of an engineering career, for to those who knew him he was the embodiment of his own high standards:

> After having followed the profession of engineering for more than fifty years, one is tempted to look back and try to assess the satisfactions of such a career . . . A successful engineer must possess and develop a high type of character. He must be meticulously honest with himself and others. He must be obedient to the laws of nature so far as he can grasp them. He must be logical, thorough, industrious, inventive, practical, firm in well-grounded opinions, yet tolerant of the views of others and able to associate comfortably with them. At the same time, he must see visions and dream dreams, and clearly visualize the embodiment of the dreams, whether in structures, machines, organizations, business or human relationships. . . . While the financial rewards may not be large, the inner satisfactions are great. The engineer feels that he has at least done a little to advance civilization and the enjoyment of life by his friends and the public in general.

6. E. H. COLPITTS: IN THE SHADOW OF WORLD WAR II

Director Hovey's activities on behalf of the Engineering Foundation were cut short by his death on April 15, 1941. Hovey's demise caused the Foundation to once again re-evaluate the position of director. The board established a special committee which delivered a "Report on the Directorship" at the June 19, 1941 board meeting. The board accepted the report's recommendation that the director be limited to two terms totaling three years. At the October 16, 1941 meeting Edwin H. Colpitts became director. Like his predecessor, Colpitts was a distinguished engineer who had a long and illustrious career including pioneering work in the development of long distance telephone transmission. It became his immediate task to guide the Foundation through the turmoil that was caused by America's involvement in World War II.

World War II Effort

The Engineering Foundation's activities to support the U.S. war effort began with the board meeting of March 19, 1942 when NRC was granted permission to utilize the Foundation's room at the NRC for the temporary wartime use of the secretary of NRC's Committee on Metallurgy. At this meeting the chairman also stated that many personnel changes were taking place in the research projects which were then being funded by the Foundation due to factors such as resignations caused by war work and the decreasing availability of laboratory space and equipment for non-war related work. As a result, it was decided to review all Foundation sponsored activities in light of their relationship to the war effort. Thus only those projects which directly contributed to the war effort would be funded and all others would be discontinued. The Research Procedure Committee was directed to consider in their funding recommendations for 1942–43 the real value of each project as a contribution to the war effort. At the board meeting of June 18, 1943 it was pointed out that government agencies controlled their own projects by either initiating them or funding existing projects of interest to the armed forces and that the Foundation projects as nearly approached aid to the effort as could any private organization's research activities.

World War II once again saw the Engineering Foundation supporting NRC activities. At the October 21, 1943 board meeting

$2,000 was approved to help support the continuation of the Office of Scientific Personnel of NRC (Item 83 in Table 2). The Office, established in May 1941, operated under the auspices of the National Academy of Sciences and the National Research Council. Up to August 31, 1943, the Office of Scientific Personnel had been supported from funds supplied by the Office of Scientific Research and Development. For support beyond that date, it became necessary to ask for aid from other sources to meet a budget. The functions of the Office of Scientific Personnel extended far beyond the efforts involved in the recruiting of scientific and technical personnel for war work. For example, in the past the Office had been called upon to render services to the U.S. Office of Education, the Bureau of the Budget, the War Manpower Commission, the War Production Board, and to other government agencies including various divisions of the army and navy. With the end of hostilities the nature of the problems presented to the Office changed and pressing matters concerned with immediate postwar years received attention. To some extent the Office continued to serve as a listening post and informal mouthpiece for the scientific community in matters of developing public interest in Washington. Through this office scientists could also contribute valuable counsel on where the capacity of science to serve the nation was materially affected.

The Engineering Foundation also took a strong stand against the passage of the Kilgore Bill which called for complete government take-over of scientific and engineering research. To forestall this bill, the Foundation met with government officials and put forward a factual statement of the advances made by American engineers and researchers as demonstrated by the comparison of military equipment of our armed forces with those of other countries. It called attention to the dangers posed by the disruption and reordering of our system of scientific and engineering education that was posed by the Kilgore Bill. Due in large measure to efforts of the Foundation and like-minded organizations, the Kilgore Bill never became law.

Despite its many activities, the Engineering Foundation soon realized that despite its best efforts it could have only a limited impact on the war effort. It also became aware that World War II had made possible massive amounts of government funding for research projects. At its June 15, 1944 board meeting it was stated that "while the Foundation is doing all that it can to aid the war effort and that while its function is to apply its income to the conduct of research projects it is also proper to conserve such money not consumed by current research for use when demands for funds will be heavier following the war." This problem of the future place and role of privately funded research in the postwar scheme of things was debated continually by the board during the war years.

The end of the Engineering Foundation's wartime activities can properly be stated to have occurred at the June 17, 1946 board meeting when the Engineering Foundation offered its services and experience to the U.S. Army and Navy to help in government research and when it also voted funds toward the expenses of sending Professor Morris S. Viteles of Pennsylvania State College as a part of the Technical Services Investigation team in Germany (Item 92 in Table 2). The team was charged with appraising German control of manpower in the technical industries, and during the next two years it successfully carried out its mission.

Covering, at least in part, the observations of the above mission the Technical Industrial Intelligence Division of the U.S. Department of Commerce issued, under date of March 19, 1947, a report by Morris S. Viteles and L. Dewey Anderson entitled "Training and Selection of Supervisory Personnel in the I. G. Farbenwerke Ludwigshafen." It described the steps taken by this one large German industrial plant to improve the quality of leadership displayed by supervisory personnel. The management of this plant, in seeking methods for securing increased production, had found that while the individual workers were, in general, satisfied with the wage scale, working conditions, social and welfare programs, there was considerable dissatisfaction with the quality of supervision. The report described the conditions met and the steps taken by the management to select and train supervisors.

Due to the problems brought about by World War II and the good job that Colpitts was doing, the board voted to suspend its standing policy in order to allow Colpitts to continue to serve on an extended basis. This action was taken at its March 16, 1944 meeting.

On October 19, 1944, Director Colpitts issued an important and forward looking letter to the Engineering Foundation Board. In this letter he speculated on what he felt would be the postwar trends and what role the Engineering Foundation would play in this changed environment that would be dominated by massive government controlled funding for continuing research.

The limitation rule on the term of the directorship was again waived for Colpitts on June 21, 1945 when the board voted to keep him in office. It should be also noted that Colpitts renewed the project visitations by the director in the fall of 1946, the visitations having been suspended during the war.

Director Colpitts, as a viable part of his concern for the continuation of the Foundation as a research funding agency, realized that its endowment would have to be increased. In an effort to augment it, he received board approval to retain George Aubrey Hastings to develop and distribute publicity releases and articles that were specifically targeted toward fund raising. This action took place at the October 17, 1946 board meeting. About the same time, Director Colpitts attended a meeting with Randall Robertson, head of the mechanics and materials section, planning division, Office of Naval Research, Washington, D.C. Robertson requested this meeting as a means of increasing the cooperation between the Engineering Foundation and his office. Due to his continuing efforts on behalf of the Foundation, Colpitts received a raise in 1947.

Hastings' publicity campaign soon generated some surprising findings. It revealed to the board that the Engineering Foundation funded projects had a much greater name recognition than did the Engineering Foundation itself. This lack of awareness concerning the Foundation was in part responsible for the slow growth in the endowment.

During 1947 Director Colpitts began a process through which the Engineering Foundation began to re-evaluate its purposes and activities. It was also decided to shift the publicity policy, in light of Hastings' findings, toward the creation of more public name recognition of the Foundation. However, this new policy was short-lived and by the June 1, 1948 board meeting it was decided to terminate Hastings' contract and instead use additional funds to distribute over a wider area copies of the Engineering Foundation's annual reports.

Edwin H. Colpitts

On March 6, 1949 Edwin Henry Colpitts died at his home in Orange, New Jersey, at the age of 77.

Born in Pointe du Butte, New Brunswick, Canada, on January 19, 1872, Colpitts was graduated from Mount Allison University in 1893 with honors in science. In 1896 he received a Bachelor of Arts degree, and in 1897 a Master of Arts degree, from Harvard, where he was an assistant in physics from 1897 to 1899.

In 1899 Colpitts joined the Engineering Department of the

American Bell Telephone Company in Boston. He transferred in 1907 to the Western Electric Company in New York as head of the Physical Laboratory, becoming assistant chief engineer in 1917. In 1924 he joined AT&T as assistant vice-president in the Department of Development and Research, and when that department was consolidated with the Bell Laboratories in 1934, Colpitts became vice-president of the Bell Laboratories.

He served on the U.S. Army's chief signal officer's staff, American Expeditionary Forces, during World War I engaged in military communications research on both sides of the Atlantic; and in World War II was an active member and head technical aide of Division 6 of the National Defense Research Committee. He gave full time to the NDRC anti-submarine warfare program, and was awarded the Medal of Merit for his "adroit and expert advice and effective guidance" of a broad program relating to the improvement and development of echo-ranging systems and attack directors.

Colpitts entered telephone research as assistant to G. A. Campbell on the problem of the loaded telephone line. He was directly

Edwin H. Colpitts (1872–1949), long distance telephone pioneer, and the third director of the Engineering Foundation (1941–48).

responsible for the experiments undertaken to verify Campbell's theoretical work, and thus provided the first conclusive demonstration of the practicality of loading. He also participated in early application of loading and in much of the development of coils, testing methods, and operating standards. One of his important early achievements was an improved battery supply repeating coil for the common battery system. This coil, in which wire is wound on a doughnut-shaped ring of magnetic wire, was the first toroidal repeating coil for exchange plant, and remained standard for twenty years. Another achievement to Colpitts' credit was the design of retardation coils for use in composite telegraph circuits. Basic features of both his repeating and retardation coils have continued to be used for many years.

Cross talk between adjacent circuits also received Colpitts' attention. He introduced the first effective method of determining experimentally the capacity unbalance between wires on an actual pole line, and so laid a foundation for the design of transposition systems, and for the balancing of toll cables.

Beginning in 1911 Colpitts became the head of the Research Department and was directly in charge of the adaptation of the newly discovered thermionic tubes to long distance telephony, both by wire and by radio. He thus played an essential part in the achievement of transcontinental telephony in 1915 and in the radio transmission of speech across the Atlantic Ocean later in the same year. Some of his patents cover the use of the vacuum tube as an oscillator, modulator and amplifier. The best known of these is the so-called Colpitts oscillator, one of two fundamental vacuum-tube oscillator circuits. It has been used so generally that it goes by the name of its inventor.

In 1941 he was elected director of the Engineering Foundation, and continued in that office until resignation due to illness on January 21, 1948, barely six weeks before his death. In 1948 he was awarded the Elliott Cresson Medal by the Franklin Institute. He was a fellow of the American Institute of Electrical Engineers, the Institute of Radio Engineers, the American Physical Society, the Acoustical Society of America, the American Chemical Society, and a member of the American Association for the Advancement of Science. In 1926 he received the honorary degree of LL.D. from Mount Allison. His technical contributions are recorded in twenty-four United States patents, and in numerous papers in professional journals.

Colpitts was married twice. His first wife, Anne Dove Penney, died in 1940. He later married her sister, Sarah Grace Penney, who survived him. He was also survived by his son of the first marriage, Donald B. Colpitts, and three brothers.

7. RESEARCH COUNCILS

Shortly after World War I, two NRC projects resulted in the formation of the American Bureau of Welding and the Advisory Board on Highway Research. As described earlier, both organizations stimulated research in their respective fields and thus contributed substantially to rapid progress in welding and highway transport. The Engineering Foundation cooperated in the establishment of the Advisory Board (now Transportation Research Board) and maintained liaison with the American Welding Bureau. These two successful organizations provided models for subsequent formation of several research councils—independent bodies formed on a nationwide basis of representatives of public, industrial, and private organizations interested in research in some broad field of engineering. Eleven such organizations are discussed below: six of them are concerned with structures and the remaining five with miscellaneous subjects including water waves, corrosion, air resources, metal properties, and buildings.

As their name implies, research councils are concerned with investigations of unsolved problems. Most civil engineering structures are made up of beams joined in some fashion to columns. The two most commonly used modern materials used in engineered structures are steel and reinforced concrete. Aside from corrosion, the two most troublesome items in steel structures are joints and buckling of columns. Welding, which came to the fore in the 1920s as a means of making structural joints, had the great advantage of economy but its acceptance met with considerable resistance. Thus it is not surprising that the first structural research council was concerned with welding. Four others dealt also with steel—columns, riveted and bolted joints, painting, and fasteners—and the remaining one was concerned with reinforced concrete.

Welding Research Council

In response to communications from the American Institute of Electrical Engineers, the American Welding Society and the American Welding Bureau of NRC, late in 1934 the Foundation took the first steps toward aiding the organization of a cooperative research to cover the broad field of welding. The purpose was to bring together all interested groups, including ASME, ASCE, AIME, other technical societies, and industries, such as General Electric, American Rolling Mill, and A. O. Smith which were dependent on welding, to coordinate their efforts. The Foundation suggested as the most important initial item of the cooperative program a critical review of the world literature in all languages since the beginning of modern welding processes, the results to be printed in suitable books. To aid this review and preparation of a comprehensive program, the Foundation offered a grant of $5,000 and the benefits of its extensive experience in organizing and conducting the "Alloys of Iron Research." The resulting Welding Research Committee, sponsored jointly by the American Institute of Electrical Engineers and the American Welding Society, was chaired by Professor Comfort A. Adams and had the following objectives: (1) assembling, digesting and publishing available information on specific subjects of broad interest in the welding field; (2) assisting research in universities on specific fundamental problems; (3) stimulating and conducting research in the welding field; and (4) correlating present and future programs of welding research.

The activities of the committee were carried on under three divisions: Literature, Fundamental Research, and Industrial Research. Industry quickly realized the importance of these objectives and cooperated with the Engineering Foundation in organizing this comprehensive research program. The Literature Division, chaired by J. H. Critchett, made many studies, translations, and abstracts of American and foreign papers and research, which were published in *Supplements of the Welding Journal.* The Fundamental Research Division, chaired by H. M. Hobart, affiliated with the American Bureau of Welding for a number of years, was transferred to the Welding Research Committee. Its main purposes were: (1) to interest the college faculties of this country in problems which relate to welding in order to familiarize the professors and students of the universities with the important process of welding; and (2) to bring to bear upon fundamental problems those scientific minds which are free from commercialism. The Industrial Research Division, chaired by G. F. Jenks, worked in three areas: (1) the organization and coordination of research activities in the field of welding required by technical organizations; (2) the organization and coordination of welding research activities of general interest to the industry; and (3) the organization and fostering of research activities in the field of inspection and testing of materials used in welding, and of welded joints and structures.

Committees were formed to investigate and report on plain

carbon steel, low-alloy steels, high-alloy steels, aluminum alloys, copper alloys, nickel alloys, methods of testing, weld stresses—causes and effects, nondestructive testing, and fatigue testing. Their reports were published and distributed all over the world in the monthly *Supplement to the Welding Journal.* Close contacts were maintained with literally dozens of universities. The committee supplied machined specimens for welding researchers when needed, and assisted in the formulation of problems, in small grants in aid, and in the establishment of fellowships. Contacts were established with the British Institute of Welding and with several departments of the U.S. government. The government contacts were particularly important during World War II when the committee made available its facilities and information to the National Defense Research Committee.

In January 1943 the Welding Research Committee was completely reorganized and renamed the Welding Research Council (Item 62 in Table 2). The principal reason for this reorganization was to bring about a simplification of structure and to enable the Welding Research Council to carry on its activities more closely related to the war effort. With the discontinuance due to the war of certain committees, the work was formulated under the following project committees:

Aircraft Welding Research Committee
Fatigue Testing (Structural) Committee
High Alloy Steels
Light Alloys
Literature Committee
Nickel Alloys
Resistance Welding
Structural Steel Committee
University Research Committee
Weldability
Weld Stress Committee

Under oversight of the above named committees, studies and investigations of broad scope relating to many phases of welding were continued. Cooperation with the aircraft industry was maintained, in part at least, through the National Committee for Aeronautics. Various research projects started under the University Research Committee were taken over and expanded under the National Defense Research Committee.

The end of hostilities curtailed or brought to an end the government support of many areas of welding research. Consequently, it became necessary to examine critically any abandoned projects and to revise the program of continuing basic research as far as possible. The Council has continued to operate and direct its research through the following committees:

Aircraft Welding Research (later consolidated with Resistance Welding Research)
High Alloys Committee
Resistance Welding Research Committee
Universities Research Committee
Fatigue Testing (Structural) Committee
Structural Steel Research Committee
Weldability Committee
Weld Stress Committee
Pressure Vessel Research Committee

The Welding Research Council grew in about ten years from its small beginnings (a contribution of $5,000 from the Engineering Foundation) to an annual budget of about one quarter million almost wholly contributed by industrial supporters. It was organized for the purpose of increasing industrial efficiency in applications of all welding processes and to this end to cooperate in industrial and scientific research. The Council has continued the fundamental analysis of many complex problems involved in the design, fabrication, and service performance of all kinds of welded structures for the wide variety of services met within industry. On the basis of these analyses it formulated and proposed research programs for which, particularly in the case of research undertaken in university laboratories, the Council furnished financial support. In addition the Council, over its first ten years, published some seventy critical digests of the literature on specific subjects and these constituted the best reference library in this field. Prior to the war the Council's predecessor obtained active association or participation in its activities of some three hundred of the ablest men of this country interested in welding research and related sciences. It had developed research centers in some thirty laboratories, mostly in universities. It shared with others credit for substantial accomplishments in the welding art and was in position to clearly indicate the research problems which war conditions made most urgent.

For the first few years after the Council was formed, its activities and importance as a research organization steadily increased. The growth of the research expenditures by the Council is illustrated below. Of the $250,000 budget for the fiscal year ending September 30, 1951, approximately $150,000 was expended by WRC and the remaining $100,000 was paid directly by participants to the laboratories where the work was done.

Fifteen years after its founding, The Welding Research Council installed H. C. Boardman as its second chairman, replacing Comfort A. Adams who was not only instrumental in the creation of the Council but also brought it through its formative years to becoming an established organization recognized as a leader throughout the world. Boardman had served as vice chairman since 1944. A. B. Kinzel replaced Boardman as the Welding Research Council's vice chairman and became chairman three years later. In 1958 Kinzel was succeeded by Walter A. Green. W. Spraragen continued as the operating head of the organization until 1960 when he was replaced by K. Koopman.

In evaluating the work of the Council it should be remembered that its most important service was not spectacular and not reducible to a dollar value except in rare instances. This most

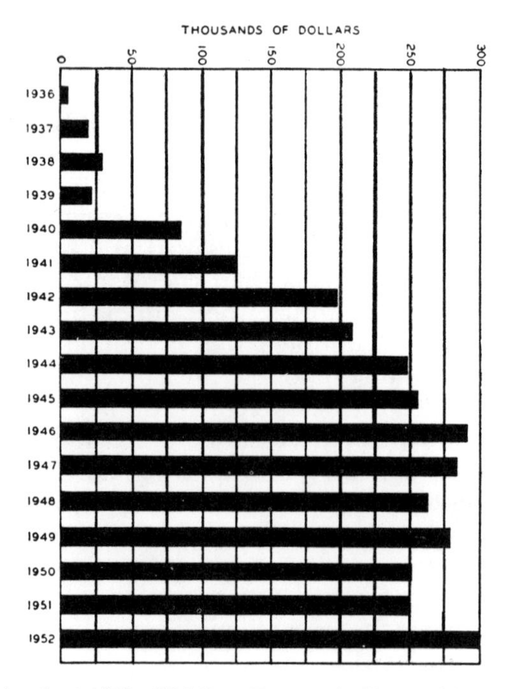

Research budget of the Welding Research Council (Project 62) for the first seventeen years.

important work consisted in filling the gaps in the knowledge of the fundamentals, clarifying the perspective, checking and possibly employing the prevailing theory, improving and solidifying the understanding, and doing the very important educational work which involves the greatest possible dissemination of the knowledge and understanding among those who can apply it to the solution of vital problems.

Structural Stability Research Council

Of the ninety projects sponsored by the Foundation during its first thirty years, a considerable number were supported solely by Foundation grants. Most of the projects in this class were established at the request of one of the founder societies with the object of securing data to fill an important gap in engineering knowledge. Others, as for example the National Research Council and the Advisory Board on Highway Research, received Foundation sponsorship and financial aid until they were organized and had demonstrated their value sufficiently to attract outside support. At this point the Foundation usually withdrew, although in some cases it continued sponsorship and sporadic financial aid.

As is evident from the data in Table 2, most of the projects sponsored by the Foundation were of the cooperative type and a number of them, notably the "Alloys of Iron Research," "Engineers' Council for Professional Development," and "Welding Research Council" have proved so successful that they became permanent organizations.

In 1944 the Committee on Research of the American Society of Civil Engineers, chaired by B. A. Bakhmeteff, decided to recommend formation of councils to organize, plan, finance, and carry on research in civil engineering with the expectation that these would become important quasi-permanent instrumentalities under Engineering Foundation sponsorship. The formation of such councils was approved by the Board of Direction of ASCE, and in 1944 the Foundation was requested to sponsor and to grant financial aid to the first of these, the Column Research Council (Item 84 in Table 2).

The sequence of events that led to the formation of the Column Research Council was initiated in October 1939 when Jonathan Jones, chief engineer of the Fabricated Steel Construction Division of Bethlehem Steel Corporation, submitted to the ASCE Structural Division a memorandum entitled "Unit Stresses in Columns with

High Yield Point Steels." At that time the structural carbon steel in common use, ASTM A7, had a minimum specified yield point of 33 ksi. Higher strength structural silicon and structural nickel steels had found limited use in long span bridges, and some proprietary steels of higher strength were used in certain applications. It was anticipated that higher strength steel would be used increasingly in both bridges and buildings.

Jones' memorandum was referred to the ASCE Committee on Design of Structural Members which had come into being in January 1940. A program was already underway to "study the physical properties of four structural alloys in relation to their behavior under various conditions of stress and shape." Tests for this program were then being carried out at the Fritz Engineering Laboratory at Lehigh University. The ASCE committee reviewed Jones' memorandum and formulated a five point program covering requirements for additional knowledge in the field of columns. At the suggestion of Jones, and with the approval of the Structural Division Executive Committee, the scope of the project was enlarged to cover all phases of compression member behavior.

In 1941 the American Institute of Steel Construction (AISC) and the American Standards Institute, as well as various city building code bodies, were considering independently the development of design formulas for steel columns. Jones saw the need to give to the ASCE program a broader base in order to avoid proliferation of column formulas. Accordingly, in December of 1941, Jones wrote:

> . . . I urged and do urge that it is a national necessity that as many as possible of the bodies who are interested in writing formulas for steel columns get together in some kind of a central group and carry on the research and analyze the results in a way that will be satisfactory to all.

Action was curtailed by the war during 1942 and 1943 but F. H. Frankland of AISC embraced Jones' idea and in April 1943 wrote to the Engineering Foundation:

> . . . many of those concerned came to the conclusion that the whole work involved in column research might properly be something to be organized under the auspices of the Engineering Foundation. It is suggested that such a proposed organization could be set up as a Column Research Council, corresponding to the Welding Research Council, into which all column research and study could be funneled for coordination.

By early 1944 the Column Research Council concept was accepted and in January of that year ASCE agreed to sponsor it under the auspices of the Engineering Foundation. An organizing committee was formed, consisting of Shortridge Hardesty as chairman and including C. A. Ellis, F. H. Frankland, S. C. Hollister, Jonathan Jones and B. G. Johnston. The proposed plan for the Council was to have it made up of official representatives of specification writing bodies together with other organizations specifically related thereto. Accordingly, in late 1944 Hardesty sent out an invitation to the prospective organizations, outlining the objectives of the proposed council and asking that they appoint representatives. Hardesty's prospectus included the 1944 Annual Report of the ASCE Committee on Design of Structural Members but, in his later words, this was ". . . not because we want the Council to follow it, but so that the scope of their work as they organized it would be set forth. In fact . . . this (the Council) is intended to be of much wider scope than the ASCE Committee had planned."

Responding to the 1944 invitation, eighteen organizations sent a total of thirty-three representatives to the first meeting of the Council, held in the board room of ASME in the Engineering Societies Building in New York City on September 25 and 26, 1945. The adopted plan of organization followed closely that proposed by the organizing committee. The executive committee, made up of nine elected members, would not be concerned with technical matters. The chairman and vice chairman of the council would also hold the same offices on the executive committee on which they would be *ex officio* members. Every member of the council would be assigned, according to his preference, to a 'category group' that would represent his primary field of interest. These groups were not to be organized as committees but would proclaim the broad scope of the council and would provide liaison with various specification writing bodies. The following category groups were proposed:

1. Railway bridges
2. Highway bridges
3. Tier buildings
4. Industrial buildings and hangars
5. Machinery
6. Derricks and cranes
7. Military structures
8. Fixed and floating marine structures
9. Towers
10. Aircraft
11. Ship structures
12. Mobile equipment

Reporting to the executive committee would be a technical board that would include representation from all category groups. Under the technical board there would be a Committee on Research and a Committee on Recommended Practice.

Shortridge Hardesty was elected the first chairman of the Council and S. C. Hollister the first chairman of the technical board. L. S. Beedle served as the Council's first secretary starting in 1947. Following its initial meeting on September 26, 1945, the Committee on Research proposed that four subcommittees be formed at once, patterned after the column project of the ASCE Committee on Design of Structural Members. The names of these subcommittees were as follows:

1. Mechanical Properties of Materials
2. Initial Eccentricities of Compression Elements
3. Local Buckling of Compression Elements
4. Columns in Structural Frames

This organization provided a basis for the recommendation to the ASCE Structural Division that the Committee on Design of Structural Members be disbanded and its work taken over by the Column Research Council through its Research Committee.

Jonathan Jones had sown the seed that led to the formation of the Council and Shortridge Hardesty went on to nurture its early growth as chairman for the first twelve years of its existence. The efforts of these two men were of paramount significance to the Council. When, due to ill health, Hardesty retired from the chairmanship in 1955, he was made an honorary member, and when he died a few months later, the following resolution was passed:

> Column Research Council records with sadness the passing on of Shortridge Hardesty. As chairman and director for twelve years, Mr. Hardesty gave fully of his love, devotion, and material assistance to the organization and continued support of Column Research Council activities. His mind not only grasped the practical problems of engineering application but instinctively went beyond to encompass the fundamental knowledge that underlies research. His influence was personal inspiration to all who worked in Column Research Council and his kindly spirit will remain with it for so long as it may continue.

B. G. Johnston became the second chairman of CRC in 1956. The terms of office, names and affiliations of CRC chairmen were as follows:

1945–55	S. Hardesty, Hardesty and Hanover, Inc.
1956–61	B. G. Johnston, University of Michigan
1962–65	E. H. Gaylord, University of Illinois
1966–69	L. S. Beedle, Lehigh University
1970–73	T. V. Galambos, Washington University
1974–77	G. Winter, Cornell University
1978	J. W. Clark, Aluminum Company of America
1979–82	J. S. B. Iffland, Iffland, Kavanaugh and Waterbury
1983–86	J. Springfield, Carruthers and Wallace Limited
1987–	S. J. Errera, Bethlehem Steel Corporation

In practice, the initial scheme of organization proved unwieldy and in 1951 the technical board and the overall Research Committee were eliminated. New task-oriented research committees reported directly to the Executive Committee, which then took on both the technical and administrative functions of the Council. From the thirty-three representatives that were at the first meeting in 1945, the membership of the Council grew to 205 by 1981, and the number of sponsoring/participating organizations from eighteen to thirty-three. The initial financial as well as administrative support came from the Engineering Foundation, which agreed in 1945 to contribute $5,000 contingent on the raising of $10,000 by the Council from other sources. In addition to the Engineering Foundation, there have been four major sources of funds in support of the Council: the Association of American Railroads, the American Institute of Steel Construction, the American Iron and Steel Institute, and the National Science Foundation.

In 1959, the Engineering Foundation Board decided to discontinue its services as a fiscal agent for the various research councils after September 30, 1959. This decision forced the Column Research Council to re-evaluate its affiliation. It established itself as an independent organization acting as its own fiscal agent. From the beginning, the Council's financial matters were handled by the chairman and the secretary. Thus, aside from the additional paper work, the Council's new independent status meant only that the headquarters moved with its chairman. In 1966, when L. S. Beedle became chairman, the Council's headquarters moved to Lehigh University. Four years later, at the end of Beedle's term as chairman, the Council appointed Beedle to the new office of director and Lehigh University became the permanent seat of the Council's headquarters. In 1976, because of the expanded scope of its coverage, the name of the Council was changed to the Structural Stability Research Council.

The objectives and purposes of the Council were first formulated by Hardesty. Over the years they have undergone several revisions. The 1980 revision, used throughout the following decade, contained five items:

1. To maintain a forum where structural stability aspects of the behavior of frames, columns, and other compression-type elements in metal and composite structures can be presented for evaluation, and pertinent structural research problems proposed for investigation.
2. To review the world's literature on structural stability of metal and composite structures and study the properties of metals available for their construction, and to make the results widely available to the engineering profession.
3. To organize, administer, and guide cooperative research projects in the field of structural stability, and to enlist financial support for such projects.
4. To promote publication and dissemination of research information in the field of structural stability.
5. To study the application of the results of research to stability design of metal and composite structures, and to develop comprehensive and consistent strength and performance criteria and encourage consideration thereof by specification-writing bodies.

Projects originating within the Council were originally planned by the research committees, which established research priorities. The appropriate committee then solicited proposals and financial support, awarded the project, guided its course, reviewed the resulting reports and approved their release for publication by CRC or other appropriate medium. One of the first projects was a critical survey of published data on the strength of structural members under compressive loading made by Friedrich Bleich at the David Taylor Model Basin with funds contributed by the Bureau of Ships. The resulting book *Buckling Strength of Metal Structures* was a major first step in the development of the Council's research program. It was also a forerunner of the future design guides. After Bleich's sudden death in February of 1950, the book was completed later that year by his son, Hans Bleich, who was employed at the time by the firm of Hardesty and Hanover. Published in 1952, the book is still used as a reference.

Over the years, the Council was connected with a great many research projects carried out at Lehigh University's Fritz Engineering Laboratory and at numerous other structural research institutions. The projects can be classified into the following categories:

1. Centrally Loaded Columns
2. Beams and Girders
3. Beam-Columns
4. Tapered Members
5. Built-Up Members
6. Members with Distributed Lateral Restraints
7. Members in Frames
8. Stiffened Plates
9. Curved Bars and Arches
10. Tubes and Shells
11. Local Buckling
12. Thin-Walled Metal Construction
13. Composite Columns

The broad coverage of the subject of compression members is evident. The results of research were used to improve the design practice through publication of successive editions of a guide for the design of compression members. The concept of the design guide was introduced in 1956 by Jonathan Jones and L. S. Beedle. The purpose was to provide an authoritative source of stability design information for engineers and specification-writing bodies. The first two editions, published in 1960 and 1966, were titled the *Guide to Design Criteria for Metal Compression Members*, and the third and fourth editions, in keeping with the changed name of the Council, were titled *Guide to Stability Design Criteria for Metal Structures*. The fourth edition was published by John Wiley & Sons in 1988 and was edited by T. V. Galambos.

Research Council on Structural Connections

Based on his discussions with G. A. Maney of Northwestern University, C. A. Ellis of Purdue University, and H. H. Lind of the Industrial Fastener Institute, W. M. Wilson of the University of Illinois wrote to E. H. Colpitts, director of the Engineering Foundation, on June 30, 1945 as follows:

> There is a need for additional work on riveted and bolted structural members. Some of us have given considerable thought to the problem and have concluded to ask the Engineering Foundation to create a Research Council on Riveted and Bolted Structural Members.

The proposal was presented to the ASCE Board of Direction at its July 16–17, 1945 meeting. The ASCE undertook to organize experimental research on riveted and bolted structural joints and requested the Engineering Foundation to sponsor the undertaking. The Foundation agreed and a temporary committee on organization was appointed with C. A. Ellis as chairman and W. M. Wilson as secretary. During the following year and a half the efforts of the temporary committee were devoted largely to organization and to studies basic to the development of a fundamentally sound research program. These preliminaries were completed early in 1947 when the permanent organization of the Research Council on Riveted and Bolted Structural Joints (Item 90 in Table 2) was established. The first meeting of the Council was held on January 15, 1947.

The following seven organizations were the initial participants in the Council:

University of Illinois
Northwestern University
Association of American Railroads
American Institute of Bolt, Nut, and Rivet Manufacturers
Bureau of Public Roads
Illinois Division of Highways
American Institute of Steel Construction

The first two listed institutions conducted the initial experiments and the remaining five provided the major portion of funding with the balance coming from the Engineering Foundation. In 1948–49, the list of sponsors was increased by the addition of American Iron and Steel Institute and the Industrial Fastener Institute. The financial support of the Council through 1957–58 exceeded $433,000 of which $24,000 was contributed by the Foundation.

After thorough consideration of the available data and consideration of the several variables requiring to be investigated, seven research projects, each under a project committee, were organized. The names of the projects and of the corresponding project committee chairmen are listed below:

Project 1—Effect of Bearing Pressure on Static and Fatigue Strength on Hot and Cold Riveted Joints—Jonathan Jones
Project 2—Effect of Rivet Pattern on Static Strength of Structural Joints—W. M. Wilson

Project 3—Strength of Rivets in Shear and Combined Shear and Tension—T. R. Higgins

Project 4—Fatigue Strength of Bolted Structural Joints—W. C. Stewart

Project 5—The Effect of Grip Upon the Fatigue Strength of Riveted Joints Designed to Fail in the Rivets—K. H. Lenzen

Project 6—The Fatigue Strength of High Strength Steel Riveted Joints—J. L. Beckel

Project 7—The Effect of Rivet Pattern Upon the Fatigue Strength of Structural Joints—L. E. Philbrook

T. R. Higgins served as the first chairman and W. M. Wilson as the first secretary of the Council. The years of service, names and affiliation of the Council's chairmen were as follows:

1947–49	T. R. Higgins, American Institute of Steel Construction
1950–52	W. C. Stewart, Industrial Fastener Institute
1953–56	E. J. Ruble, Association of American Railroads
1957–60	E. L. Erickson, Bureau of Public Roads
1961–64	R. B. Belford, Industrial Fasteners Institute
1965–70	J. L. Rumpf, Temple University
1971–73	W. A. Milek, American Institute of Steel Construction
1974–79	E. R. Estes, Old Dominion University
1980–82	F. J. Palmer, Copperweld Corporation
1983–88	T. S. Tarpy, Stanley D. Lindsey & Associates
1989–	M. I. Gilmor, Canadian Institute of Steel Construction

During the fiscal year 1948–49, investigations were started at Purdue University and at the University of Washington and during 1956–57 at Lehigh University. State highway departments of Washington and Pennsylvania contributed funds for the investigations at the University of Washington and at Lehigh University, respectively.

The project "Fatigue Strength of Bolted Structural Joints," in progress at the Universities of Illinois and Washington, aroused international interest in the use of high strength bolts as fasteners in structural members. In 1951 the Council published specifications, the "Assembly of Structural Joints Using High-Tensile Steel Bolts." The ASTM and the American Standards Association specifications for these bolts and nuts were issued during the 1950s and the fasteners became readily available. By 1955, thousands of copies of the Council's specifications were distributed, and the actual use of bolts instead of rivets for field erection was reported in a number of buildings throughout the country. The use of high strength bolts found rapid acceptance. In a few years the high strength bolts completely replaced rivets in new civil engineering structures and loose rivets in existing structures.

On February 9, 1956 the Council held a brief ceremony at which a plaque was presented to W. M. Wilson "in recognition of his pioneering research on the development of the high strength bolt as a fastener for structural steel." Professor Wilson was the first to observe the excellent fatigue life of high strength bolts in loose fitting holes and his early work on such bolts was an important

Fatigue testing machine with a riveted structural joint undergoing testing at Northwestern University.

stimulus in the formation of the Research Council on Riveted and Bolted Joints.

In 1974 the *Guide to Design Criteria for Bolted and Riveted Joints* by J. W. Fisher and J. H. A. Struik was published as a part of the Council's program. T. R. Higgins, chairman of the subcommittee on specifications which guided the preparation of this book, wrote in the Foreword to the first edition:

> . . . because of the accumulation of . . . much new information concerning the behavior of joints assembled with [high strength] bolts, largely the result of the Council's research program, it was felt that a succinct digest of this material, within a single publication, would be a valuable aid to specification-writing bodies as well as to those engaged in the design and investigation of such connections.

The second edition of the *Guide*, authored by G. L. Kulak, J. W. Fisher and J. H. A. Struik, was published in 1987.

Throughout the years, the Council continued its affiliation with the Engineering Foundation which has provided the services as the Council's fiscal agent. In 1979 the name of the Council was changed to Research Council on Structural Connections.

Reinforced Concrete Research Council (RCRC)

The investigation of reinforced concrete columns (Item 47 in Table 2) carried out in the 1930s at the University of Illinois and at Lehigh University showed conclusively that the so-called straight-line theory could not adequately predict the strength of centrally loaded reinforced concrete columns and, therefore, was not suitable for design purposes. The new procedure for column design adopted as a result of this investigation was based on inelastic response of the component materials, i.e. of steel and concrete. Work in the United States as well as abroad indicated that any rational theory for the response of reinforced concrete members to loads must take into account the inelastic properties of the component materials regardless of the type of loading. In other words, the findings for concentrically loaded columns suggested strongly that similar consideration should be applicable to beams and beam-columns.

Earlier experimental and analytical investigations of reinforced concrete beams had indeed shown that the prediction of beam strength requires the knowledge of the inelastic as well as elastic properties of steel and concrete. On the other hand, reliable test data on beam-columns covering the practical ranges of variables were not available. To secure test data needed for the development of a rational design procedure for reinforced concrete based on strength, the Executive Committee of the ASCE Structural Division proposed the establishment of a research council. A letter dated February 24, 1948 from J. M. Garrelts of Columbia University to B. A. Bakhmeteff, chairman of the ASCE Committee on Research, outlined the situation as follows:

> On behalf of the Executive Committee of the Structural Division, I present to the Committee on Research, A.S.C.E., this application for support of Fundamental Research on Reinforced Concrete. We request that the American Society of Civil Engineers assume sponsorship over this Research and apply to the Engineering Foundation for a suitable Grant in Aid.

After a discussion of the technical factors involved, the letter continued:

> On October 2, 1947, the Executive Committee of the Structural Division, A.S.C.E., requested Society approval for the organization of a Council on Research in Reinforced Concrete. At its meeting in Jacksonville, Florida, October, 1947, the Board of Direction of A.S.C.E. concurred in the recommendation of the research committee and approved the organization of such a council.
>
> Mr. A. J. Boase, the Chairman of the A.S.C.E. Committee on Masonry and Reinforced Concrete, has taken steps to set up such a council and has enlisted the support of some of the interested organizations. The personnel for such a council will consist of authorities on Reinforced Concrete and representatives from interested organizations.

Fatigue testing machine and a bolted structural joint undergoing test at the University of Illinois.

> The function of the Council will be to plan a program of study and experimental research in reinforced concrete, direct the execution of that program, and to interpret the results in the form of a code for the design of concrete structures. Such a code with a scientific basis may reduce the cost of reinforced concrete, may permit its use in longer span structures than now considered possible, and will eliminate most of the uncertainty in design.

A description of two specific investigations proposed by the ASCE Committee on Masonry and Reinforced Concrete followed. The letter was concluded thusly:

> Two of the organizations interested in this investigation have already agreed to contribute $5000 each to the Council for the first year's work. It is believed that additional funds will be forthcoming from other bodies interested in this problem when the work of the Council gets under way.
>
> Mr. Boase has submitted the following list of personnel for the Council. Additional members will be named as the work of the Council indicates the advisability of enlarging the personnel.
>
> Proposed Council on Reinforced Concrete:
>
> Mr. Robert Blanks, director of research, Bureau of Reclamation, Denver, Colorado—chairman
> Prof. Jewell M. Garrelts, Columbia University—secretary
> Prof. Clyde T. Morris, Ohio State University
> Prof. Frank E. Richart, University of Illinois
> Mr. F. R. Smith, chairman, Comm. of Masonry, AREA
> Mr. Raymond Archibald, chairman, Bridge Specification Committee AASHO
> Mr. Albert E. Cummings, research engineer, Raymond Concrete Pile Co., N.Y.
> Mr. Harold D. Jolly, chairman, Engineering Practice Committee, Reinforcing Steel Institute
> Mr. Arthur J. Boase, Portland Cement Association

The council (Item 99 in Table 2) began functioning in 1948 with R. F. Blanks as chairman and J. M. Garrelts as secretary. The list of RCRC chairmen follows:

1948–58	R. F. Blanks, Bureau of Reclamation to 1951 Ideal Cement Company after 1951
1958–64	T. Germundsson, Portland Cement Association
1964–65	R. H. Corbetta, Corbetta Construction Company
1965–68	D. McHenry, Portland Cement Association

1968–80 C. P. Siess, University of Illinois
1980–89 E. Cohen, Ammann & Whitney

The first project undertaken by the Reinforced Concrete Research Council (Item 99 in Table 2) was to determine the effect of combined bending and direct stress on reinforced concrete structural members. The program was planned by A. J. Boase and testing began in the Talbot Laboratory of the University of Illinois early in 1949. A total of 120 columns, ninety square and thirty round, were tested. The major variables were concrete quality, amount of reinforcement and eccentricity of load. At the RCRC meeting held in Chicago on September 9, 1949, principal investigator E. Hognestad reported that the tests of square columns were completed and the results were in a reasonable agreement with values computed using the Jensen stress block. On the completion of all tests, an improved flexural theory was developed by means of which the behavior under load, mode of failure, and ultimate loads

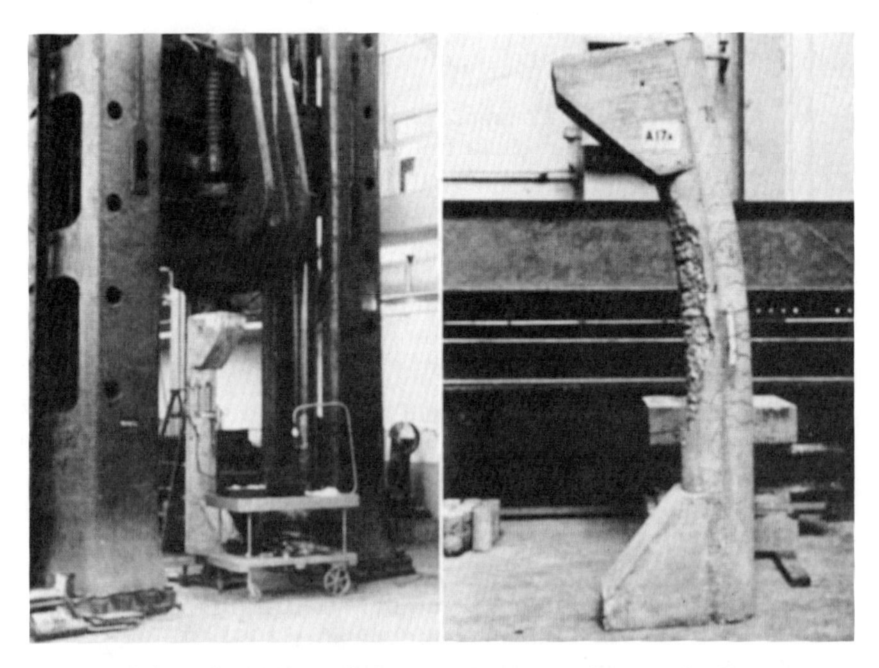

A reinforced concrete column in a testing machine at the University of Illinois and the appearance of this column after test to failure.

were predicted with satisfactory accuracy. The final report was completed in October 1950 and the results were published in the "University of Illinois Bulletin 399."

The second major RCRC project was an investigation of shear and diagonal tension in reinforced concrete beams. The purpose of this work was to study the strength and basic behavior of reinforced concrete beams designed to fail in diagonal tension. It involved the manufacture and testing of sixty reinforced concrete beams with various percentages of longitudinal and web reinforcement as well as three grades of concrete. The project was started in July 1950. In 1951 the Council authorized an extension of the investigation involving forty-four additional beam tests. Analytical studies of the results indicated that two equations are required to predict the shear strength of a reinforced concrete beam: one for the load at which diagonal cracks form and the other for the load at which the compression zone of concrete fails.

To determine the accuracy of design procedures for rigid frame bridges and to compare elastic and plastic theories of design, a project entitled "Analysis of Skewed Rigid-Frame Bridges" was carried out at Virginia Polytechnic Institute. The study included tests of an aluminum model followed by tests of a 1/10-scale reinforced concrete model.

Early in its life, RCRC provided general oversight for research in reinforced concrete at the Bureau of Reclamation laboratory at Denver. Included were beam and pull-out tests, large beam tests, installation of SR-4 gages on reinforcing steel in the foundation of a 5,000,000 lb testing machine, reinforced concrete rigid-frame tests, and stepped-column tests.

During the fiscal year 1950–51, the Council voted to sponsor projects at Ohio State University, at the Treat Island Waterway Experiment Station, Eastport, Maine, and at Lehigh University. The one at Ohio State was started in 1948 and was aimed at determining by photoelastic methods the stress distribution in simple reinforced concrete beams in bending. Several progress reports were issued, mostly on the perfection of the technique. The work was discontinued when it was realized that errors in stress measurements were too great to permit reasonable corrections.

The work at Treat Island was a series of exposure tests on reinforced concrete beams. The purpose of the tests was to determine the relationship between various levels of stress in main reinforcing bars and deterioration caused by weathering. Exposure tests of twenty beams of air-entrained and thirty of non-air-entrained concrete were started in November 1951. The specimens were inspected in June 1952, June 1953, and June 1954 when the first series of tests was completed. The specimens were subjected to about 1,000 cycles of submergence in sea water which were caused by rising tides and approximately 300 cycles of freezing and thawing. Air-entrained concrete was far superior to non-air-entrained concrete under the exposure conditions at this location. The tests of non-air-entrained concrete beams resulted in the definite conclusion that the rate of deterioration of beams reinforced with A-305 bars stressed to 40 or 50 ksi is only 60% of the rate of deterioration of beams with old-style bars stressed at 20 ksi. A second program of exposure including sixty-four additional specimens, all with air-entrained concrete, was started before the

Testing a reinforced concrete beam in shear at the University of Illinois.

beginning of winter in 1954–55. The first inspection was made in July 1955 and others followed in July 1956, 1957, and 1958.

The investigation sponsored by RCRC at Lehigh University was concerned with tests and analyses of full-scale precast, prestressed concrete bridge beams subjected to repetitive loading. Several pilot beam tests were made prior to RCRC sponsorship which started in October 1951. During the winter and spring of 1952, a loading frame complete with its auxiliary hydraulic load controller was designed and fabricated for use in these fatigue tests. The first 38-ft. prestressed concrete test beam was pretensioned. It was poured late in August and testing started in the fall of 1952. It was subjected to 1,300,000 applications of design live load, including impact, without suffering damage. No slip of strands or appreciable permanent deflection occurred, and all cracks closed upon removal of the live load. Cracking progressed and strand stress variation due to live load increased during the first part of the cyclic loading. The beam was then subjected to 100,000 cycles of 1.5 times the design live load. It was finally tested to destruction, failing at a load corresponding to the dead load plus 4.2 times the live load, including 30% impact. The second 38-ft. test beam was post-tensioned using ungrouted strands. It was made of two I-shaped beams bolted together. This test beam was subjected to 1,000,000 applications of simulated 36-ton truck load followed by the test to destruction in July. It performed as predicted; its failure load was equal to the dead load plus 330% live load, including 30% impact. As in the first beam, failure was by crushing of the concrete.

The second phase of Lehigh studies of prestressed concrete bridges was concerned with the lateral load distribution in a multiple-beam system. An initial theoretical study was followed up with a test of a three-lane prestressed concrete bridge at Centerport, Pennsylvania, and by testing of a prestressed concrete bridge in the laboratory. Also in the laboratory, tests were made on twenty-two prestressed concrete beams 12 ft. long to determine the bond characteristics of 7/16-in. strands.

The third phase of the work at Lehigh included: (1) a study of the ultimate shear strength of prestressed concrete beams, (2) a study of the ultimate design criteria for slip of the strand and (3) a study of the capacity of beams to resist repetitive loading. The first item involved analytical studies and twenty beam tests as a check on the analysis, the second pull-out tests of strand and four tests of beams, and the third involved illustrative examples of design calculations and comparisons with existing test data. An additional testing program was carried out to better define the fatigue properties of prestressing strand.

The investigation of the effects of combined bending and direct stress completed at the University of Illinois in 1950 was followed up in 1951 by tests of columns subjected to sustained eccentric loads. Forty-four column tests were reported; of these thirteen were made with fast static loading, twelve with slow loading and nineteen with sustained loading. The test findings indicated that the ultimate strength of an eccentrically loaded column under sustained loading is only about ten percent below that for fast static loading.

A study of shearing strength of reinforced concrete slabs subjected to a centrally located concentrated load was authorized and started at the University of Illinois in 1952. The object was to secure data on the strength of reinforced concrete slabs subjected to combined shear and flexure. Twenty slabs and nine beams were tested. The major finding of the investigation was that the intensity and type of flexure that is combined with a shearing force has a considerable influence on the shearing strength of the slab. It thus appeared that shearing stresses alone have little meaning. Shear and flexure in slabs should be regarded as a combined stress

Test of a full scale pretensioned 38-ft. concrete beam containing forty 5⁄16-in. bonded strands at Lehigh University.

problem analogous to combined axial load and flexure in columns.

An investigation to determine shear strength of reinforced concrete frame members without web reinforcement was carried out at the University of Illinois under RCRC sponsorship in 1953–1955. The tests included thirty-three knee frames subjected to combined axial compression, shear, and moment; and thirty-eight stub beams subjected to moment and shear. In all knee frames the axial load was equal to the external shear since the two lengths of any one knee frame were of equal length. The major variables were the length of shear span, the strength of concrete, and the percentage of tensile reinforcement. Experimental and analytical studies established conclusively that the strength of a reinforced concrete frame member without web reinforcement is affected by shear through the formation of diagonal tension cracks. If the percentage of longitudinal reinforcement is small or the shear span is very long, a member without web reinforcement fails in flexure without prior formation of diagonal tension cracks; in such case, the strength of the member is unaffected by shear. If, however, a diagonal tension crack forms, the strength of the beam is usually lower than that corresponding to flexural failure. Depending on its make-up, such member may fail either simultaneously with the formation of the diagonal tension crack or it may fail at a higher load. Analytical expressions were developed for predicting the diagonal tension cracking load and shear compression strength. The diagonal tension cracking load was expressed in terms of the nominal shearing stress, and the shear compression strength was expressed in terms of the shear moment capacity.

Extensive investigations of short concentrically and eccentrically loaded columns led to the development of generally applicable analytical expressions for predicting the ultimate strength of short reinforced concrete columns. The experimental studies have shown that for short columns the effect of lateral deflections on the strength is small and that short columns are not in danger of buckling. Accordingly, the analytical expressions disregarded both deflections and buckling. On the other hand, it was known that the strength of long columns may be reduced significantly by both of these factors. To investigate the effects of these factors on column strength, an analytical investigation was carried out at the University of Illinois under the sponsorship of RCRC during the period 1953–56. Theoretical analyses were developed for hinged and for restrained columns. Both concentric and eccentric loads were considered. The treatment of columns loaded concentrically was based on Engesser's tangent modulus theory and the treatment of

Test of a symmetrical reinforced concrete knee frame without web reinforcement at the University of Illinois.

eccentrically loaded columns was based on the principles advanced by von Karman. Both analyses utilized the stress-strain relationship for concrete determined by Hognestad from tests of short eccentrically loaded columns. The analyses were compared with the test results of six earlier experimental studies reported in the literature. The test data, including forty-eight concentrically and seventy-nine eccentrically loaded long hinged columns, and the analyses were in satisfactory agreement, thus indicating that the analyses with their basic assumptions yield reliable results. A practical design procedure based on the results of these theoretical analyses was recommended.

An investigation of hipped-plate construction was started in 1954 at Syracuse University, based on plans developed by an ASCE technical committee. Included were a reinforced concrete model and several small-scale aluminum models. Two new investigations were started in 1955: one at Cornell University dealing with investigation of medium-strength steel for concrete reinforcement and the other at the University of California, Berkeley, concerned with the strength of concrete under combined stresses. All three of these investigations were completed during the fiscal year 1956–57.

Two new projects were started at the University of Illinois: one concerned with the effect of moment ratio on the shear strength of reinforced concrete beams started in 1955–56 and the other on multiple panel reinforced concrete floors for buildings started in 1956–57. The first study involved tests of unsymmetrical knee frames without web reinforcement with the ratios of axial force to shear varying from 0 to 6. It covered the entire range of failures from that caused by shear in the absence of axial load to that caused by eccentric compression. The observed diagonal tension cracking loads were found in good agreement with the results of the earlier RCRC tests of symmetrical knee frames. On the other hand, the shear compression strength was found to increase with axial load considerably faster than indicated by the earlier tests.

The second Illinois project had the cooperation of the Headquarters of U.S. Air Force, General Services Administration, Public Building Service, and the Office of the Chief of Engineers, U.S. Army in addition to RCRC. The study involved tests of five specimens, each representing a square floor of nine square panels, and consisting of a floor slab, portions of sixteen columns supporting the slab, and, in some cases, column capitals or floor beams between the columns. All specimens were 15 × 15 ft. in plan. Slab thickness, slab reinforcement, and the presence or absence of floor beams and capitals were the principal variables. Each specimen was subjected to numerous tests with vertical loads to determine its deformation characteristics and strength. The project continued for several years and generated sufficient information to provide valid comparisons of the behavior of various types of slabs used in buildings and to place the design of such slabs on a common basis.

In response to the 1959 basic changes in the policy of the Engineering Foundation Board, RCRC relinquished its affiliation with the Engineering Foundation and became affiliated with ASCE which assumed the function of RCRC's fiscal agent.

Tests to determine the effect of long-term loading (sixteen months) on the strength of reinforced concrete columns at the University of Illinois.

Steel Structures Painting Council

The Steel Structures Painting Council (Item 122 in Table 2) was organized in 1950 at the Mellon Institute of Industrial Research in Pittsburgh through the efforts of the American Institute of Steel Construction. The Council was composed of representatives of about twenty associations and organizations concerned with manufacture, specification, and use of paints and other coatings for the protection of structural steel surfaces. Its objectives were (1) to review the literature on the art of cleaning and painting steel structures, (2) to prepare codes and specifications covering the cleaning and painting of steel structures, and (3) to carry on research on surface preparation and on the application of paint and other coatings to make the protection of steel structures more economical and effective.

The Council established a research fellowship at the Mellon Institute. Joseph Bigos filled the position of senior fellow and director of research of the Steel Structures Painting Council through 1957 when he was succeeded by John D. Keane. Through the fiscal year 1957–58, the Engineering Foundation contributed $3,000 while $88,000 was contributed by other participants.

Among the many continuing activities supervised by the Council were the following: An ongoing literature survey; paint tests; exposure tests to investigate painting over welds, painting of rusty steel, protection of steel structures against brine drippings; weathering versus shop painting tests; investigation of chemical versus mechanical treatment of surfaces; and testing of heat-resistant paints. Inspections were made of several of the projects, and many reports were prepared by the director.

During the first half of the 1950s, the Council published the *Steel Structures Painting Manual* in two volumes. Volume I—*Good Painting Practices*—described the critical selection factors for various industries. Volume II—*Systems and Specifications*—gave detail specifications for surface preparation, pretreatment and paint application, and for paint and paint systems. The manual was the first publication containing authoritative information on good painting practices for structural steel and the first presentation of specific recommendations and guides for such painting. The research at Mellon Institute on cleaning and painting, carried out in close cooperation with technical representatives of industry and other interested associations, succeeded in the promulgation of specifications for surface preparation, pretreatment and paint application, and for paint and paint systems. The specifications made it possible, for the first time, to specify painting of a steel structure simply by reference to a system number. The manual and the specifications received enthusiastic acceptance from users and highly favorable reviews from the technical press both in the United States and abroad.

Fasteners Research Council

Rapid technical advances and increasingly sophisticated mechanisms in the aerospace, automotive, electronics, and related industries made desirable the formation of an organization to encompass the entire field of mechanically fastened joints. The Fasteners Research Council (RC-A-64-4 in Table 2), organized in August 1963, was a scientific and educational organization dedicated to the advancement of mechanical fastening technology through basic and applied research. It concerned itself with basic phenomena in the application of mechanical fasteners. A $3,300 Engineering Foundation grant to the Council aided the development of several basic fasteners research projects.

With the exception of adhesives, there were two basic methods of joining materials: one was by welding and the other was by mechanical fasteners such as bolts, screws, rivets, nails, etc. By far, mechanical fasteners constituted the largest field, since fasteners were used to join all types of materials, organic as well as inorganic. The materials engineering explosion compounded the problems of mechanical fastening systems and the need for basic and applied research has multiplied.

The Engineering Foundation became the sponsor of the Fastener Research Council in May of 1964. The American Society for Metals, the Society of Automotive Engineers, the Industrial Fasteners Institute and the American Institute of Steel Construction joined the Foundation as co-sponsors in the early months of the Council's existence.

Extraordinarily little was known about the basic theory which underlies mechanical fastening assemblies. Many of the developments have proceeded empirically—the joint or the clamping effort worked, or the new technique proved better, without an understanding of the fundamental and underlying reasons. The increasing reliance upon mechanical fastening systems and the need for improvements coincided with the recognition that research on basic theory and basic understanding must be undertaken. The Fastener Research Council stepped into this vacuum and initiated various technical activities aimed at correcting these deficiencies.

Council on Wave Research

On the basis of discussions with B. A. Bahkmeteff and M. P. O'Brien, Professor J. W. Johnson of the University of California at Berkeley wrote a letter on May 23, 1949 to the chairman of ASCE Hydraulics Division, L. G. Straub, in which he proposed the formation of a nationwide body for studying waves and related problems. At the request of the Hydraulics Division, the formation of a Council on Wave Research (Item 105 in Table 2) was approved by the ASCE Board of Direction in 1950 and, in response to an ASCE request, the Engineering Foundation provided a $1000 grant for organizational purposes. The detail objectives of the Council were stated as follows: (1) to encourage, sponsor, and coordinate research on wind-generated waves and related problems, such as coastal erosion and protection, structural design in coastal areas, and waves on inland waterways; (2) to review the results of both laboratory and field studies in this field, and the effects of engineering works; (3) to develop design methods, criteria, and procedures for the use of practicing engineers; (4) to publish the findings of the Council; and (5) to recommend the assignment to appropriate agencies of certain projects and services, the continuation of which is advisable. The organization of the Council was accomplished rapidly. At the outset, it was composed of representatives of twelve universities, six large industrial companies, three research institutes working in oceanography, thirteen departments of the government, and the Illinois State Division of Waterways. The University of California at Berkeley was the seat of the Council. The officers of the Council were from the staff of the University: Chairman Morrough P. O'Brien, dean of engineering, Secretary J. W. Johnson and Executive Engineer R. L. Weigel. The first meeting of the Council was held on November 9, 1951 in Houston, Texas at the conclusion of the Second National Conference on Coastal Engineering sponsored by the Council.

The principal activity of the Council was the sponsoring and organizing of the Coastal Engineering Conferences to bring together workers in the field of wave and shore erosion research and to summarize knowledge of coastal engineering. The Council was not organized in time to assume full responsibility and sponsorship for the First Annual Conference on Coastal Engineering, which was held at Long Beach, California, in October 1950. This conference was organized by the Departments of Engineering and the Division of Extension of the University of California. The editing and publication of the proceedings were, however, in the hands of the Council. The volume was published early in 1951.

The Second Conference on Coastal Engineering was held at Houston, Texas, on November 7–9, 1951. Thirty-one papers, on foundation problems, basic design, design of coastal structures, case histories, and some specific engineering problems, were presented. This conference was sponsored and its organization was handled by the Council with the cooperation of the Southwest Research Institute. The volume of proceedings of this second conference was published in May 1952.

The Third Conference on Coastal Engineering was held at Massachusetts Institute of Technology, Cambridge, Massachusetts, on October 22–24, 1952, and included three inspection trips on October 25. Cooperating groups were the local and regional sections of the ASCE and ASME. Thirty papers were presented, of which twelve described shoreline and wave conditions on the east coast of the United States. Three papers describing conditions in Denmark, France, and Spain were presented by engineers from those countries. The proceedings of this conference were published in June 1953.

The fourth conference, dealing primarily with wave and shore problems on the Great Lakes, was held at Chicago on October 29 and 30, 1953, with inspection trips on October 31. A total of 137 persons registered at the conference. Twenty-four technical papers were presented and discussed. The proceedings of this meeting were published in 1954.

The fifth conference was held at Grenoble, France on September 8–11, 1954 at the University of Grenoble. Thirty-six papers were given and printed in the proceedings that were published in 1955.

The sixth conference on Coastal Engineering was held in Florida on December 1–7, 1957, with an attendance of 173. The registrants (University and state, county, and municipal engineers) also were given an opportunity to inspect beaches and coastal engineering works at all the principal beaches and inlets on the Atlantic and Gulf coasts of Florida. Fifty-four papers, including five on hurricanes, were presented at the technical sessions. The proceedings were published in August 1958. Foreign engineers, who prepared and presented seventeen papers, came from Ceylon, Denmark, England, France, India, Mexico, Netherlands, Portugal, and Venezuela.

By the time of the Seventh Conference on Coastal Engineering, held at Hague in the Netherlands on August 22–29, 1960, the interest in these conferences had grown greatly. The conferences were unique in that a large variety of research workers were brought together from all over the world for an exchange of information. Participating individuals included engineers, oceanographers, meteorologists, geologists, geographers, and applied mathematicians. The important product of these international symposia was that the exchange of information between these workers revealed deficiencies in the basic knowledge. This spotlighting of the gaps in existing knowledge greatly assisted the Council on Wave Research in its task of preparing the comprehensive report "Deficiencies in Research on Water Gravity Waves."

The world-wide nature of interest was clearly reflected by the choice of locations for the next three conferences. They were held in Mexico, Portugal, and Japan.

In addition to the regular conferences on Coastal Engineering, the Council was involved in several other activities. It organized a conference on water-wave problems as related to naval architecture and engineering held at Stevens Institute of Technology, Hoboken, New Jersey, on October 25–27, 1954. It was sponsored jointly by the Council on Wave Research and the Society of Naval Architects and Marine Engineers.

An early, major undertaking of the Council was the compilation of a glossary of terms and a list of standard symbols used in this field of engineering. The glossary was published in September 1952. In addition to the list of terms and symbols, photographs and sketches have been incorporated to illustrate the more difficult concepts. The council also compiled tables of functions for gravity waves which were published in February 1954.

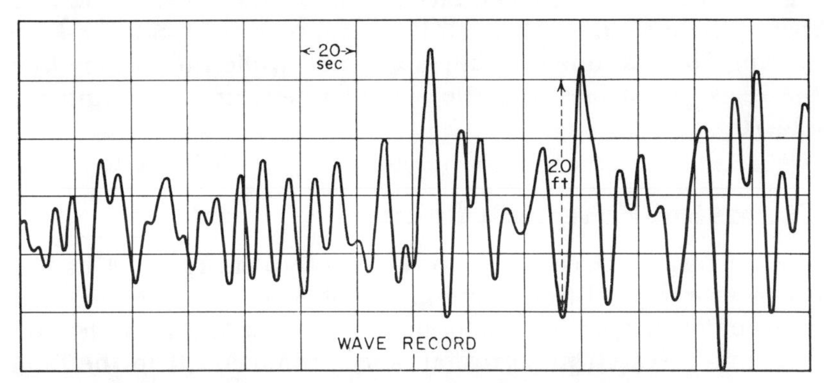

Wave record made at Oceanside, California, on June 24, 1949 illustrating the variability of surf. Waves vary both in height and period, and the alignment of the coast and the condition of the bottom may cause wide variation in the surf along the shoreline.

To keep Council members informed of current activities, a series of circulars was prepared, including a compilation of the research activities of the Beach Erosion Board of the Department of the Army, Corps of Engineers. In addition, the Council prepared a list of urgent research problems, including those pertaining to wave motion, man-made structures, sediment movement, shoreline processes, and others.

In October 1958 the Council collaborated with the Waterways and Harbors Division of the ASCE in sponsoring a three-day conference at Princeton University on problems concerning littoral drift. Two task committees of the Coastal Engineering Committee of the Waterways and Harbors Division were the ASCE sponsors of this conference. Discussions held at the conference resulted in the development of a sound program in this field of activity.

From time to time the Council has received requests for information on the facilities that might be available in this country for research on particular wave problems. To provide such information the Council has canvassed the laboratories that are known to have facilities for wave research to obtain an up-to-date description of such equipment. These data were assembled into a brochure distributed as requests for such information were received.

As the Council was primarily a coordinating agency, limited cash contributions and the income from the sale of the annual proceedings were sufficient to meet the cost of publishing the annual volume and the incidental expenses in connection with the office of the Council at the University of California. The income from the registration fees for the annual conferences was used to cover the cost of the meetings. Through the fiscal year 1957–58, the Engineering Foundation contributed $11,500 while the income from the registration fees and the sale of proceedings amounted to $24,700.

Corrosion Research Council

The Corrosion Research Council (Item 110 in Table 2) had its origin in a request from a group of engineers and scientists that AIME officially sponsor the formation of an organization for fundamental research in corrosion, and recommend it to the Engineering Foundation for support.

Such an organization was considered to be necessary because "up to now, the attack on problems of corrosion control has been mostly by empirical methods based on a valuable but still inadequate store of fundamental knowledge regarding corrosion processes and reactions. . . . Two things, therefore, are needed: increased fundamental scientific data; and a concerted team effort by scientists in the several disciplines of learning which have a bearing on corrosion—electrochemistry, physical metallurgy, solid-state physics, nuclear physics, surface chemistry and physics, reaction kinetics, and others."

The request was approved by AIME on March 17, 1954, and a proposal for sponsorship and sufficient support to organize the Council came before the Foundation Board at the semiannual meeting on June 17, 1954. After prior study and approval of the request by the Foundation's Research Procedure Committee, the project was approved by the board, and a grant of $1,000 for organizing purposes was authorized.

An organizing committee of seven experts in corrosion with H. H. Uhlig of MIT as chairman was appointed, and the first meeting was held on September 28, 1954. Between that date and September 21, 1955, when the Council was organized and the organizing committee discharged, the committee prepared a brochure describing the proposed Council and its aims, a letter for transmitting the brochure to executives in industry, and a set of by-laws and rules of administration. It also invited proposals for the first research project. The campaign for funds began in May 1955, and by the end of the fiscal year $14,350 had been subscribed by industry. The first chairman of the Council was R. M. Burns of Stanford Research Institute. F. L. LaQue of the International Nickel Company became chairman in the second year of the Council. E. H. Robie was the first executive director. In October 1959, Robie was appointed assistant to the chairman and F. T. Sisco became executive secretary.

The first research project, "Fundamental Reactions at the Surface of Metal Single Crystals in Selected Environments," was started at the National Bureau of Standards soon after January 1, 1956. The second research project of the Council, also at the Bureau of Standards, the "Relation of Stress to Corrosion," was started in January 1957. The following additional four projects were started before the end of 1961:

Mechanism of the Dissolution of Metals at the University of Pennsylvania, supported by the CRC. Started October 1959, Council support ended September 30, 1961.

Stress Corrosion in Face-Centered Cubic Metals at the University of Wisconsin, supported entirely by CRC. Started January 1961.

Effect of Nitrogen on Stress-Corrosion Cracking of Stainless Steel at the Virginia Polytechnic Institute, supported entirely by CRC. Started July 1961.

Anodic Protection Studies at the Rensselaer Polytechnic Institute, supported entirely by the CRC. Started October 1961.

Through 1957–58, the Foundation contributed $12,000 while $131,000 was obtained from numerous other sources.

Research Council on Air Resources Engineering

A Coordinating Committee on Air Pollution (RC-A-60-1d in Table 2) was established by the Engineering Foundation in 1960. The committee arranged a "Research Symposium to Advance Management and Conservation of the Air Resource" in 1962 in cooperation with the American Society of Civil Engineers, New York University, and other organizations. The symposium clearly indicated the need for (1) better diagnostic tools, and (2) a combined attack on the air pollution problem by the engineering profession. As part of the development of the study "Measurement of Stack Emissions," a workshop meeting on "Stack Sampling and Monitoring" was held in November of 1963. A study on the "Evaluation of Carbon Dioxide Concentration as an Indicator of Air Pollution in Urban Atmosphere" was carried out at the University of California, Los Angeles. The study, underwritten by the Engineering Foundation, was sponsored by the Research Council on Air Resources Engineering of the American Society of Civil Engineers often referred to as Air Pollution Research Council. The Council, chaired by W. T. Ingram, was active at least from 1962 through 1972.

Metal Properties Council

Industrial progress in the United States required the establishment of a permanent organization to prepare and analyze data on the properties of metals for code and specification-writing bodies. In response to this need the Foundation Board at its May 7, 1964 meeting approved the establishment of the Metal Properties Council (RC-A-64-2 in Table 2) as an activity of the Engineering Foundation and appropriated $25,000 to partly finance the first year's work. The Council was sponsored by the American Society of Mechanical Engineers, the American Society for Testing and Materials, and the American Society for Metals. It served as a vehicle for cooperative procurement of data on the properties of metals, analysis and condensation of such data into a usable form, and dissemination of the resulting authoritative technical information.

The purposes of the Metal Properties Council were:

1. To continually determine data on metal properties in cooperation with technical societies and government agencies responsible for the presentation of code regulations and specifications;
2. To conduct a program for obtaining such data by financial support and/or by direct sponsor contribution of reliable data;
3. To generate, collect, analyze, and organize the data so secured;
4. To prepare reports to be distributed to responsible committees, the scientific and educational communities, and the general public; and
5. To maintain an information center on similar activities abroad.

On January 1, 1966 A. O. Schaefer became the director of the Metal Properties Council. On August 1 of the same year, the Council moved to its permanent headquarters on the fourteenth floor of the United Engineering Center.

Building Research Advisory Board

The Building Research Advisory Board (BRAB) was established late in 1948 (Item 102 in Table 2) for the stimulation and correlation of building research. It was organized as a branch of the Division of Engineering and Industrial Research of the National Research Council. During the first several months of its existence, it undertook a survey of modular coordination for the U.S. Housing and Home Finance Agency. This early work, and the experience of the board and its committees during this period, made it clearly evident that a broad program, over a period of five years, should be undertaken to stimulate research in the building industry. This program was formulated and adopted early in October 1949. The building industry and the Foundation were requested to finance the project to the extent of $30,000 for the first year, with an eventual budget of $100,000 annually being anticipated.

The initial objectives of BRAB were to survey the problems of the building industry, to conduct conferences and forums where ongoing research could be discussed and evaluated, to maintain a continuing analytical classification of building research, to encourage coordinated field trials of the results of research, to act in an advisory capacity to government agencies now carrying on research in the building field, and to develop a suitable program of publication and information.

The board was made up of thirty leaders of the building industry, including representatives from universities, foundations, trade organizations, Government bureaus, publishers, consulting engineers, and industrial companies. Frederick L. Hovde, president of Purdue University, served as the first chairman and William H. Scheick as the first executive director of the board. BRAB's income for the first year was $35,700, of which the Foundation contributed $1,000. This did not include the large contributions by experts in the building industry who gave freely of their time for consultation and in attending meetings.

On January 11 and 12, 1950, the board held its first research-correlation conference on the topic "Weather and the Building Industry." Registrants for the conference sessions numbered 310. Some thirty individuals from the fields of weather science and building technology participated in the program. The proceedings of the conference were published in April. On June 30, 1950, negotiations were completed with the Housing and Home Finance Agency for a contract under which BRAB conducted a survey of housing research in educational institutions, foundations, and industry. The contract was carried out by project staff organized by BRAB to do the work. The topics of the next three conferences were "Fire Resistance of Non-Load-Bearing Exterior Walls," "Design of Laboratories for Handling Radioactive Materials," and "Moisture Resistant Characteristics of Paint and Paper."

BRAB activities generally involved conferences, preparation of reports, and formulation of policies for the building industry. Most of these activities were either carried out or advised by committees of experts drawn from the industry at large. Through the fiscal year 1954–55, the Engineering Foundation contributed $4,500 while the total from cash contributions and dues was about $50,000. The income from contracts with the Federal Construction Council and the Federal Housing Authority for advisory studies was not included in these figures since the income from these contracts was for self-sustaining projects subject to separate accounting.

8. FRANK T. SISCO'S TERM

Director Colpitts' activities on behalf of the Foundation were cut short by illness. He resigned at the executive committee meeting of January 21, 1949.

E. H. Colpitts' resignation caused the board to once again re-evaluate the position of the director. It was decided to terminate the director's position involving coordination of the Engineering Foundation's general administrative activities including the supervision of the grant projects. Instead, it created the position of technical director, whose sole primary responsibility would be the supervision of the grants program. At the January 21, 1949 executive committee meeting, Frank T. Sisco was appointed to serve in this newly created position. Sisco had long been associated with the activities of the Engineering Foundation. He had been the editor of the important "Alloys of Iron Project" which was one of the primary activities funded by the Foundation during the 1930s and '40s.

Technical Director F. T. Sisco had a strong interest in developing the Engineering Foundation's sponsorship of research activities. He, however, was appointed at a time when the Foundation was undergoing a protracted period of self-evaluation in regards to its purposes and activities. There were many board members who felt that the Engineering Foundation should return to its earlier emphasis on the humanistic aspects of engineering such as the relationship of the engineering profession to society in general. This process of self-evaluation worked at times against Sisco's proposals. The self-evaluation process was institutionalized when the board proposed the creation of the Engineering Study Committee of the UET at the March 3, 1949 board meeting of the Foundation. This committee was to be composed of one member appointed by the UET from each of the Founder Societies.

The background for the creation of this committee can be found in the minutes of the board meeting:

> Mr. Swasey established the Engineering Foundation in the hope that his donations would serve as a nucleus to attract numerous similar gifts. Mr. Swasey's vision was correct as far as research was concerned. However, we all know that research today has taken on the proportions of a riptide, but the Engineering Foundation through no fault of its own has been largely bypassed. Mr. Swasey's hope that this expansion in research would be accomplished by his great conception has not been realized . . . Mr. Swasey also envisioned means other than research for advancing the profession as well as the possibility that times and trends

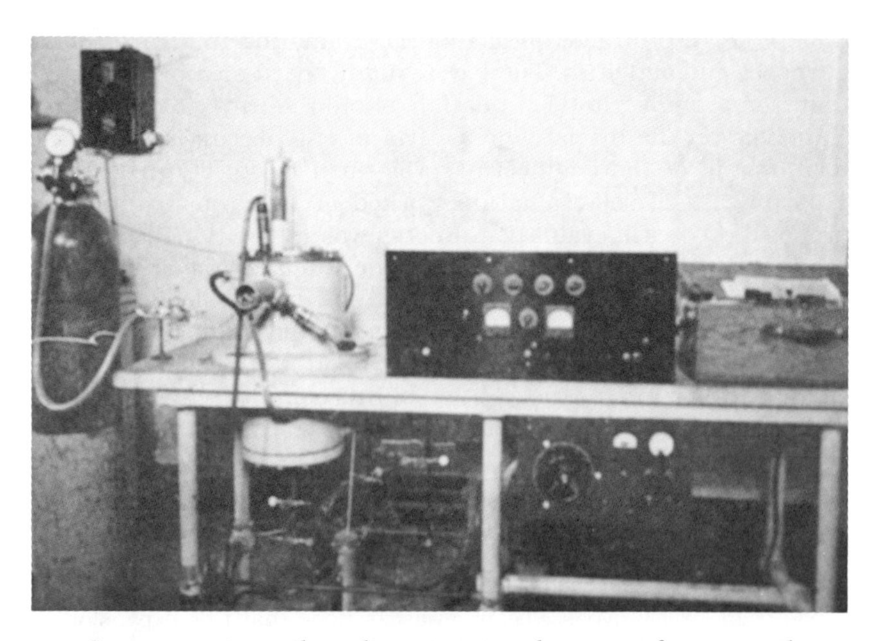

Geiger counter with scaling circuit and vacuum furnace with temperature controlling equipment for diffusion experiments at MIT.

might change the relative emphasis on the several components of his plan. Accordingly, the terms of his gifts provided that the income is to be applied not only to research but also to "the advancement in any other manner of the profession of Engineering." So far in the administration of the Engineering Foundation emphasis has been placed almost exclusively on research. Not only has this applied to the allocation of funds, but the founder societies have also leaned over backwards to name men with a research background to the Board. The recently retired director was a research man. Financial Status—Out of an endowment of $1,000,000 only $27,000 is available for research grants annually. It is submitted that this amount cannot be of much significance in an overall volume of research expenditures now running to hundreds of millions of dollars per year. It is further submitted that having assumed such proportions, research no longer needs stimulation by the Engineering Foundation and that it is now time to consider the shifting of emphasis to the second item of Mr. Swasey's objectives: the advancement of the profession. It is the committee's opinion that the funds available could be made to serve a succession of worthwhile projects designed to advance the profession by means other than research.

It was against such a background that Technical Director Sisco assumed office.

The self-evaluation process of the Engineering Foundation continued during the early months of Sisco's tenure as technical director. His primary interest was in the field of research projects. To deflect criticism from those individuals who felt that the Engineering Foundation had placed too much emphasis on funding research projects and not enough on the other aspects of the engineering profession, he wrote the following letter to the board on April 15, 1949:

In view of the fact that a few engineers have recently questioned whether the Engineering Foundation was doing enough to advance the engineering profession in comparison with its aid to engineering research, I have divided the projects into two groups, details of which are given in Tables I and II and the summary is likewise divided in this manner. Since its establishment thirty-four years ago, the Foundation has sponsored ninety-six projects of which sixty-nine can be classed as engineering projects and twenty-seven can clearly be classed as projects for the advancement of the engineering profession. Sponsorship of monetary grants of approximately $676,000 has been responsible for carrying out projects in research or for the advancement of the profession representing a total cost of nearly $9,000,000. Certainly the reputation that the Engineering Foundation has acquired for the advancement of engineering is well deserved and I feel sure Mr. Swasey would be proud of the record that the Foundation has made. You are, of course, as familiar as

I am with the fact that many of the Engineering Foundation's projects have received international acclaim and that hundreds of technical papers and twenty five or more text books have been published as the direct result of the Foundation activities which have been read and used by engineers all over the world. Such a fact is important, but it cannot be readily shown in a statistical report.

The tactic of compiling reports and writing formal letters to the board proved to be an effective one, and Sisco utilized it throughout his association with the Engineering Foundation.

One of the key elements of Sisco's career as technical director was his willingness to undertake the required visits to the research projects that were funded by Foundation grants. He was far more active than Colpitts had been in this regard and he also prepared detailed analytic reports for the board at the conclusion of each annual cycle of visits. It also appears that Sisco possessed an orderly mind with a flair for administration. Although he was not initially charged with many of the Engineering Foundation's administrative duties, he soon assumed *de facto* control of these activities which he immediately consolidated and streamlined. He also undertook to purge the Foundation's project files of much extraneous material.

Sisco encouraged the renewal of the somewhat dormant policy of involvement by the Engineering Foundation in the activities of other research oriented organizations. With the consent of the board, Sisco became the representative of the Foundation on the board of the Division of Engineering and Industrial Research of NRC in 1949.

During the early years of Sisco's tenure as technical director, the board also began to develop new policies which would shape its ongoing and future relationships with the various research councils. At its June 16, 1949 meeting the board approved "that the Engineering Foundation continue to propose or endorse the policy of project programs being planned on national council lines, as an excellent means of interesting industry, thus drawing research and industrial personnel together in greater understanding and better cooperation."

Sisco also continued to refine the Foundation's contingency fund policy in relation to research grants. The following minimum ratios were adopted by the board at its June 16, 1949 meeting: EF one-to-three outside for the first year of a project, one-to-four for the second and one-to-five for the third year. It adopted the following statements: (1) that the Foundation reduce its grant proportionately, depending on the percentage collected of the required outside support, (2) that nonpayment of the granted funds be made if less than two thirds of the specified outside funds are raised.

During the early 1950s, Sisco continued to refine the contingency fund aspects of the Engineering Foundation's grant policy. At its May 11, 1950 meeting, the Research Procedures Committee accepted his position that contributions in kind and a reasonable figure for overhead could be included in the contingency funding, while depreciation and obsolescence of equipment along with heat, light, and power could not be counted in deciding if a project met the Foundation requirement for outside funding. Sisco based this belief on the supposition that expenses for such items as heat and equipment wear represented essential contributions to any serious project by the sponsoring organization and also represented a bare minimum commitment that was the first step in starting a project.

Sisco displayed a strong interest in maintaining and increasing the involvement of the Foundation in international engineering affairs. His editorial background caused him to be appointed to the NRC's Committee on International Publications.

Since research was his chief priority, Sisco continued to seek board approval for the revision of the Engineering Foundation's grant policy. The Engineering Foundation had long emphasized the need for cooperative research in its grant policy. So it was natural that Sisco would also emphasize such research in his revision of the grant policy, which was adopted by the Research Procedures Committee at its May 11, 1950 meeting: "In the future the Engineering Foundation would actively encourage the organization of such cooperative projects and that it would favor such projects in assigning its funds available for research."

During 1950 Sisco continued to display his earlier interest in scientific publications in his role as the primary representative of the Engineering Foundation. At the October 19, 1950 board meeting he reported that he had attended a two day conference on primary publications that was designed to discuss ways and means of expanding scientific and technical publications facilities in order to take care of the recently increased number of invaluable reports of research in the face of rapidly rising publication costs.

During the early 1950s Sisco continued to serve as the Engineering Foundation's *de facto* ambassador to the general engineering and scientific communities. He fulfilled this function by his attendance at many and varied international conferences and by his participation in the deliberations of engineering boards. During the same period Sisco also extended his ambassador role to include his annual visitations to the currently funded projects. At the board meeting of June 21, 1950 Sisco stated in his semiannual report that he made a special effort to talk about the Engineering Foundation to the heads and deans of college engineering departments during the course of each project visitation. His goal was to make these influential individuals more familiar with the aims and activities of the Engineering Foundation and the catalytic effects that its sponsored research projects had exerted on the recent growth of engineering in the U.S. Sisco concluded his report with his strong belief that such dignified publicity would stimulate interest in the work of the Engineering Foundation and would pay dividends in the future in goodwill and possibly in added industry and trade association support. "Such efforts should be continued," Sisco reported, "since the outbreak of the Korean emergency has caused a slowdown in applications for Engineering Foundation support."

At the October 19, 1950 board meeting he described the recently published book on soil mechanics which resulted from one of the Engineering Foundation's most important civil engineering projects (Project 38 in Table 2). As Sisco described it, this new volume presented a summary of this project's research findings, which were compiled between 1924 and 1948. The book—*Subsurface Exploration and Sampling of Soils for Engineering Purposes* by M. Juul Hvorslev—was specifically cited at a recent conference as a particularly good example of a modern engineering text. Finally, it should be noted that under Sisco's direction the board also voted $3,500 for the Council on Wave Research for the specific purpose of printing a publication on coastal engineering, "with the understanding that all profits from the book shall be used by the Council for its operations and that this income shall be considered by the Engineering Foundation in connection with the application by the Council for any future grants from the Foundation."

The decrease in the number of applications and the corresponding rise in the amount of unallocated funds available for grants, greatly disturbed Sisco. At the June 21, 1950 board meeting he stated "that the Engineering Foundation has played and can continue an important role in catalyzing research in the broad and important field of engineering that is not at present occupied by either pure scientific, industrial, or governmental research and it is strongly recommended that the Engineering Foundation Board members consider the present tendency toward a reduced number of projects and increased amounts of unallocated funding carefully with the possibility of discussing this situation with the secretaries of research committees of the Founder Societies. Such a discussion may result in suggestions for increasing the usefulness of the Foundation to the engineering arts and sciences."

Aware of the continuing efforts of the board members to undertake a process of self-evaluation of the Engineering Foundation's direction and purposes, Sisco countered by basing his activities on the successful precedents of the Engineering Foundation's research projects. He sincerely believed that the decline in the number of applications was solely the result of a growing lack of awareness of the Engineering Foundation, and he felt that an increased program of publicity and public relations oriented publications would overcome this deficiency. The foundation of his plan was the creation of a bimonthly engineering news magazine similar in concept to the widely read *Popular Research Narratives* which had been published by the Engineering Foundation between 1921 and 1929. Initially proposed at the October 18, 1951 board meeting, this idea was not accepted by the board, which was beginning to seriously question the Engineering Foundation's continuing emphasis on funding research projects.

During 1951 the Engineering Foundation rendered an important service to the U.S. Government when Technical Director Sisco worked as the editor of the State Department's *Engineering Newsletter.* This publication was widely distributed in both English and foreign language editions and thanks to Sisco's editorial skills its format and contents selection were greatly improved. Sisco also received board approval to serve as secretary for a commission on technical assistance, which had been set up by the Committee on International Relations of the Engineers' Joint Council. The objective of the commission was to lend technical assistance to the Department of State, the United Nations, and UNESCO.

Despite the refusal of the board to authorize and fund his proposed engineering news magazine, Sisco continued to advocate an increased number of public relation publications as a means of increasing professional awareness of Engineering Foundation activities. He won a partial endorsement when the board at its June 19, 1952 meeting voted to continue its existing authorization of the preparation and publication of suitable articles on Foundation activities.

During 1952, Technical Director Frank T. Sisco continued to be concerned about the decline of the number of suitable grant applications and the corresponding rise in the amount of unallocated money. He reported:

> It is generally acknowledged that in the last ten years the Foundation has done a noteworthy job in organizing and financing engineering research that within a few years has proved to be worthy of extensive industrial and trade association support and also in acting as a coordinating agency in those fields of engineering research that lacked such facilities. The technical director feels, however, that much remains to be done along these lines and he would appreciate greatly any assistance that the members of the Board and the research committees of their respective societies could give to him during the coming months.

In the early 1950s an Engineering Foundation funded project made a notable contribution to the development of the peaceful uses of atomic energy when the "Diffusion in Steel" research project (Item 97 in Table 2) completed work on the qualities of radioactive iron alloys. This research would play an important role in the development of steel components of nuclear electrical generating reactors.

Period of 1952 Through 1957

The continuing efforts of Sisco on behalf of the Engineering Foundation were recognized by the board at its October 1952 meeting, by its action to upgrade his position to director with full authority over the Foundation's administrative activities. However, he was not given a raise. Among the first duties that he was assigned in his new position, was the rewriting and simplification of the Foundation's rules of administration and the preparation of articles on the achievements of specific research projects. He was also designated to prepare a discussion of engineering research that would stimulate a more aggressive research attitude on the part of the engineering profession. To aid him in this task, the board created a Publicity and Public Relations Committee.

To further increase the publicity activities of the Engineering Foundation, it was suggested, that in cases where research councils or projects had become so successful that they no longer requested Foundation funding, that the Engineering Foundation make token grants to them for the purpose of associating itself with these continuing research efforts.

Despite his increased role in the administrative affairs of the Engineering Foundation and his growing involvement in the Foundation's public relations campaign, supervision of the research grants program continued to be Director Sisco's primary responsibility. At the meeting of the Research Procedures Committee that was held on May 13, 1953, he stated that "the increasingly important role that the Foundation is playing in engineering research is becoming evident. A few years ago the Foundation was supporting between eleven and thirteen projects and the Engineering Foundation annual expenditure of $35,000 to $40,000 was catalyzing industry funds of between $200,000 and $300,000. At present the Foundation is supporting seventeen or eighteen projects and the same Engineering Foundation's total expenditure is attracting well over $500,000 in outside support." Director Sisco attributed this increase in both the number of projects and the total amount of outside support to the Foundation's upgraded public relations activities which would be further enhanced by an eight-page booklet describing the aims and activities of the Engineering Foundation. Director Sisco believed that this new brochure would also be an effective fund raising tool.

Due in large part to the success of Sisco's public relations activities, it was voted at the board meeting of May 13, 1953 to increase his salary to $4,000 per year. At Director Sisco's instigation the board also continued to expand its publicity and public relations activities. At its June 17, 1954 meeting the board voted to add to its future annual reports a brief history of the Foundation as another means of raising awareness of its purpose and activities. The effectiveness of the eight-page pamphlet was noted at this meeting since its distribution had been directly responsible for two recent high quality grant proposals. An even greater enlargement of the Foundation's public relations and publicity activities was also noted by Director Sisco's announcement that three descriptive articles on Foundation funded projects were under preparation. These articles would focus on the achievements of the Column Research Council, the High Temperature Steam Generation project, and the Committee on Evaluation of Engineering Education. All of these treatises would be written by Mary T. Jessup of the ASCE editorial staff, who would also be employed to assist the director in his public relations activities.

Director Sisco continued to be aware of criticism that the grants policy of the Engineering Foundation continued to focus primarily on research at the expense of projects to advance the engineering profession. In his semiannual report for the period between 10/1/53 and 3/1/54 he wrote that the Engineering Foundation funded twenty projects ". . . sixteen of which are definitely concerned with some phase of engineering research and the other four are focused on advancement of the engineering profession, especially the advancement of engineering education. It is estimated that during the current fiscal year industry and others will contribute between $400,000 and $450,000 to these twenty Foundation sponsored projects, all of which have made excellent progress." To his critics who questioned the Foundation's continuing emphasis on research, Sisco pointed out that the Engineering Foundation funded two year project "Evaluation of Engineering Education" was almost completed. Director Sisco also continued his campaign to make the activities of the Engineering Foundation better known to both the engineering profession and the general public. To aid him in this task he had earlier asked the heads of all Foundation sponsored projects to submit brief articles to him on the progress. However, he soon found that the summaries submitted by the project directors were too technical in nature to be used for publication. He was forced to reject these summaries and he asked the project heads to resubmit them to him in a more simplified form. Sisco's public relations activities were enhanced by a decision of the Public Relations Committee to open the Engineering Foundation's files to feature story writers.

The efforts of the Foundation to increase awareness of its activities was noted at the October 21, 1954 board meeting, when it was formally announced that Jessup was preparing 1,000 word descriptive articles on Foundation sponsored projects for possible publication in the journals of the Founder Societies. At the same meeting, the board also received a summary of the ongoing projects from Director Sisco: "It is evident that the comparatively large volume of the important engineering research is carried on annually under Foundation sponsorship with Foundation funding support that amounts to less than ten percent of the total cost. This catalytic effect which developed over the years from Foundation sponsorship and early financial support is unique in the history of research and education/foundations."

Sisco also summarized the two year study of engineering education (Item 106 in Table 2) which he considered to be among the most important undertaken by the Engineering Foundation during recent years. "This two year study of engineering education was made by a committee of ASEE and 122 institutional committees appointed by deans of all engineering colleges having accredited curricula and ended with a noteworthy report. The committee made one recommendation which if implemented should have far reaching consequences: that curricula should be divided broadly into engineering-general and engineering-scientific. These two branches should have a common stem consisting of 100 semester hours of mathematics, basic science, engineering science, and humanities, with a little essential technology. The other thirty-eight semester hours should consist of design, analysis, further science and technology, and electives." This report marked a high point in the Engineering Foundation's long-term sponsorship of projects related to the improvement and reform of engineering education.

During the remainder of 1954 Sisco continued to demonstrate his interest in the reform of the Foundation's administrative procedures. At the December 20, 1954 board meeting he secured approval of new rules of administration which he had recently formulated to streamline the Foundation's daily activities. In his semiannual report for the period between October 1, 1954 and March 31, 1955, Director Sisco was also able to report the continued success of his public relations campaign. Two journals had published the first of the 1,000 word articles on Foundation funded projects that had been prepared by Jessup, and a general article on the Engineering Foundation had also appeared in the April 1955 issue of the *ASTM Bulletin.* Sisco was also able to report that the increased number of Foundation projects would soon wipe out the surplus that had been accumulated since World War II. This point can be considered as the apex of the Foundation's traditional policy of emphasizing support for research projects. From this point onward a continuing process of self-examination and evaluation would cause the board to begin the development of a policy that would represent a radical break with its past.

The first evidence of the process by which the Foundation would drastically alter its priorities for funding appeared at the October 13, 1955 board meeting. At this meeting the executive committee of the board, which had been inactive for several years, began to once again function. In its capacity as a planning body, the executive committee would become the major catalyst for change in the Engineering Foundation's policies.

The year 1956 began with a significant addition to the endowment of the Engineering Foundation when it was reported at the May 3 board meeting that a gift of a hundred shares of General Motors Corporation stock was received from Professor O. W. Boston of the University of Michigan. The market value of these shares was estimated at $4,412.50. At this meeting, Director Sisco tendered his semiannual report for the period between October 1, 1955 and March 31, 1956, which noted that the projects that were sponsored by AIME were receiving more than fifty percent of the Foundation's available funding. This dominance of AIME projects was the continuation of a trend that had become apparent since 1950. Director Sisco attributed this success to the fact that AIME made extensive efforts to publicize the activities and purposes of the Engineering Foundation to its members on a regular basis. This activity had recently inspired ASCE to do likewise and as a result it was increasing the number and quality of the projects that it sponsored for Engineering Foundation funding.

In spite of Sisco's success in revitalizing and upgrading the Foundation's traditional research project grant program, the harbinger of drastic changes that would undo much of his success was also apparent at the May 3 board meeting. At that meeting the executive committee voted to consider the following items:

1. Should or can the Foundation search for means to increase its endowments.
2. Are there specific types of projects in which the Foundation's support has been or can be particularly effective and should the Foundation tend to concentrate its support on such projects.
3. Appointment of a planning committee to (a) study the possibility of obtaining additions to the endowment fund, and (b) study future policy for the most effective allocation of available funds.

The result of these proposals was a recommendation that the chairman be empowered to appoint a committee of six members of which there shall be one member from each founder society and at least two of these four shall be present members of the Engineering Foundation's Board and that there shall be two other members of the committee who shall represent other research viewpoints and further that the director of the Engineering Foundation shall be an *ex-officio* member of the committee and further that it be the duty of the committee to study and recommend to the board future aims and policies. This committee was directed to make a preliminary report by October 11.

Before the committee's report could be delivered an event occurred that significantly increased the endowment of the Engineering Foundation. For over a decade the Foundation had been receiving funding from the interest on the remainder of the estate of Edwin McHenry which had not been assigned to support his immediate heirs. When the designated heirs had passed away all of the principal of the McHenry estate would be available to support Engineering Foundation projects. The estate was set up as a trust dedicated to the memory of Mrs. Blanche H. McHenry for the benefit of the Engineering Foundation. It was estimated that well over $400,000 would be available. By the autumn of 1956 all of the designated heirs had passed away and the McHenry estate was now totally available to help support the activities of the Engineering Foundation. When all of the legal issues had been resolved a total of $408,357.63 was made available to the Foundation.

The report of the planning committee was delayed and other matters came to the board's attention. At its May 16, 1957 meeting, the board voted to turn down a request to make a grant to support research in Israel and it thus established a policy on foreign research that stated: "It is the policy of the Engineering Foundation that it does not support projects located outside of North America."

The board also continued to modify its grant policy in an effort to both secure more publicity for the Engineering Foundation and to better serve the needs of the engineering profession. At this same meeting, it approved the following modification to its grant policy: "As a condition of approval of all Foundation grants, all results of research be made available for publication and that no work supported in part or in whole by the Foundation shall remain unpublished for more than two years."

Despite the increased number of high quality applications for grant funding, the Research Procedures Committee remained concerned about a perceived lack of interest in the Foundation's grant program. To address this problem, it approved the following statement at its October 10, 1957 meeting: "It was decided that it might be well for the committee to take attitude and to seek out more research projects rather than wait for them to come to the Engineering Foundation." As one means for the committee to take some initiative it was suggested that consideration be given to the establishment of a modest scholarship program.

Period of 1957 Through 1959

The long delayed report of the Planning Committee caused the board to take further action as a part of its ongoing self-evaluation and self-examination process at the March 13, 1958 meeting of the executive committee. At this meeting it was suggested by Chairman E. M. Barber that a further study of the Engineering Foundation's program be required to determine whether its efforts and limited funds could be more profitably utilized in other fields of endeavor. To accelerate this process he requested that each member of the dual Executive-Planning Committee write a letter within two or three weeks expressing his thoughts on possible concepts of the Engineering Foundation's program. He was particularly concerned about the relevance of the Engineering Foundation's policy of providing support for specific projects. Barber expressed his thoughts in the foreword to the 1956–1957 Annual Report:

> Successive Engineering Foundation Boards have focused attention on different activities as seemed best suited to the times and to the needs of the engineering profession. Also they have devised a variety of procedures for sponsorship and support of the selected activities to obtain the greatest value from the funds available. For several years past the two principal activities have been (1) sponsorship and support of research councils to deal with specialized engineering subjects requiring technology, drawn from several of the commonly recognized technical disciplines; and (2) sponsorship and support of specific research projects, often, but not exclusively, on recommendation of one or more of the Founder Societies.
>
> In carrying out these activities the Foundation Boards have followed a 'seed money' philosophy whereby worthy projects may be given rather substantial support to organize and demonstrate their worth. Later, continuing support may be made contingent on the ability of the project to obtain the major portion of its funds from other sources, and, still later, the Engineering Foundation's support may be reduced to a mere token of continuing approval.
>
> The pattern of our technology is changing rapidly, however, and the changes create new challenges and new opportunities for the Engineering Foundation to fulfill its purpose and to be of greater service to the engineering profession. For example, there are notable increases in governmental and industrial research and in the availability of funds to support research projects with specific and worth-while objectives. These conditions tend to minimize the value of Engineering Foundation's support for specific researches, excepting possibly support for the organization of engineering society research committee projects.
>
> Another notable condition is that important new engineering subjects more often than not fail to fall neatly into the commonly accepted scope of any one of the recognized technical disciplines but involve a mixture of many disciplines. With the large research funds now available the technology of these subjects grows rapidly but often very nonuniformly and without the perspective and balance that could be obtained by a better coordination of the contributions to the progress from all of the relevant technical disciplines.
>
> Under these conditions there is growing need for critical reviews of important subjects to be prepared on a collaborative basis by qualified representatives of the several relevant technical disciplines and for these reviews to be made available to engineers, industry, and the public. Where such reviews have been conducted on an appropriately broad

basis, they have stimulated forward surges in the subject reviewed: various aspects of the subject have been put into their proper perspective, the need and opportunity for specific research have been highlighted, and potential applications have been stimulated.

It is suggested that the Engineering Foundation Board explore the possibility of initiating a program of critical reviews of important engineering subjects, the reviews to be prepared in the form of symposia with published proceedings. To insure that the subjects receive well-rounded treatment, the reviews should be composed of coordinated contributions from selected representatives of the Founder Societies. The role of the Engineering Foundation should be mainly to supply the initiative and the staff work and to serve as a focal point for the combined action of the Founder Societies. As an example of a subject, 'water resources' might warrant this broad treatment. This is a many-faceted subject that involves more or less of the technology of each of the Founder Societies; thus, water conservation is largely in the field of civil engineering, while purification, underground supplies, treatment, etc., involve mechanical, electrical, chemical, and mining engineering technology.

Critical reviews are a needed engineering research activity that is almost uniquely suited to the organization, finances and purposes of the Engineering Foundation. Note that the Engineering Foundation Board includes two representatives of each Founder Society and that it is almost the only financially self-sustaining organization in which the societies unite in technical activities. The united action that is implicit in a critical review program would be an intangible value added to the very real value of the reviews themselves.

During the 1956–57 budget year the Engineering Foundation provided sponsorship and 'seed money' support for twenty-eight projects. This compares with sixteen projects given similar support during 1950, when the board's current procedures were evolved. During 1956–57 the major portion of the available funds went to the support of specific research projects. Division of the available funds by twenty-eight shows that the average support per project is only a fragment of what is required to conduct a year of research. It is for this and analogous reasons that I have devoted so much space to consideration of new ways for Engineering Foundation to fulfill its purpose. In doing this, I believe that I reflect the mood of the entire Board whose members, although not having agreed on any particular new method, are questioning and searching for new and useful methods.

By the board meeting of May 16, 1958, the Executive-Planning Committee could report "that it is the sense of the board that a basic objective of the board in the administration of its funds is the stimulation of investigation and research in new areas as contrasted to continuing financial support." As a result, the Research

Procedure Committee was directed to formulate ways of implementing the above stated policy and report back to the board by October of 1958.

Before the Research Procedure Committee could comply with the board's directive it had another item of business to consider. The Ford Foundation, which was the wealthiest foundation in America, was developing a new division of science and engineering and it asked for the cooperation of the Engineering Foundation in formulating its rules of administration. Unfortunately, the concept of the Ford Foundation's Engineering Division differed so greatly from the practices of the Engineering Foundation that the proffered help was of little use.

This same meeting also recorded a major change as the American Institute of Chemical Engineers became the fifth Founder Society. This action further broadened the scope of the Engineering Foundation's support. The number of trustee representatives on the board was increased from four to five and of Founder Societies representatives from eight to ten.

The growing success of the research councils that had been initially funded and sponsored was recounted in the "Semi-Annual Report" of the director, Frank T. Sisco, for the period October 1, 1957 to March 31, 1958: "In general the research councils are adequately financed and the income is correctly budgeted to insure the greatest return for the smallest cash outlay. The amount of cooperative effort that goes into the work of the research councils is very high. If a monetary value could be placed on the aid rendered these councils by many of this country's most noted engineers, it would probably amount to ten times the available contributed funds." Sisco also summarized the general status of the Engineering Foundation's research projects and ironically produced what would prove to be a valedictory for the traditional grant program of the Engineering Foundation.

During the years 1950–1955 the amount of unallocated funds available to the Engineering Foundation Board for the sponsorship and support of engineering research increased from about $15,000 to $45,000. This increase was due to two factors: (1) a slight increase in the endowment from the $1,065,000 to $1,141,000 and a somewhat higher return from this endowment; and (2) a paucity of worthwhile new projects. As discussed earlier, in 1952 the board recognized the problem and appointed a public relations committee with the task of making the Engineering Foundation better known to the engineering profession. The new committee did two things to help solve the problem. It prepared a new brochure and it also secured the services of Jessup, the news editor of ASCE, to prepare occasional press releases and articles concerning the achievements of the various funded projects. These efforts were successful. Between 1950 and 1952 only fourteen to seventeen applications for research support were submitted annually while twenty-seven to thirty-two were submitted annually for the period 1955 to 1958. During the same period support for worthy projects increased from approximately $35,000 to $75,000 annually.

The Research Procedure Committee continued to take a more aggressive attitude in seeking new projects, rather than restrict itself to reviewing applications which were submitted for its consideration. Several projects were suggested. One such suggestion was a National Science Foundation project on documentation in engineering which would receive ten percent of available Engineering Foundation grant funds. The Engineering Foundation could also serve as coordinating agency for the five founder societies and help to prepare critical reviews of the state of existing knowledge in engineering and in this work the Foundation would work closely with the research committees of the founder societies. Another possible field could be holding symposia particularly in the areas where the interests of the societies overlap. Ironically, this last statement would prove to be a precursor of the future direction of the Engineering Foundation and the swan song of the Research Procedure Committee.

On September 19, 1958 the Planning Committee in conjunction with the Research Procedure Committee delivered one of the most important documents in the development of the Engineering Foundation. This joint report altered completely the direction of the Foundation's activities and it also represented the culmination of the process of self-evaluation that had characterized the Engineering Foundation since the end of World War II. The report stated that

> there was unanimous agreement that the income of the Foundation is so small relative to the vast funds available from other sources for research, that it is insignificant in supporting research. The proper use of the income would appear to be rather stimulate research. It is believed that a radically new approach to our broad policy and method of operation is necessary and that this can be achieved only by wiping the slate clean. It is therefore recommended: (1) That no new grants of funds for the period beyond October 1, 1959, be made. If any special grants are necessary properly to fulfill and terminate existing obligations, they should be made out of the budget for the current fiscal year. (2) That for

a suitable period of time, perhaps as long as three years, as much of the Foundation's income as necessary be devoted to formulating and instituting a new long range program. (3) That as a first step, the executive committee, acting as a planning committee formulate the broad outlines of such a program. (4) That an executive head be employed to develop the program in close collaboration with the board and the executive committee. (5) That until the new program develops projects for properly employing current income any excess of income over that required under item 2 will be accumulated.

The joint committee report was accepted by the board at its October 16, 1958 meeting.

It is ironic that just as the Engineering Foundation Board was about to change the direction of the organization's grant policy, an opportunity arose to fund a new and exciting area of scientific research. At the October 16, 1958 meeting, the board voted to turn down a request to fund pioneering work in the field of radio astronomy. Despite the strong support of Director Frank T. Sisco, the board voted to refuse this request for funding due to the impending change in its grant policy.

The Director's annual report for the period between October 1, 1957 and March 30, 1958, marked both the conclusion of the Engineering Foundation's traditional policy of supporting research projects on a continuing basis and final summation of the value of the cooperative research endeavors of the many research councils which the Engineering Foundation had helped to originate. One research council in particular, the Council on Riveted and Bolted Structural Joints was cited for the immense contribution that it had made to the development of modern American cities—In 1952 the Council prepared its now famous specifications for the assembly of structural joints using high tensile strength steel bolts. The research work of this Council on the properties of riveted and bolted structural joints which resulted in these specifications and in many technical papers has had remarkable consequences for the construction industry. During the building boom of the mid-1950s the face of New York City was virtually changed. In New York and other metropolitan centers, the sound of the riveting hammer was seldom if ever heard. The field erection of these new block square twenty to forty story office buildings, apartment houses, and hotels was carried out almost entirely by the use of high tensile strength steel bolts. The Council continued its research program and investigated the variables that affected the static and dynamic properties of riveted and bolted joints. During the 1950s, its research budget was approximately $50,000 per year.

During the planning sessions of the executive committee, many suggestions were discussed. Among the most notable and prophetic of these suggestions was one made by board member Estill I. Green. Although he only served for one term, Green proposed that the Engineering Foundation might sponsor, organize and at least partially underwrite international conferences on selected engineering subjects. Green's idea soon gathered a great deal of board support provided that these events would be onetime affairs and that the Engineering Foundation avoid encouraging or supporting their perpetuation as annual events.

The wide board support that Green's proposal gained was indicative of the direction in which the Engineering Foundation was moving. It was becoming increasingly obvious that the board was favoring allocation of grant money only to start new fields of research or to underwrite specific events such as conferences. The traditional approach of supplying long-term support for specific research projects had been almost completely discarded. To further bring about this change-over in policy, the Research Procedure Committee was directed to review all current projects and to designate those projects which, in its opinion, should continue to receive Engineering Foundation sponsorship without any commitment of financial support.

The Engineering Foundation continued rapidly on its course of policy change. It further severed its links to the research councils by announcing that after September 30, 1959, it would no longer serve as fiscal agent for any of them. It is ironic that Estill I. Green, whose proposal had done so much to inspire change, resigned from the Engineering Foundation Board at the October 22, 1959 meeting. To satisfy critics that it was abandoning its commitments, the board voted at this meeting to contribute $100 to any council or project which requested it. This money would be available until the new grant policy was totally in place.

The October 22, 1959 board meeting also marked the end of Sisco's tenure as director. Although no particular reason was ever recorded or stated for this action, in retrospect one can surmise that Sisco realized that the new direction of the Engineering Foundation's policy held little place for him. Sisco's entire career as director of the Engineering Foundation had focused on the support and enhancement of the Foundation's traditional policy of initial and continuing support for specific research projects. At the time of his resignation, the board took the unusual action that $6,000 be appropriated out of the current budget to pay Sisco one year's salary in four annual installments of $1,500 each on

January 1, 1960, 1961, 1962, and 1963. It was also emphasized by the board that this action should not be considered as establishing a precedent since every future case of this kind would require individual consideration. It was also made part of the record that the unusual condition would be attached that each payment would be contingent on the recipient having not engaged in any activity which in the opinion of the board is prejudicial to the interests of the Engineering Foundation.

Frank T. Sisco

Frank Thayer Sisco had a great influence on the development of the Engineering Foundation during his tenure as technical director and later director. The continuation and enhancement of the Foundation's traditional policy of supporting long-term basic engineering research reached its final stage during his years at the helm. Through his regular visits to the research projects that the Foundation was then funding, he was able to establish an effective

Frank T. Sisco (1889–1965), long time editor of the "Alloys of Iron" publication series and the technical director (1949–53) and the fourth director (1953–59) of the Engineering Foundation.

dialogue not only with the researchers themselves, but also with the deans and administrative heads of the engineering departments of the cooperating universities. Through these meetings and through his accurate evaluations of the projects he was often able to discern new and exciting areas of future research which could then become the focus of Engineering Foundation grant support. He also served both as a tireless publicist for the activities of the Foundation and its ambassador to the national and international engineering community. Sisco was also an efficient and effective administrator whose reports to the board often convinced others to support his position by their thoroughness and masterful use of statistics.

Sisco's professional career had prepared him well for his achievements as the head of the Engineering Foundation. Born on April 4, 1889 in Lawrence, Kansas, he was raised by his paternal grandparents at Clinton, Iowa. In 1908 he enrolled at the University of Illinois where he showed an aptitude for technical subjects and where he also led an active social life. Forced by the death of his grandfather to abandon his studies after only two years, he gained employment with a succession of steel companies.

During the next decade, Sisco filled responsible positions at various chemical and metallurgical laboratories. Among the most important of these jobs were those of chief chemist and superintendent of the Hess Steel Company of Baltimore, Maryland, which he held between 1916 and 1920 and chief chemist for the U.S. Navy's short-lived armor plate plant at South Charleston, West Virginia, in 1920–21. Between 1922 and 1923 he was employed as chief chemist and engineer of tests for the American Steel Wire Company of Waukegan, Illinois. He joined the U.S. Army Air Corps laboratories in 1923. One of the most distinguished periods of his career was his tenure between 1925 and 1929 as the chief of the Metals Branch of the U. S. Army Corps at Wright Field which included the metallographic and radiographic laboratories, the experimental heat treatment laboratory and the experimental foundry.

The Metals Branch at Wright Field pioneered in many outstanding developments in aircraft metallurgy which occurred between 1925 and 1930. It was primarily responsible for the development of an Al-Mg-Ni alloy for pistons and air-cooled cylinder heads and for the development of a successful sodium-cooled valve. It carried on the original research work which led to the use of welded chromium molybdenum steel tubing for aircraft fuselages and wings—a development which was responsible for the rapid abandonment of plywood and fabric construction. It also played a vital role in making the duralumin propeller practical.

From his work at all of these positions Sisco gained both the practical and theoretical knowledge of engineering and metallurgical principles along with the administrative skills which would later enable him to serve so effectively as the head of the Engineering Foundation.

In 1930 he relocated to New York City where he entered a new phase of his career, one largely devoted to editorial and executive endeavors. He played an important role in the direction of the "Alloys of Iron" project as its editor (1930–43), consultant (1943–45), editorial director (1945–50) and director (1950–61). This monumental project ultimately resulted in the publication of sixteen monographs. His editorial work was aided by the advice and help of his wife, the former Anneliese Gruenhaldt, whom he married in 1932. In 1949, when Sisco was appointed technical director of the Engineering Foundation, all of the editorial work on manuscripts of the "Alloys of Iron" monographs was turned over to Mrs. Sisco.

Besides the pivotal role that Sisco played in the development of the Engineering Foundation's "Alloys of Iron" project, his editorial and administrative skills were also utilized on behalf of AIME. Between 1934 and 1937 Frank T. Sisco served as the secretary of the Iron and Steel Division of AIME and in 1940 he became its chairman. Between 1941 and 1945 he held the position of secretary of AIME's Metals Division. He also regularly wrote a two page column in AIME's publication *Mining and Metallurgy*. The columns displayed not only his knowledge of engineering history but also his profound faith in engineering's future achievements. The columns attracted a wide readership. Sisco also continued to write scholarly papers and articles, and by the time of his death he was credited as the author of thirty-three technical papers.

After his service as director of the Engineering Foundation, Sisco served on the Engineering Index Board—first as president and later as honorary member—and as the secretary of the Corrosion Research Council until his death.

A complex man of many interests which included music, fiction, and crossword puzzles, Sisco had a profound influence on the engineering community. He was remembered by his contemporaries as a cheerful advisor who was unselfish in his devotion to the pursuit of engineering excellence through research. He died in New York on January 12, 1965.

References

2.1 R. Eliassen and J. C. Lamb, III, "Mechanisms of the Internal Corrosion of Water Pipe," *Journal, American Water Works Association*, Vol. 45 (1953), pp. 1281–1294.

2.2 R. Eliassen and J. C. Lamb, III, "Mechanism of Corrosion Inhibition by Sodium Metaphosphate Glass," *Journal, American Water Works Association*, Vol. 46 (1954), p. 445.

2.3 R. Eliassen and J. C. Lamb, III, "Corrosion Control with Sodium Metaphosphate Glass," *Journal, New England Water Works Association* (Mar. 1955), pp. 32–68.

2.4 R. Eliassen and R. T. Skrinde, "Experimental Evaluation of 'Water Conditioner' Performance," *Journal, American Water Works Association*, Vol. 49, No. 9 (1957).

2.5 A. J. Romeo, R. T. Skrinde, and R. Eliassen, "Effects of Mechanics of Flow on Corrosion," *Proceedings, ASCE*, Paper 1702 (Jul. 1958).

2.6 R. E. Davis, "Physical Properties of Concrete," *Proceedings, ASCE*, Vol. 54, No. 5, part III (May 1928), pp. 119–214.

2.7 R. E. Davis, "Flow of Concrete Under Sustained Compressive Stress," *Proceedings, American Concrete Institute*, Vol. 24 (1928), pp. 303–326.

2.8 R. E. Davis and H. E. Davis, "Flow of Concrete Under Sustained Compressive Stress," *Proceedings, ASTM*, Vol. 30, Part II (1930), pp. 707–731.

2.9 R. E. Davis and H. E. Davis, "Flow of Concrete Under the Action of Sustained Loads," *Journal, American Concrete Institute* (Mar. 1931), pp. 837–902.

2.10 R. E. Davis, H. E. Davis, and J. S. Hamilton, "Plastic Flow of Concrete Under Sustained Stress," *Proceedings, American Society for Testing Materials*, Vol. 34, Part II (1934), pp. 354–386.

2.11 R. E. Davis, H. E. Davis, and E. H. Brown, "Plastic Flow and Volume Changes of Concrete," *Proceedings, American Society for Testing Materials*, Vol. 37, Part II (1937), pp. 317–330.

2.12 J. W. Kelly, "Some Time-Temperature Effects in Mass Concrete," *Proceedings, American Concrete Institute*, Vol. 34 (1938), pp. 573–586.

2.13 R. E. Davis, E. H. Brown, and J. W. Kelly, "Some Factors Influencing the Bond Between Concrete and Reinforcing Steel," *Proceedings, American Society for Testing Materials*, Vol. 38, Part II (1938), pp. 394–406.

2.14 R. E. Davis, "Autogenous Volume Changes of Concrete," *Proceedings, American Society for Testing Materials*, Vol. 40 (1940), pp. 1103–1112.
2.15 F. H. Morton, "Creep of Tubular Pressure Vessels," *Trans. ASME*, Vol. 61 (Apr. 1939), pp. 239–245.
2.16 *Nature, Occurrence, and Effects of Sigma Phase*, Symposium, American Society for Testing Materials, Philadelphia, Special Technical Publication No. 110 (1950).
2.17 *Corrosion of Materials at Elevated Temperatures*, Symposium, American Society for Testing Materials, Special Technical Publication No. 108 (1951).
2.18 R. F. Miller and J. J. Heger, *Strength of Wrought Steels at Elevated Temperatures*, American Society for Testing Materials, Special Publication (1950).
2.19 T. N. Armstrong and R. J. Greene, "Nickel-Chromium-Molybdenum Steel Valve Casting After 50,000 Hours of Service at 900 F," *Trans. ASME*, Vol. 72 (1951), pp. 751–754.
2.20 J. L. Ham, "An Introduction to Arc-Cast Molybdenum and Its Alloys," *Trans. ASME*, Vol. 73 (1951), pp. 723–732.
2.21 M. E. Holmberg, "Experience with Austenitic Steels in High Temperature Services in Petroleum Industry," *Trans. ASME*, Vol. 73 (1951), pp. 733–742.
2.22 A. M. Hall and E. E. Fletcher, "An Investigation of the Role of Aluminum in the Graphitization of Plain Carbon Steel," *Trans. ASME*, Vol. 73 (1951), pp. 743–760.
2.23 D. N. Frey, "The General Tensional Relaxation Properties of a Bolting Steel," *Trans. ASME*, Vol. 73 (1951), pp. 755–760.
2.24 W. C. Stewart and W. G. Schreitz, "Thermal Shock and Other Comparison Tests of Austenitic and Ferritic Steels for Main Steam Piping," *Trans. ASME*, Vol. 75 (1953), pp. 1051–1072.
2.25 W. F. Simmons and H. C. Cross, *Report on the Elevated-Temperature Properties of Stainless Steel*, ASTM Special Technical Publication 134.
2.26 J. R. Kattus, *Properties of Cast Iron at Elevated Temperatures* (ASTM: May 1957).
2.27 *Relaxation Properties of Steels and Super Strength Alloys at Elevated Temperatures*, ASTM STP No. 187.
2.28 *Wrought Medium-Carbon Alloy Steels*, ASTM STP No. 199.
2.29 G. A. Hawkins, H. L. Solberg, and A. A. Potter, "The Viscosity of Water and Superheated Steam," *Trans. ASME*, FSP-57-11 (Oct. 1935), pp. 395–400.

2.30 A. A. Potter, H. L. Solberg, and G. A. Hawkins, "Investigation of the Oxidation of Metals by High-Temperature Steam," *Trans. ASME*, RP-59-10 (Nov. 1937).
2.31 G. A. Hawkins, H. L. Solberg, and A. A. Potter, "The Viscosity of Superheated Steam," *Trans. ASME* (Nov. 1940), pp. 677–688.
2.32 Hawkins, et al., "Corrosion of Unstressed Specimens of Alloy Steel by Steam at Temperatures up to 1800 F," *Trans. ASME*, Vol. 65 (May 1943), pp. 301–308.
2.33 Sibbitt, Hawkins, and Solberg, "The Dynamic Viscosity of Nitrogen," *Trans. ASME*, Vol. 65 (Jul. 1943), pp. 401–405.
2.34 C. J. Slunder, A. M. Hall, and J. H. Jackson, "Laboratory Investigation of Superheater-Tubing Materials in Contact with Synthetic Combustion Atmospheres at 1350 Deg F," *Trans. ASME*, Vol. 75 (1953), pp. 1015–1019.
2.35 J. H. Jackson, et al., *Trans. ASME*, Vol. 75 (1953), pp. 1021–1035.
2.36 A. M. Hall, D. Douglass, and J. H. Jackson, "Corrosion of Mercury-Boiler Tubes During Combustion of a Heavy Residual Oil," *Trans. ASME*, Vol. 75 (1953), pp. 1037–1049.
2.37 V. Paschkis and G. Stolz, Jr., "Quenching as a Heat Transfer Problem," *Journal of Metals* (Aug. 1956), p. 1074.
2.38 V. Paschkis and G. Stolz, Jr., "How Measurements Lead to Effective Quenching," *Iron Age* (Nov. 22, 1956), pp. 95–97.
2.39 F. G. Keyes, "A Summary of Viscosity and Heat-Conduction Data for He, A, H2, O2, N2, CO, CO2, H2O, and Air," *Trans. ASME*, Vol. 73 (1951), pp. 589–596.
2.40 F. G. Keyes, "Measurements of the Heat Conductivity of Nitrogen/Carbon Dioxide Mixtures," *Trans. ASME*, Vol. 73 (1951), pp. 597–603.
2.41 F. G. Keyes, "Thermal Conductivity of Gases," *Trans. ASME*, Vol. 77 (1955), p. 13.
2.42 E. L. Carpenter and L. Holdridge, "Cottonseed Oil Industry, Its History, Economics, Processes and Problems," *Mechanical Engineering*, Vol. 53, No. 5 (May 1931), pp. 353–359.
2.43 R. W. Morton, "Cottonseed Pressure-Cooking Research," *Mechanical Engineering* (Oct. 1940), p. 731.
2.44 J. Leahy, "Processing Oil Seeds and Nuts," *Southern Power and Industry*, Part 1 (Jan. 1941), Part 2 (Mar. 1941).
2.45 P. B. Bucky and A. S. Toering, "Mine Roof and Support Design," *Engineering and Mining Journal* (Apr. 1935).
2.46 P. B. Bucky, A. G. Solakian and L. S. Baldin, "Centrifugal

Method of Testing Models," *Civil Engineering* (May 1935), pp. 287–290.
2.47 P. B. Bucky, "Roof Control Problems in High Speed Mechanization," *Coal Age* (Jan. 1938).
2.48 A. M. Gaudin, "Grinding as a Chemical Reaction," *Mining Engineering*, Vol. 7 (1955), pp. 561–562.
2.49 R. J. Charles and P. L. deBruyn, "Energy Transfer by Impact," *Mining Engineering*, Vol. 8 (1956), pp. 47–53.
2.50 R. J. Charles, "Energy-Size Reduction Relationships in Comminution," *Trans. AIME*, Vol. 208 (1957), p. 80.
2.51 S. R. Keim, "Fluid Resistance to Cylinders in Accelerated Motion," *Proceedings, ASCE*, Vol. 22 (Dec. 1956), pp. 1113–1 to 1113–14.
2.52 I. D. Bakalar, "On the Temperature-Dependence of Self-Diffusion in Alpha Iron," *Journal of Metals*, Vol. 188 (1950), p. 1213.
2.53 F. S. Buffington, "Effect of Uniaxial Compressive Stress on Self-Diffusion in Alpha Iron," *Journal of Metals*, Vol. 188 (Oct. 1950), p. 1213.
2.54 M. Cohen, C. Wagner and J. E. Reynolds, "Calculations of Interdiffusion Coefficients When Volume Changes Occur," *Trans. AIME*, Vol. 197, pp. 1534–1536.
2.55 E. Reynolds, B. L. Averbach and E. Cohen, "Self-Diffusion and Interdiffusion in Gold-Nickel Alloys," *Acta Metallurgica*, Vol. 5 (Jan. 1957), p. 29.
2.56 E. O. Hovey, *Movable Bridges*, 2 volumes (New York: John Wiley and Sons, 1926–27).
2.57 E. O. Hovey, *Steel Dams*, (American Institute of Steel Construction: 1935).
2.58 B. G. Johnston, "History of Structural Stability Research Council," *Journal of the Structural Division*, Proceedings of the ASCE, Vol. 107, No. ST8 (1981), pp. 1529–1550.
2.59 Letter from F. H. Frankland to the attention of J. Arms, Secretary of the Engineering Foundation dated April 1, 1943.
2.60 F. Bleich, *Buckling Strength of Metal Structures* (New York: McGraw-Hill Book Co., 1952).
2.61 *Guide to Design Criteria for Metal Compression Members* (Column Research Council of Engineering Foundation: 1960).
2.62 Column Research Council, *Guide to Design Criteria for Metal Compression Members*, Second Edition (New York: John Wiley and Sons, Inc., 1966).

2.63 B. G. Johnston, *Guide to Stability Design Criteria for Metal Structures*, Third Edition (New York: John Wiley and Sons, Inc., 1976).

2.64 *Guide to Stability Design Criteria for Metal Structures*, ed. by T. V. Galambos, Fourth Edition (New York: John Wiley and Sons, Inc., 1988).

2.65 *Specifications for Assembly of Structural Joints Using High-Tensile Steel Bolts* (Council on Riveted and Bolted Structural Joints, 1951).

2.66 W. M. Wilson, W. H. Munse, and M. A. Cayci, "A Study of the Efficiency Under Static Loading of Riveted Joints Connecting Plates," University of Illinois, Engineering Experiment Station, Bulletin 402.

2.67 W. C. Stewart, "Work of Research Council on Riveted and Bolted Joints," ASCE, Convention Preprint, Paper No. 48 (Sep. 1952).

2.68 F. Baron and E. W. Larson, Jr., "The Effect of Grip on the Fatigue Strength of Riveted and Bolted Joints," American Railway Engineering Association Bulletin 503 (Oct. 1952).

2.69 T. R. Higgins and W. H. Munse, "How Much Combined Stress Can a Rivet Take?", *Engineering News Record*, Vol. 149 (Dec. 4, 1952), pp. 40–42.

2.70 H. H. Nicholson, "A Million and a Half High-Strength Bolts Speed Up Steel Erection," *Civil Engineering* (Aug. 1954).

2.71 J. W. Carter, K. H. Lenzen, and L. T. Wyly, "Fatigue Failure in Riveted and Bolted Single Lap Joints," *Proceedings, ASCE*, Vol. 80, Separate No. 469 (Aug. 1954), pp. 469–1 to 469–35.

2.72 F. Baron and E. W. Larson, Jr., "Comparative Behavior of Bolted and Riveted Joints," *Proceedings, ASCE*, Vol. 80, Separate No. 470 (Aug. 1954), pp. 470–1 to 470–19.

2.73 R. A. Hechtman, et al., "Slip of Joints Under Static Loads," *Proceedings, ASCE*, Vol. 80, Separate No. 480 (Sep. 1954), pp. 480–1 to 480–18.

2.74 J. W. Carter, J. C. McCalley, and L. T. Wyly, "Comparative Test of a Structural Joint Connected with High-Strength Bolts and a Structural Joint Connected with Rivets," AREA Bulletin 517 (Oct. 1954).

2.75 M. H. Bell, "High-Strength Bolts in Structural Practice," *Proceedings, ASCE*, Vol. 81, Separate No. 651 (Mar. 1955), pp. 651–1 to 651–36.

2.76 W. H. Munse, "Bolted Connections—Research," *Proceed-*

ings, ASCE, Vol. 81, Separate No. 650 (Mar. 1955), pp. 650–1 to 650–18.

2.77 J. R. Fuller, T. F. Leahey, and W. H. Munse, Jr., "A Study of the Behavior of Large I-Section Connections," *Proceedings, ASCE*, Vol. 81, Separate No. 659 (Apr. 1955), pp. 659–1 to 659–26.

2.78 J. W. Fisher and J. H. A. Struik, *Guide to Design Criteria for Bolted and Riveted Joints* (New York: John Wiley & Sons, Inc., 1974).

2.79 G. F. Kulak, J. W. Fisher, and J. H. A. Struik, *Guide to Design Criteria for Bolted and Riveted Joints*, Second Edition (New York: John Wiley & Sons, Inc., 1987).

2.80 E. Hognestad and I. M. Viest, "Some Applications of Electric SR-4 Gages in Reinforced Concrete Research," *Journal, American Concrete Institute* (Feb. 1950), pp. 445–454.

2.81 E. Hognestad, "A Study of Combined Bending and Axial Load in Reinforced Concrete Members," Bulletin 399, University of Illinois Engineering Experiment Station (Nov. 1951).

2.82 E. Hognestad, "Inelastic Behavior of Tests of Eccentrically Loaded Short Reinforced Concrete Columns," *Journal, American Concrete Institute* (Oct. 1952), pp. 117–139.

2.83 E. Hognestad, "What Do We Know about Diagonal Tension and Web Reinforcement in Concrete?", University of Illinois Engineering Experiment Station, Circular 64, Urbana, Ill. (Feb. 1952).

2.84 K. G. Moody, et al., "Shear Strength of Reinforced Concrete Beams, I—Tests of Simple Beams," *Proceedings, ACI*, Vol. 51 (1955), pp. 317–332.

2.85 K. G. Moody, et al., "Shear Strength of Reinforced Concrete Beams, II—Tests of Restrained Beams without Web Reinforcement," *Proceedings, ACI*, Vol. 51 (1955), pp. 417–434.

2.86 R. C. Elstner and others, "Shear Strength of Reinforced Concrete Beams, III—Tests of Restrained Beams with Web Reinforcement," *Proceedings, ACI*, Vol. 51 (1955), pp. 525–540.

2.87 K. G. Moody and I. M. Viest, "Shear Strength of Reinforced Concrete Beams, IV—Analytical Studies," *Proceedings, ACI*, Vol. 51 (1955), pp. 697–730.

2.88 D. H. Pletta and D. Frederick, "Model Analysis of a Skewed Rigid Frame Bridge and Slab," *ACI Journal*, Vol. 26 (Nov. 1954), pp. 217–230.

2.89 T. B. Kennedy, "Tensile Crack Exposure Tests of Stressed

Reinforced Concrete Beams," *ACI Journal*, Vol. 27 (Jun. 1956), pp. 1049–1063.

2.90 A. Roesli, A. C. Loewer, Jr., and W. J. Eney, "Machine to Apply Repeated Loads to Large Flexural Members," Bulletin, American Society for Testing Materials (Feb. 1954), pp. 50–53.

2.91 R. E. Walther, "Investigation of Multi-Beam Bridges," *ACI Journal*, Vol. 29, No. 6 (Dec. 1957); Proceedings, Vol. 54, pp. 505–526.

2.92 R. E. Walther, "Shear Strength of Prestressed Concrete Beams," *Proceedings, Third International Congress of Prestressed Concrete* (Berlin: May 1958).

2.93 I. M. Viest, R. C. Elstner, and E. Hognestad, "Sustained Load Strength of Eccentrically Loaded Short Reinforced Concrete Columns," *ACI Journal*, Vol. 27 (Mar. 1956), pp. 727–755.

2.94 E. Hognestad, "Shearing Strength of Reinforced Concrete Column Footings," *ACI Journal*, Vol. 25, (Nov. 1953), pp. 189–208.

2.95 JoDean Morrow and I. M. Viest, "Shear Strength of Reinforced Concrete Frame Members Without Web Reinforcement," *ACI Journal*, Vol. 28 (Mar. 1957), pp. 833–869.

2.96 B. Broms and I. M. Viest, "Ultimate Strength Analysis of Long Restrained Reinforced Concrete Columns," *Journal of the Structural Division, ASCE*, Vol. 84, ST1 (Jan. 1958), pp. 1510–1 to 1510–38.

2.97 B. Broms and I. M. Viest, "Ultimate Strength Analysis of Long Restrained Reinforced Concrete Columns," *Journal of the Structural Division, ASCE*, Vol. 84, ST3 (May 1958), pp. 1635–1 to 1635–30.

2.98 B. Broms and I. M. Viest, "Design of Long Reinforced Concrete Columns," *Journal of the Structural Division, ASCE*, Vol. 84, ST4, (Jul. 1958), pp. 1694–1 to 1694–28.

2.99 J. W. Baldwin, Jr. and I. M. Viest, "Effect of Axial Compression on Shear Strength of Reinforced Concrete Frame Members," *ACI Journal* (Nov. 1958), pp. 635–654.

2.100 *Proceedings of the First Conference on Coastal Engineering*, ed. by J. W. Johnson (Council on Wave Research, 1951).

2.101 *Proceedings of the Second Conference on Coastal Engineering*, ed. by J. W. Johnson (Council on Wave Research, 1952).

2.102 *Proceedings of the Third Conference on Coastal Engineering*, ed. by J. W. Johnson (Council on Wave Research, 1953).

2.103 *Proceedings of the Fourth Conference on Coastal Engineering*, ed. by J. W. Johnson (Council on Wave Research, 1954).
2.104 *Proceedings of the Fifth Conference on Coastal Engineering*, ed. by J. W. Johnson (Council on Wave Research, 1955).
2.105 *Proceedings of the Sixth Conference on Coastal Engineering*, ed. by J. W. Johnson (Council on Wave Research, 1958).
2.106 *Proceedings of the First Conference on Ships and Waves*, ed. by J. W. Johnson (Council on Wave Research, 1955).
2.107 *Proceedings of the First Conference on Coastal Engineering Instruments*, ed. by Robert L. Weigel (Council on Wave Research, 1956).
2.108 R. L. Weigel, *Waves, Tides, Currents, and Beaches: Glossary of Terms and List of Standard Symbols* (Council on Wave Research, 1952).
2.109 R. L. Weigel, *Gravity Waves—Tables of Functions* (Council on Wave Research, 1954).
2.110 J. Kruger, "Effect of Illumination on the Oxidation of Copper Single Crystals in Water," *Journal of Applied Physics*, Vol. 28 (1957), pp. 1212–1213.
2.111 J. Kruger, "Some Basic Corrosion Research at NBS," *Industrial and Engineering Chemistry*, Vol. 50 (Mar. 1958), pp. 55A–56A.
2.112 "Weather and the Building Industry," *Proceedings of the Research Correlation Conference, Jan 11 and 12, 1950*, BRAB Conference Report No. 1 (Washington D.C.: National Research Council, 1950).
2.113 "Fire Resistance of Non-Load-Bearing Exterior Walls," *Proceedings of the Research Correlation Conference, November 11, 1950*, BRAB Conference Report No. 2 (Washington, D.C.: National Research Council, 1951).
2.114 "Laboratory Design for Handling Radioactive Materials," *Proceedings of the Conference, November 27–28, 1951*, BRAB Conference Report No. 3 (Washington, D.C.: National Research Council, 1952).
2.115 "Condensation Control in Building," *Proceedings of the Conference, February 26–27, 1952*, BRAB Conference Report No. 4, (Washington, D.C.: National Research Council, 1952).
2.116 M. Juul Hvorslev, "Subsurface Exploration and Sampling of Soils for Civil Engineering Purposes," Waterways Experiment Station, Corps of Engineers, U.S. Army, Vicksburg, Mississippi, November 1949. Reprinted by Engineering Foundation in 1962 and 1965.

2.117 "Annual Report 1960" (New York: Engineering Foundation).
2.118 "Annual Report 1961" (New York: Engineering Foundation).
2.119 "Annual Report 1962" (New York: Engineering Foundation).
2.120 "Annual Report 1963" (New York: Engineering Foundation).
2.121 "Annual Report 1964" (New York: Engineering Foundation).
2.122 "Annual Report 1965" (New York: Engineering Foundation).
2.123 "Annual Report 1966" (New York: Engineering Foundation).
2.124 "Annual Report 1967" (New York: Engineering Foundation).
2.125 "Otis Ellis Hovey, Hon. M. Am. Soc. C. E.," *Trans., ASCE* (1941), pp. 1537–1543.
2.126 "Edwin H. Colpitts 1872–1949," *Bell Laboratories Record* (Apr. 1949), p. 150.
2.127 "Colpitts, Edwin Henry," *Who's Who in Engineering* (1948).
2.128 "Edwin H. Colpitts, Phone Pioneer, 77," *New York Times* (Mar. 7, 1949).
2.129 E. H. Robie, "Frank T. Sisco," *Mining Engineering* (Mar. 1965), p. 99.
2.130 "Sisco, Frank T.," *Who's Who in Engineering* (1964), pp. 1718–19.
2.131 "The Reinforced Concrete Column Investigation—A Review," *ACI Journal* (Jan.–Feb. 1934), pp. 5–9.
2.132 C. F. Scott, "A Brief: Summarizing the Results of the Investigation of Engineering Education and Related Activities," Society for the Promotion of Engineering Education, Proceedings of the 42nd Annual Meeting, 1935.

CHAPTER 3

Change in Direction: Re-evaluation of the Role of the Engineer in Society 1959–75

1. FOUNDATION AT THE CROSS ROADS

The resignation of Director Frank T. Sisco marked a watershed in the development of the Engineering Foundation. The previous history of this organization had been noted by its continuing sponsorship of specific research projects and cooperative national research councils. However, during the 1960s, the direction of the Engineering Foundation was greatly altered by a new emphasis on other areas of activity.

The new direction of the Engineering Foundation was summarized in the foreword to the 1958–59 annual report prepared by Chairman W. M. Peirce under the date of October 22, 1959.

> For many years the policy of the Engineering Foundation's Board has been to make small appropriations to a considerable number of projects, most of such appropriations being contingent upon the sponsor's obtaining considerably greater support from other sources. Here and there larger appropriations have been made where necessary to 'get a project off of the ground'. A second element of the policy has been to continue token support for many approved projects for long periods of years, even though major and adequate support had become available from other sources.
>
> In a letter to the board in 1941, Dr. Buckley challenged this policy raising questions which have persisted in the minds of some members up to the present time. Many efforts to gradually modify this policy have failed and the distribution of income to a large number of projects has been continued, even though it has been obvious that the amounts so available are by present day standards insignificant even as seed money.
>
> During the chairmanship of Mr. Barber, the board faced up to this

situation and undertook to examine its traditional policies. The board delegated this task to the Executive Committee acting as a planning committee. This group has held several long meetings in the course of which many points of view and many ideas have been critically examined, and some progress made in formulating concrete proposals for reorienting Foundation policy.

First it appeared to be unreasonable for the busy men who constituted the Engineering Foundation Board to spend their time parcelling out its small income in driblets for the direct support of research. (This view has been independently expressed by each of the several individuals with whom we have consulted.) Even if concentrated on a very small number of projects, the funds available would be of doubtful significance in direct support of research. Today the problem is not to find money for research but to define important problems and find competent men to carry them out.

The committee then considered avenues by which the Foundation could make effective use of its limited income and its unlimited access to the top men in the engineering professions. Many suggestions were received both from within the board and from qualified people outside the board whose opinions were solicited. Some of the suggestions related to the improvement of engineering education. Others related to surveys of areas crossing boundaries between the various branches of engineering with a view to locating gaps in engineering knowledge and then finding ways to fill these gaps. Another suggestion that was logically related to the previous one is that when the Foundation, either through such surveys just suggested or through other channels, finds an area in which worthwhile engineering research projects might be carried out, it brings together a group of experts drawn from the founder societies to delineate the project and to determine where it might be carried out and that the Foundation then takes it upon itself to seek the necessary financial support. If necessary, the Foundation could contribute the initial support to get a project underway. The amount required to do this would, under today's conditions, be substantial in most cases, and the Foundation might have to limit its financial support to a very small number, perhaps even a single project at a time.

As its work progressed, the planning committee came to feel that among the types of activity just described can be found the basis for an appropriate Engineering Foundation program. They were convinced, however, that such a program cannot fully be developed nor could it be administered directly by the board without the aid of an executive head of broad background and a high order of ability. The committee felt that a large part of the Foundation's income could be wisely and effectively spent in the employment of such a man, and that the first problem is to find such a man. The board concurred in this view and the search for such a man is being actively pursued.

Since the committee did not feel that continuance of the old policy would bring an adequate return from the funds spent, it recommended discontinuing this policy feeling that it was better to let the income accumulate for whatever period necessary than to spend it inefficiently simply to avoid a break in the activities of the Foundation. The board followed the recommendation.

In taking this action, the board was fully mindful of the many worthwhile achievements under projects that the board had helped to initiate. It has been a source of gratification to the board that its sponsorship of such organizations as the several research councils has been considered of continuing value and it was the definite hope of the committee in making its recommendations that the Foundation could continue to be of service in ways other than direct financial support of any project in which the Foundation has had a part. Ways in which the Foundation can be of such service are the subject of continuing study and at its October 22, 1959 meeting, the following resolution was adopted:

> The Engineering Foundation Board wishes to make clear its policy regarding its relations with projects which it has supported in the immediate past. It is the desire of the board to continue to lend its endorsement to projects, which in its opinion, advance the cause of engineering.
>
> While the Board feels that it should not make long continuing substantial contributions to any single project, it is glad to continue its endorsement of projects which have become creative and self supporting where it is desired and is willing to consider token financial support where this seems advisable in order to give weight to its endorsement.

To realize the changes that were outlined in the above report, the board appointed Harold K. Work as the Engineering Foundation Director. Unlike his predecessor, Director Work was employed on a part time basis at a salary of $600 per month. He was expected to work an average of six days per month.

One of Director Work's first tasks was to discontinue the Research Procedure Committee. In its place a new research committee was formed. It was composed of a chairman and two vice chairmen, all three members of the Foundation Board; the director and chairman of the Foundation as ex-officio members; and the chairmen of the research committees of the five founder societies or individuals actively interested in engineering research.

The duty of the committee would be to seek out areas in which engineering research is needed, then make a study as to how and where such research might be carried out. It would then prepare an estimate of the cost involved and develop a well formulated plan for presentation to the board or Executive Committee for approval. The Research Committee would also be delegated with the task of drafting revisions to the Rules of Administration in accordance with the above provisions. It would also seek means of radically increasing the endowment of the Engineering Foundation.

The Research Procedure Committee in its last action gave $100 in token grants to all existing projects. The dispensing of these token grants continued through 1965.

During 1960 the Executive Committee continued to function as a planning committee evaluating new ways in which the Engineering Foundation could serve the engineering profession. On May 6, 1960, it reported to the board "that these prolonged studies leave the committee convinced that stimulation and support of engineering research remains one of the primary functions of the Engineering Foundation and should be the first program."

Due to these changes, the period of the 1960s was a period of adjustment for the Engineering Foundation as it carried on its process of re-evaluation of its function and activities. Old projects, some of which had been receiving Foundation support for more than a decade, were phased out while new activities such as the sponsorship of conferences gained increasing prominence.

One of the new ideas which won the support of the Engineering Foundation Board was the development of a plan to transform the exhibit area of the then Engineering Center Building into a Hall of Engineering Progress. On March 8, 1963, the board of the Engineering Foundation sent a letter in support of this idea to the board of the United Engineering Trustees, Inc. In this letter the Engineering Foundation pledged to make over $100,000 available to support this idea if it would win the approval of the UET Board. However, the idea of an Engineering Hall of Progress did not win the support of the UET Board and it was never developed.

At the same time, the board of the Engineering Foundation also

> resolved that the activities of the Engineering Foundation office will include the new function of serving as the Engineering Foundation Office of Research Cooperation—having the purpose of serving the research activities of the technical societies particularly in the interdisciplinary areas. The services of the Offices of Research Cooperation are to be made available to the extent that the research groups of the societies find them useful and will include, but not be limited to, maintaining and providing

1. Current information on the title, scope and personnel of active engineering society research programs and committees.
2. Assistance in developing liaison between groups and in organizing cooperative or joint projects where these are considered desirable by the several groups concerned.
3. Current information on fund raising plans and programs of engineering society research groups.
4. Information including mailing lists, sample brochures, letters on methods that have been proven successful in fund raising for research and assistance in developing and conducting fund raising.
5. Custodial service for funds collected for individuals or joint research projects. However, it should be noted that the Engineering Foundation will not ordinarily provide funds for the continuing support of research projects.

2. HAROLD K. WORK

The career of Harold K. Work was marked by distinguished achievements as an administrator. As director of the Engineering Foundation, he was confronted between 1959–1965 with the difficult task of implementing the many major policy changes that were brought about by the Foundation's ongoing program of self examination. During his tenure as director, Work successfully supervised the establishment of the Engineering Foundation's research conferences and he also played a pivotal role in the successful efforts to bring about the creation of the National Academy of Engineering.

Harold Knowlton Work was born on May 21, 1901, at Hartford, Connecticut. He received his A.B. from Columbia University, New York City, in 1923 and a degree in chemical engineering from this same institution in 1925. He earned a Ph.D. in chemistry from the University of Pittsburgh in 1929.

Between 1925 and 1929, Work served as a research fellow at the Mellon Institute of Pittsburgh. After receiving his Ph.D., he remained in Pittsburgh serving as a division head in the research laboratories of Aluminum Company of America until 1934, when he was employed as a chemical engineer at the jobbing division of this same firm. In 1936, Work became the manager of research and development of Pittsburgh's Jones and Laughlin Steel Corporation, a position he held until 1948 when he was promoted to become director of research.

Leaving the employment of Jones and Laughlin in 1949, Work

re-entered academic life as the director of the Research Division of the School of Engineering and Science of New York University, where he served until 1966. In 1957 he was appointed as associate dean of the School of Engineering and Science of New York

Harold K. Work (1901–77), first secretary of the National Academy of Engineering and fifth director of the Engineering Foundation (1959–65).

University. As associate dean, Work was able to serve as a spokesman for the restiveness of many research engineers who felt that they did not have adequate representation. As quoted in the 1989 volume *The Making of the N.A.E. : The First 25 Years* by Lee Edson, Work wrote: "We should accept the fact that the NSF is for scientists and then try to set up a National Engineering Foundation to meet the need for basic engineering research. This should be supervised and operated by engineers and should fill the position that the National Institutes of Health do for medicine. This would seem to be a way to meet the basic needs of engineers while the N.S.F. could continue to do the excellent job that it has done for scientists."

After serving as director of the Engineering Foundation between 1959 and 1965, Work became the first appointee to the position of secretary of the National Academy of Engineering. He served in this capacity until he resigned in 1968 in order to accept a position under the auspices of the United Nations as an advisor to the Institute of Standards and Industrial Research of Iran. He also became an advisor to the World Bank.

Work played an active role in the affairs of many important professional scientific and engineering organizations. He served as chairman of the Industrial Research Institute in 1944–1945 and successively between 1945–1949 as national treasurer, vice president and president of the American Society for Metals. Work was elected vice president of the American Society for Engineering Education and served in this office between 1954–1956. He simultaneously served this same organization as chairman of its Engineering Research Council. Work had served as chairman of the Pittsburgh Section of the American Chemical Society in 1946. He was also an active member of the American Association for the Advancement of Science, the American Institute of Chemists, the American Iron and Steel Institute, the American Institute of Chemical Engineers, the American Society for Testing and Materials, the Engineers Joint Council, the New York Academy of Sciences, and the American Institute of Mining, Metallurgical and Petroleum Engineers.

He also played a prominent role in such governmental advisory agencies as the National Research Council, the National Center for Atmospheric Research, and the President's Conference on Technical and Distribution Research for the Benefit of Small Business.

On a more local level, he was a member of the New York State Council on Industrial Research and Development.

Work received numerous honors including the Illig Medal from Columbia University and the Robert W. Hunt Prize from AIME.

In 1928, Work married Miss Margaret Leal. This marriage provided him with both love and commitment and it produced four children. Harold K. Work died on June 12, 1977.

3. FOUNDATION CONFERENCES

The primary focus of the Engineering Foundation during the 1960s was the establishment of its continuing series of engineering conferences. As a part of the early re-evaluation of its purposes, the board of the Engineering Foundation had decided to sponsor a series of international conferences where engineers and other scholars could exchange information on a given subject.

It was also decided to organize these conferences along the pattern of the well established and highly successful Gordon Research Conferences which were held annually in many fields of science. The Engineering Foundation Research Conferences were planned to utilize the facilities and administrative expertise of the Gordon Research Conferences but they were to be completely independent from them. At its March 8, 1962 meeting, the board of the Engineering Foundation selected two important conference topics, one that was linked to its traditional funding subject, and one that was on the cutting edge of the rapidly evolving American aerospace technology. The study of comminution was a subject that had been the basis for several previously funded Engineering Foundation projects, while the properties and structure of composite materials was a new area of research. Since composite structures, with the light weight and high strength were being increasingly used in aircraft and space programs, a conference on composite structures would be highly relevant to the engineering community. To begin the organization of these conferences, $5,000 was voted for Director Work's use as seed money. However, in the middle of March the chairman-designate of the conference on comminution and thickening informed the Conferences Committee that he would not be able to carry out his conference as scheduled and suggested rescheduling for summer of 1963. A new topic, "Engineering Research for the Developing Countries," was suggested by the Engineers Joint Council and the Council became a cosponsor of the conference.

The first Engineering Foundation Research Conference was held at Tilton School, New Hampshire, in July 1962 on the topic of "Engineering Research for the Developing Countries." As was the case for almost all Engineering Foundation conferences, the location was selected as one that would provide surroundings conducive to informal meetings with emphasis on discussions by all participants. Sessions were limited to mornings and early evening hours, leaving ample time for recreation combined with opportunity for pairs or small groups to continue the discussions. The duration of the first as well as of most subsequent conferences was four and one half days with the participants arriving on Sunday afternoon and leaving on Friday noon. The conference was orga-

Attendees of the First Engineering Foundation Conference. Entitled "Engineering Research for the Developing Countries," it was held at the Tilton School in New Hampshire in July 1962.

nized by T. O. Paine of General Electric Company who served as chairman. The second conference held at the same location in August was organized and chaired by W. J. Harris, Jr. of Battelle Memorial Institute on the topic of "Composite Materials." The attendance was eighty-seven professionals at the first conference and fifty-two at the second. The resolutions committee of the first conference reported that:

> We live in the most catastrophically revolutionary age that men have ever faced. This revolutionary age places upon the United States a major and urgent responsibility to share its strength in education, research facilities, and ability to transmit knowledge with the less developed nations. As a result, the engineering and scientific community has a responsibility to channel technical and educational assistance to those countries that need it.

The early research conferences of the Engineering Foundation were resounding successes. In response, the board of the Engineering Foundation voted at its October 9, 1962 meeting that the Engineering Foundation should attempt to arrange three conferences for the summer of 1963 and that $10,000 should be allocated to cover possible expenses in connection with the organization and administration of these proposed conferences. The board also adopted the sensible policy of reviewing the previous year's conferences and further acted to implement several recommendations that had been developed by these meetings.

Four conferences were held in August, 1963 on the following topics: "Technology and the Civilian Economy," "Comminution," "Engineering in Medicine," and "Urban Transportation Research." The facilities of the Proctor Academy in Andover, New Hampshire, were used for all four conferences as well as for the 1964 and 1965 conferences. The average attendance of the 1963 conferences was fifty-four. The limited number of attendees at the first six conferences was typical of all Engineering Foundation conferences even though the average has increased over the years. The limited attendance was considered essential since the principal objective was to develop a thorough discussion of the subject on hand.

In response to recommendations that were developed by the conference on composite materials, the Engineering Foundation's board voted to retain a competent writer on a full or part time basis to prepare an annotated bibliography on composite materials and to arrange for its publication. It appropriated $10,000 for this purpose. It was further recommended that the Engineering Foundation also encourage centralized publication of scientific and engineering studies in the field and to develop intersociety programs relating to composite materials. Unfortunately, the publication did not materialize and the appropriation was deauthorized in 1969.

The Research Conference Committee also developed specific guidelines for the Engineering Foundation Research Conferences which were accepted by the entire board at its September 2, 1962 meeting. These are the essentials of that document:

1. Recognizing the importance and complexity of the interaction of science, engineering and national affairs, the Engineering Foundation should sponsor a small number of conferences, perhaps two or three each year for the next five years, as a continuing experiment in the stimulation of research on engineering studies in fields that are of vital importance but which are receiving limited national attention.

2. These conferences should also include those meetings which could fall within the scope of individual professional societies.

3. These conferences should be organized by small steering committees with full responsibility for program and attendance. One year is the desirable lead time for the organization of a conference.

4. The conferences should be ad hoc with the same subjects not being treated annually except in special cases. However, adequate provision should be made by the Engineering Foundation for follow-up programs of sponsored research and specific professional/society activities or appropriate programs that will enhance the contribution of Foundation conferences to the field of engineering.

5. The Research Committee should have a continuing responsibility for selecting conference topics from those proposed with due regard for the general principles of selection. This committee will also have responsibility for follow-up and evaluation.

6. Recognizing that conferences organized by the Engineering Foundation will generally involve senior industrial and uni-

versity engineers, the conferences should last for less than one week, and should be scheduled at conference sites that minimize travel time for participants. However, the conferences should provide complete opportunities for informal association of the conferees.

7. While the conferences should continue to be off-the-record, provision should also be made for duplication of data presented in slides when acceptable to the speaker.

8. The Engineering Foundation should request that both the Department of Defense and the Office of Science and Technology of the Executive Office of the President indicate problems of critical national importance on which conferences of the type such as was held on composite materials can be developed in the future.

During succeeding years, the board of the Engineering Foundation attempted to implement all of these recommendations. Relevancy to current engineering research interest in engineering remained the overall focus of the conferences. As the 1960s progressed, the research conferences further evolved until they became the Engineering Foundation's primary activity. The number of conferences per year and their attendance are shown below

Attendance at Engineering Foundation Conferences

Year	Number of Conferences	Attendance Total	Attendance Average	Year	Number of Conferences	Attendance Total	Attendance Average
1962	2	139	69.5		193	13618	
1963	4	204	51.0	1976	32	2748	85.9
1964	4	268	67.0	1977	28	2208	78.9
1965	3	207	69.0	1978	29	2361	81.4
1966	5	311	62.2	1979	25	2019	80.8
1967	10	708	70.8	1980	27	1752	64.9
1968	12	668	55.7	1981	20	1593	79.7
1969	16	1079	67.4	1982	21	1471	70.0
1970	16	1184	74.0	1983	19	1697	89.3
1971	16	1594	69.3	1984	18	1355	75.3
1972	26	1685	64.8	1985	16	1354	84.6
1973	26	2059	79.2	1986	21	1699	80.9
1974	18	1202	66.8	1987	23	1773	77.1
1975	28	2310	82.5	1988	25	2056	82.2
sub-totals	193	13618	70.6	totals	497	37704	76.0

based on calendar years. Additional information on the 497 individual conferences may be found in Table 3 in Appendix A which includes the date, title, location, chairman's name and affiliation, and attendance for each conference.

At its meeting of May 9, 1963, the board of the Engineering Foundation voted $10,000 to cover expenses in connection with the Engineering Foundation Research Conferences for 1964. It also approved four conferences for the coming year: "Technology and the Civilian Economy II," "Engineering in Medicine II," "Technological Challenges for the US in World Markets, 1964–1974," "The Building Construction System—A Challenge to Innovation." Each of these early conferences would eventually give rise to important reports. It was also decided to establish a policy that all future Engineering Foundation Research Conferences should be completely self-supporting. As a result the $10,000 that was voted at this meeting to support the organization and administration of the Engineering Foundation Research Conferences was considered only as a contingency.

Looking back from our present perspective of the early 1990s when the problems of the urban mass transit are an acute and daily reality for officials in almost all of America's major metropolitan areas, it becomes evident how perceptive the report that was generated by the Engineering Foundation's Research Conference on urban transportation appears to be. Among this report's recommendations was a call for a thorough analysis of the various components that compose urban mass transportation and a call for well-funded research in this area.

> The Urban Transportation problem is not receiving the financial support or recognition that its importance warrants. No city or metropolitan area in the country can be complacent about its transportation system. The situation ranges from getting by to chaotic. With increasing population pressure there is no reason to expect that its present status will not deteriorate. The close relationship between land use and urban transportation should be particularly analyzed. . . . All of this has proceeded while planners, engineers and economists have continued to witness the rapid changeover of urban travellers to the more expensive but more comfortable and convenient private automobile. In view of these simple facts, it is difficult to imagine that the potential impact of research directed rather singularly at the styling, power, and air conditioning aspects of bus and rail transit vehicles can hope to compete with private automobile transport particularly as the nation becomes more affluent. Technological research should be construed and pursued as

something concerned with more than just power, speed, or comfort. It should be more properly considered as seeking ideas and innovations which would lead to the improvement of the means of designing, constructing, and/or operating vehicle systems; it should be thought of as the development of new control and guidance mechanisms which in turn would also lead to more efficient operations with better service and complete changes in running urban mass transit systems.

Diversity was another hallmark of the activities of the Engineering Foundation during the 1960s and early 1970s. This diversity was the result of a policy change which shifted the focus of the Engineering Foundation's funding support from long term engineering research to conferences and later short term research grants. This change in direction came about due to the desire of the Engineering Foundation to better focus its limited resources

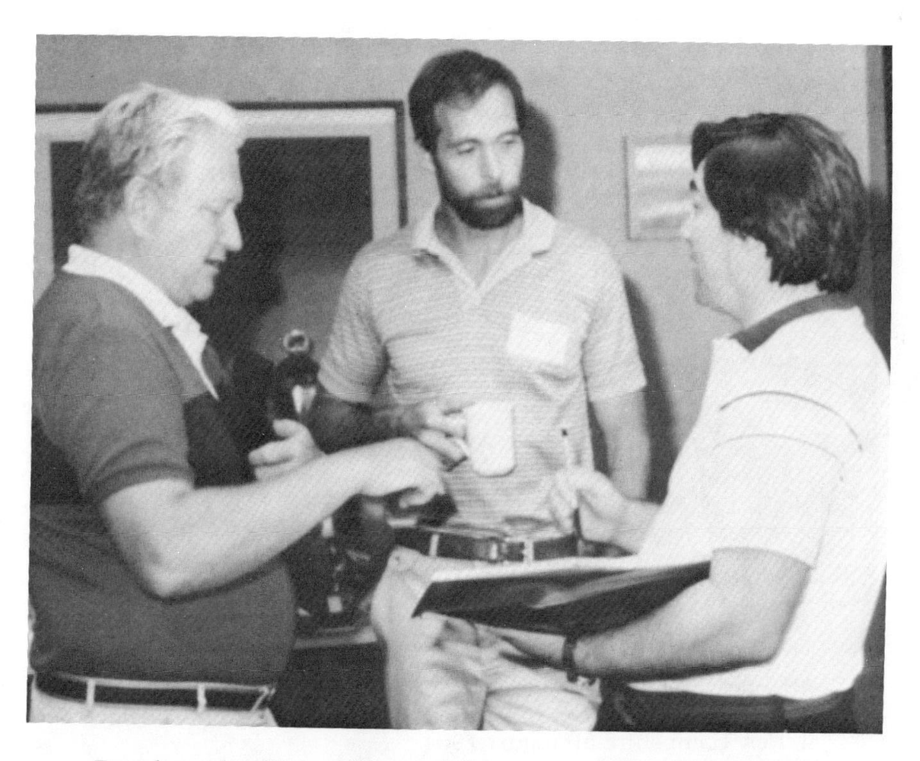

Board member Richard Bryers, left, was one of a number of board and committee members who participated in Engineering Foundation conferences in 1984.

on the identification of the problems that faced society during this period to find ways in which engineering could best be applied to the solution of them. As a result, the Engineering Foundation came to increasingly perceive itself as a catalyst for the development of cooperative research strategies by the various components of the engineering community which could then be focused on individual problems. As Director Work stated in his report for 1963, "The Foundation has turned more to projects designed to advance the engineering profession generally than to support specifically detailed engineering research activities."

The Engineering Foundation initially utilized its research conferences and grants to identify areas of concern. Quite often these areas of concern were strikingly similar to modern problems which

Attendees of the conference "The Building Construction System—A Challenge to Innovation" held at the Proctor Academy in New Hampshire in August 1964.

continue to confront society during the 1980s. Through its conferences and research grants, the Engineering Foundation became a pioneer in focusing the attention of the engineering community on such areas of concern as urban mass transportation, the technology gap between the third world and the developed countries of the west, and air pollution. It also pioneered in its sponsorship of the application of engineering to medicine and thus played a major role in providing support for the new field of biomechanics. Finally it should be noted that due to the fact that many of its conferences during the 1960s and early 1970s were international in scope, they brought together engineers and scientists from many nations which helped to develop new areas of international cooperation.

The Engineering Foundation's research conferences continued to receive additional financial support during 1972. At the October 11 board meeting, it was announced that $100,000 had been received from the National Science Foundation to support these scholarly meetings. In total 1685 attended the twenty-six research conferences that were held during 1972. The success of the research conferences induced the board to seek additional staff members and rental space to handle conference related matters.

The board of the Engineering Foundation took several actions at its May 16, 1973, meeting to implement its plan to augment the Foundation's ability to organize and administer the research conference program. William E. Reaser was retained to serve as assistant director of conferences, and additional funding to support the research conferences was received from not only the National Science Foundation but also the U.S. Department of Transportation and the National Cancer Institute.

During 1975 the Engineering Foundation's conferences became increasingly successful. It was reported at the November 6 board meeting that the total registration for these scholarly meetings was up over 57% from the previous year. More than triple the number of foreign scientists and engineers attended the 1975 conferences than had attended the 1974 conferences. As a result, it was decided by the board to hold a limited number of future conferences at foreign locations. The board also noted that due to other commitments, William E. Reaser resigned his position as assistant director of conferences.

4. CONFERENCES COMMITTEE AND SANDFORD S. COLE

Early in 1962, Research Committee formed a subcommittee that was to be responsible for planning and organizing Engineering Foundation Research Conferences. W. C. Schreiner, Research Committee chairman, assumed the chairmanship of the new subcommittee. The chairmen of the first two conferences, T. O. Paine of General Electric Co. and W. J. Harris, Jr. of Battelle Memorial Institute became members along with Director Work. The subcommittee was expanded in 1963 by the addition of L. K. Wheelock and board member R. A. Kinckiner. It soon became apparent that the conferences were filling a real need and that the two to four conferences held during each of the first few years did not begin to saturate the market. However, it was also realized that further expansion of the conference program would require additional staff, particularly since the program during its early years was to a large extent the result of work by two volunteers: T. O. Paine and W. J. Harris, Jr. Harris served on the conference committees from their formation in 1962 to 1969, during which time he made substantial contributions to the development of the conference program. The expansion was embarked on in 1966 by the appointment of board member Sandford S. Cole as director of conferences.

The appointment of Cole bore fruit almost immediately: while only five conferences were held in 1966, their number doubled in 1967 and starting with 1969 there were never fewer than sixteen conferences per calendar year. Thanks to this growth, the conferences became the principal activity of the Foundation for years to come. In 1970, the committee in charge of conferences changed its name to Conferences Committee and became independent from and of equal status with the Research Committee. At the same time, the official name of the conferences was shortened to Engineering Foundation Conferences. The number of conferences continued to increase till 1976 when thirty-two were held. After settling into the high twenties for the remainder of the 1970s, the number started a gradual but steady decline which prompted a significant realignment of the staff. On January 1, 1984, Harold A. Comerer was appointed director of the Engineering Foundation responsible for all activities of the Foundation staff and Cole was appointed consultant to the Conferences Committee with responsibility for conferences development and evaluation, and such other tasks as may be assigned to him by the Conferences Committee and the board. Comerer had served since May 13, 1976 as the assistant director of Conferences and had extensive prior experience in publishing and other related areas.

Sandford S. Cole

Sandford S. Cole played a large role in the development of the Engineering Foundation Conferences. Under his leadership, these scholarly meetings grew in number, size, and scope. He also helped to initiate the concept of sponsoring conference series in such emerging fields as biotechnology and under his leadership these conferences quickly became recognized as the premier international meetings in their fields.

Cole was born on November 24, 1900 at Cuba, N.Y. After receiving his elementary and secondary education at public

Sandford S. Cole (1900–86) played a large role in the organization and development of the Engineering Foundation Conferences. Director of Engineering Foundation Conferences 1966–84.

schools, he entered Alfred University from which he graduated in 1923 with a B.S. degree in Ceramic Engineering. In 1933 Cole received a Master of Science in Ceramic Engineering from Alfred University and in the following year he received a Ph.D. in the same field from Pennsylvania State University.

Cole began his distinguished research career as an industrial fellow of the Mellon Institute of Pittsburgh, Pa. in 1923. He held this position until 1934 when he entered the employ of the National Lead Company. He held many responsible positions in this firm's research facilities and by the time of his retirement in 1965 he had risen to become assistant manager of research. His technical specialties were silica refractories, conversion of quartz to cristobalite, conversion of anatase to rutile, chalk-resistant titanium dioxide, and smelting titaniferous ores. During the course of his career Cole was the author of twenty-eight publications and his research resulted in fourteen patents.

Throughout his professional career, Sandford S. Cole was active in several engineering organizations. In 1964 he served as the president of the Society of Mining Engineers of AIME and in 1965 he served as vice president of the AIME. He also served for several years on the board of the AIME. He was also a board member of the Engineering Foundation, the Engineers Joint Council and a trustee of Alfred University. Cole was a fellow of the American Ceramic Society and the American Institute of Chemists and a member of Sigma Xi, Phi Kappa Phi, Keramos, the National Institute of Ceramic Engineers, and the American Chemical Society.

Between 1966 and 1984 he made many notable contributions to the advancement of engineering education in his role as director of the Engineering Foundation Conferences. Through his efforts, financial assistance was made available to graduate students to enable them to attend Engineering Foundation Conferences where these graduate students could meet the best engineering talents in their own chosen fields. In recognition of Cole's achievements in this area, the Engineering Foundation established a formal program of Engineering Foundation Fellowships to carry on the tradition that he had begun.

Cole married Frances Halderman in 1925. This couple's marriage was long and happy, producing three sons. In 1981 Sandford and Frances Cole established a chair in applied mathematics at Alfred University and made provision for the establishment of a similar position in chemistry. Sandford S. Cole died on July 21, 1986.

5. NATIONAL ACADEMY OF ENGINEERING

Perhaps the most significant of the Engineering Foundation's achievements during the 1960s was the pivotal role that it played in the establishment of the National Academy of Engineering. Through the utilization of its funds and organizational abilities, the Engineering Foundation was able to serve as catalyst in the creation of this important institution. Since 1960 the Engineering Foundation as a member of the Engineers Joint Council, had worked to create a National Academy of Engineering. At its March 8, 1962 meeting, the Engineering Foundation Board voted to continue the efforts to aid the National Academy of Sciences in establishing a National Academy of Engineering and to provide funding for this activity. The Engineering Foundation director, Harold K. Work, was delegated to assist in this project.

Work's advocacy of the National Academy of Engineering is evident in the following excerpts from his article "The Question of Establishing a National Academy of Engineering" which appeared in *Journal of Engineering Education*, Vol. 51, No. 9, May 1961:

> . . . it would help to have some concept as to what the Academy might be like. This includes objectives, organization, method of operation, etc.
>
> As objectives, the following are presented for consideration:
>
> 1) Render a national service which is not being rendered now.
> 2) Develop a clearer understanding of the attributes of an engineer.
> 3) Raise the standards of the engineering profession so that the profession may better serve mankind.
> 4) Honor a substantial group of engineers.
>
> An obvious question can be immediately raised namely could the National Academy of Sciences be alerted to take care of the needs of the engineering profession? If so, that should be given serious consideration. If, on the other hand, a separate academy is required, every effort should be made to see that such an academy cooperates closely with the National Academy of Sciences so as to best serve the national interest.
>
> It is interesting to note that such a dual academy arrangement exists in Sweden. There the older Academy is involved in distributing the Nobel Prizes in Physics and Chemistry and has a distinctly pure science flavor. The newer academy is largely directed at engineering and its divisions reflect this, for they are almost all engineering in character. Should a decision be made that an Academy of Engineering is needed, a whole new series of decisions will then have to be made.
>
> 1) How should the Academy be initiated?

2) Should it be autonomous and self-perpetuating or should it be allied to existing engineering organizations?
3) Is a Congressional Charter necessary or desirable?
4) How large should the Academy be and should it be initiated at full strength or is a period of growth desirable?
5) What should be the criteria for election to the Academy?
6) How should the Academy be subdivided? Should this be along conventional engineering groupings? Should some of the newer interdisciplinary groupings be used such as power, materials, and communications? Should other divisions such as production, design management, research, and teaching be used?
7) How should the Academy be financed?
8) The housing of the Academy presents problems. Should it be located in Washington or New York? If in Washington, should an attempt be made to find an imposing or modest home? Is such a home necessary immediately or can this wait until the Academy is well underway?

This gives a sample of the problems that face us in deciding whether or not to form a National Academy of Engineering and how to do it.

Coming back to the question of whether or not an Academy should be established this depends to some extent on how such individuals regard the current status of the engineering profession. There seems to be a wide spectrum of opinion on this point. At the one extreme is the engineer who is completely satisfied with everything about the profession. At the other extreme, we have those who view with alarm the future of the profession—who feel that if present trends continue the engineer will soon join the horse and buggy. In between these two extremes are many thoughtful engineers, who while proud of the past record of the profession are concerned as to whether or not the engineer has evolved sufficiently to meet his obligations in the nuclear and space age.

The reason for the wide spread in thinking of the current status of engineering is it is believed in no small degree a matter of timing or where the effect appears first. At the universities, for example, there is much concern over the trend away from engineering toward science. The same situation seems to be developing in medicine as was mentioned in the *New York Times* a few weeks ago. The trend appears more deep seated than the usual variations in popularity of one profession with respect to another. Concern over this trend has not yet extended very far outside of the universities, but that is to be expected since the effect is immediate in the universities and more remote for those outside. There must be a reason for the trend. It is suggested that one of the factors is the determined effort to advance pure science. Huge sums of money from the National Science Foundation and a concerted effort by the country as a whole to advance American basic science have resulted in a skilled and highly successful campaign.

> The question is now raised as to whether or not the overcorrecting of the past deficiencies of American basic science has affected the applied sciences of engineering and medicine. Now I am sure that there will be some that will challenge this point of view, who will contend that pure science is good and there can never be too much of it. This I do not agree with. Some years back, Dean Hollister pointed out at an Engineering Manpower Commission meeting why it was so difficult to increase the supply of engineers. He showed the number of people in the total population with intellect to be professional people, lawyers, doctors, engineers, and the like is small. Any attempt to increase one group will generally be accomplished at the expense of the others. It is my impression that this balance has been disturbed.
>
> We must then face the question as to whether this is bad or good. Without going into detail, it is my contention that we need the surgeon, the engineer, and other applied scientists who have been trained to attack specific projects against deadlines within economic limits. This in no way detracts from the pure scientist who shuns specific objectives, deadlines, and cost in order to achieve his objectives. But we need both in proper proportion. If the premise is agreed to that the engineering profession and the good of mankind can be harmed if things are out of balance then steps should be taken to correct the situation. The formation of a National Academy of Engineering might contribute to this, but it must recognize that this is merely one step of several that are required to solve the problems facing the profession.

It should be noted that Director Work's article advocating the establishment of a National Academy of Engineering was published during the time when the U.S. was engrossed with what it perceived as a race for scientific and space supremacy with the U.S.S.R. Sputnik and the early failures of the U.S. satellite programs had brought forth a massive government programming of financial support for scientific research and education. In this process many felt that the traditional role of engineers in both general society and the technology community was becoming redundant. By advocating the establishment of a National Academy of Engineering, Director Work was combatting these feelings and also preparing the ground for an increased awareness of the value of engineering to the solution of national problems.

Other individuals who were involved in the Engineering Foundation also became advocates for the creation of a strong umbrella organization that could represent the interests of engineers and they supported the efforts of Director Work to bring such an entity into reality. Among the most prominent of these individuals was Augustus B. Kinzel.

Kinzel was a member of the board of the Engineering Foundation between 1943 and 1955, and again between 1960 and 1964. He served as chairman in 1945–48 and was also president of the Engineers Joint Council. In that capacity, Kinzel soon became aware that EJC had a fundamental weakness due to the fact that it was composed of representatives of fourteen major engineering societies each of whom felt that they could not take any formal actions without securing the approval of the boards of their respective societies. This difficulty was compounded by the fact that the board of EJC met on an irregular basis and that almost all board members served for only a single one year term. The end result of all of these factors made it almost impossible for President Kinzel to obtain any concerted action from EJC.

Kinzel's conclusions concerning EJC were shared by two influential members of the Engineering Foundation Board, Antoine M. Gaudin and Eric A. Walker. Both Gaudin, as the Foundation's chairman during 1960–61, and Walker, who served on the board between 1962 and 1966, supported the actions of Director Work and the initiatives of Kinzel to create a strong umbrella organization that would enable the views of the American engineering community to be heard on issues of national importance.

As president of EJC, Kinzel was able to begin an investigation of the feasibility of establishing an engineering academy. He obtained from the Engineering Foundation the commitment of financial support for the formation of an exploratory committee. He also secured permission for Director Work to serve as the secretary of this new committee.

The exploratory committee was initially composed of representatives from EJC, the Engineering Foundation, the Engineers Council for Professional Development, and the National Academy of Sciences. The committee began its work by examining the failure of previous attempts to establish a national academy of engineering. It soon approved a concept and an organizational structure of the academy that was in line with the proposal that had been earlier stated by Director Work. The exploratory committee was then dissolved and a new body was appointed. It was composed of representatives from EJC and was chaired by Walker. Opposition soon formed to its work, particularly from those scientists who believed that engineers do not devote enough of their energies to conducting original research. Their arguments were countered by the new committee which stated that engineers are also devoted to research although it is more often of an applied rather than of a fundamental nature. Chairman Walker assured concerned individuals that the proposed academy would emphasize creative engineering contributions as a requirement for membership. Walker's committee soon won the firm endorsement of the Engineering Foundation Board which had its own ideas concerning the future shape of the academy.

The board of the Engineering Foundation hoped that the proposed National Academy of Engineering would provide a mechanism or structure that would assure a broad representation of the profession. This representation would be composed of engineers whose careers had been distinguished by major achievements. The board also perceived that the proposed National Academy of Engineering would function as a part of a broader National Academy of Sciences complex. In his May 9, 1963 report to the board of the Engineering Foundation, Director Harold Work emphasized the formation of the proposed National Academy of Engineering by stating that "The role that the Engineering Foundation is playing in the formation of the National Academy of Engineering is similar to the one it played in the formation of the National Research Council." Director Work also stated that the Foundation staff had prepared a comprehensive proposal for the establishment of the National Academy of Engineering within the framework of the National Academy of Sciences. This proposal, he also stated, had been accepted by the National Academy of Sciences. Director Work believed that this proposal's acceptance was a significant achievement because it would insure that there would be no cleavage or separation between pure and engineering research. Throughout 1963 and 1964, the board of the Engineering Foundation continued its strong support for the establishment of a National Academy of Engineering. At its May 7, 1964 meeting, the Foundation board voted $10,000 to support the ongoing studies of the proposed National Academy of Engineering. When the Academy became fully functional all remaining funding that had been pledged by the Foundation to support the studies was transferred to the newly established organization.

By December of 1964 the National Academy of Engineering had been organized. During the first nine months of its existence, the Engineering Foundation's New York office served as the Academy's headquarters until a permanent home for it could be readied in the National Academy of Sciences Building in Washington, D.C. It was appropriate that the Engineering Foundation should render this service since, according to its 1965 annual report, it had expended over $30,000 during the previous four years to fund the studies that had led to the Academy's establishment. It was also significant that Harold K. Work, whose initiatives as director of the Foundation had played such a critical role in this process, was selected to serve as the first secretary of the National Academy of Engineering.

Perhaps the best summary of the National Academy of Engineering's functions was contained in the January 1965, Vol. XV, No. 1 issue of the *News Report of the National Academy of Sciences, National Research Council:*

> The objects and purposes of the National Academy of Engineering as formulated by the Committee of Twenty-five and set forth in the Articles of Organization are:
>
> 1. To provide means of assessing the constantly changing needs of the nation and the technical resources that can and should be applied to them; to sponsor programs aimed at meeting these needs and to encourage such engineering research as may be advisable in the national interest;
> 2. To explore means for promoting cooperation in engineering in the United States and abroad with a view to securing concentration on problems significant to society and encouraging research and development aimed at meeting them;
> 3. To advise the Congress and the executive branch of the government, whenever called upon by any department or agency thereof on matters of national import pertaining to engineering;
> 4. To cooperate with the National Academy of Sciences on matters involving both science and engineering;
> 5. To serve the nation in other respects in connection with significant problems in engineering and technology;
> 6. To recognize outstanding contributions to the nation by leading engineers.

To emphasize its role as a link between the engineering technology and research science, the National Academy of Engineering adopted a representation of a viaduct as its corporate symbol. During its first year, NAE's expertise was successfully directed toward such diverse areas of investigation as the design of buildings to withstand earthquakes, environmental engineering, support for developing nations, and the creation of a special radio program series for the Voice of America. The National Academy of Engineering also made vital contributions to such ongoing NRC activities as the formulation of patent policy, the upgrading of civil defense and emergency disaster planning, and construction research. By 1966 NAE was further expanding the scope of its efforts and entering the realm of public policy through the creation of a Committee on Public Engineering Policy which soon became known as COPEP. COPEP was formed to serve as a potential source of expertise for advising the Federal government on public policy questions which in some way involved engineering. COPEP almost immediately played a prominent role in the creation of the Congressional Office of Technology Assessment.

In 1968 NAE began to serve as an important outside advisory review body for NASA. As such, it had a great influence on the early development of the Space Shuttle program. The NAE Aeronautics and Space Engineering Board addressed such critical design issues as the future shuttle's size and payload requirements, sequential or concurrent booster/orbital development, and technological requirements for the shuttle's structure, materials, and thermal protection system.

During the turbulent years of the Vietnam war NAE was a divided organization with one faction hoping to lessen NAE's involvement in defense and aerospace projects, while instead focusing its activities on environmental issues. This divergence of opinion also caused conflict between NAE and the National Academy of Sciences. However, during the 1970s, these two research bodies did successfully complete many important cooperative projects including an influential multivolume report on the adverse health effects of air pollution, including the relationship of automotive exhaust

Poster session at an Engineering Foundation Conference.

emissions to air quality and the cost and benefits ratio that would result from the installation of pollution devices on America's new cars.

During the 1980s NAE continued to be a vital component of America's research activities. It also became a major broad based supporter of engineering education. In 1985 it issued a nine volume study entitled, *Engineering Education and Practice in the United States*. The NAE also played a major role in stimulating the growth of engineering education in industry.

The Engineering Foundation's catalytic role in the creation of the National Academy of Engineering ranks with its earlier funding of the National Research Council as one of its most important achievements and its greatest service to both the engineering community and the nation.

6. ENGINEERING INFORMATION SERVICES

Despite its activities in international engineering affairs, the Engineering Foundation did not neglect the needs of the American engineering community during this period. Realizing that the recent development of the electronic computer had made it possible to greatly increase both the speed and range of the interchange of technical information between engineers and scholarly institutions, the Engineering Foundation took a catalytic role in the development of such systems. At its September 26, 1962 meeting, the board of the Engineering Foundation approved a grant of $10,000 toward the cost of the installation of an electronic data processing center for UET. Of even greater importance was the action taken at the February 7, 1963 meeting of the Executive Committee of the board of the Engineering Foundation which resolved that the Information Systems Committee of the Engineers Joint Council endorse the joint proposal of the Engineering Index and the Engineering Societies Library for a $20,000 study by Helles Associates of the long term objectives of both the Index and the Library and further resolved that this endorsement be officially transmitted to the entire board with a request that the Engineering Foundation finance this study. This committee also recommended that the board vote $70,000 in additional grant support for the development of information services within the engineering profession. All of these requests were approved by the board.

The Engineering Foundation continued to support the development of engineering information services. At the April 9, 1964 meeting of the Engineering Foundation's Research Committee, an additional grant of $20,000 was allocated to the Engineering Index and this action was later approved by the entire board. It was also decided during this period to endorse studies and conferences which would be concerned with the impact of computers on engineering education and practice. In May of 1963 the Engineering Foundation made its initial grant to the development of Engineering Information Services. A revolutionary concept, the Engineering Information Services would utilize computers to increase the availability of technical literature to the engineering profession. The initial grant of May 1963 was given to the Joint Council, a consortium which was composed of the Engineers Joint Council and the Engineering Index, Inc. The Engineers Joint Council contracted to provide a professional staff for one year to carry out the development stages of a six-point program designed to improve the engineers' utilization of technical literature and application of information processing systems. The Engineering Index, Inc. was contracted to provide an analysis of its future role in a technical information center servicing the engineering profession. The actual analysis would be performed by the Battelle Memorial Institute. The ultimate goal of this project would be the creation of an Engineering Information Center in the United Engineering Center Complex.

At its October 10, 1965 meeting, the board of the Engineering Foundation voted to support the establishment of the Engineering Information Center as a coordinated service for the engineering profession as has been proposed by the Engineers Joint Council and which was then under development through the cooperative efforts of the Engineering Societies Library, the Engineering Index, Inc., and the Engineers Joint Council. The Executive Committee was authorized to spend up to $10,000 to support the implementation of this resolution. The Engineering Foundation also prepared and distributed a prospectus on the proposed Engineering Information Center. A great deal of the Engineering Foundation's activities on behalf of the Engineering Information Center was predicated on the anticipated approval of a $230,000 grant to the Engineering Index Inc. from the National Science Foundation. This grant would enable the Engineering Index, Inc. to put into effect the long planned full scale development of the Engineering Information Service.

The board of the Engineering Foundation viewed the creation of the Engineering Information Center as one of its most important activities during the 1960s. In its 1964 annual report, the Engineering Foundation termed the creation of the Engineering Information Center as a "major turning point" in the Foundation's history. It went on to further state that "The Foundation has over the past fifty years assumed leadership functions in engineering research and in cooperative efforts to advance the engineering profession. . . . One such area is information handling. Currently engineering organizations are vigorously attacking engineering information needs and it appears that the profession will require a fully automated information system. Such a system would be of inestimable value to the engineering profession, to industry and to the nation as a whole. The information system as presently envisioned in its ultimate form would, in addition to traditional library functions, provide new services to match the spectrum of information user requirements. Below is described one likely modern operation.

> A document received by the system center will be fed into an electronic scanner, translating it into English if necessary and reducing its contents to a mechanically retrievable form. Abstracts and index-term lists will be prepared manually and transmitted electronically for storage at a central processor. Interest profiles of customers will be made and these customers will automatically receive abstracts of pertinent documents. To obtain previously published information, a customer will dial or type his inquiry into a network connected to a central processor. Title lists and abstracts will be transmitted to the customer in some readable form, such as via closed-circuit TV. Wanted documents are then ordered from the network specifying urgency of delivery. The documents will be transmitted by mail, messenger or by electronics as urgency requires.

Unfortunately, the system that is described above was virtually impossible to create during the 1960s. Technology and software development had not yet reached the level of sophistication and flexibility needed to make the Engineering Information Center practical reality. Only during the late 1980s has technological reality begun to catch up with the Engineering Foundation's prophetic vision of the 1960s.

In a sidelight to the Engineering Foundation's support for the creation of an Engineering Information Center was its support of the efforts of the Engineering Societies Library to modernize and upgrade its operations. During 1985 the Foundation made available a total of $35,000 to the Library for the development of surveys and analyses, including literature searches, reference questions, publications from which photoprints were supplied, and personal interviews with users of the reading room.

After an extensive study of the automation techniques that were then in current use by other specialized libraries, a large part of the Foundation's grant was allocated to initiate automated serials control in the Engineering Societies Library. This improvement was felt to be essential not only to better enable the Engineering Societies Library to meet the needs of researchers but it was also due to the part that the Engineering Societies Library would play in the creation of the planned United Engineering Information Center. By the end of 1964 the Engineering Societies had made considerable progress through the installation of a system which included the data processing of all titles, CODEN for titles, holdings, call numbers, and the subjects of all periodicals. Printouts of all of this information were available to aid both the library staff and library users.

During the spring of 1965, the boards of the Engineering Index, Engineers Joint Council, and United Engineering Trustees, Inc. established a tripartite committee which was composed of the presidents and vice presidents from each to develop and bring into being a united engineering information system and center. The Engineering Foundation, as was reported in its 1966 annual report, provided $40,000 to support the work of the tripartite committee. The majority of this funding was utilized to subsidize a study by Information Dynamics Corporation. When it was completed in 1966, this study established that

1. Engineering management is firmly convinced that improvements in the nature of engineering information and data systems can contribute to improvements in engineering practice.
2. There is industrial acceptance of the concept of a united engineering information system established by the engineering societies.
3. Industry has expressed its willingness to contribute financial support and staff assistance.

By 1966 the Engineering Foundation was able to state in its annual report that

> The Engineering Index has continued to progress steadily in all technical phases of its plastics and electrical-electronics projects, supported by a $75,000 grant from the Foundation. The overall project was designed to offer improved information services and to provide a tested, modern base for the engineering and abstracting programs to be afforded by the United Engineering Information Center, currently under consideration by the Tripartite Committee. The initial products of the program are the

pilot bulletins, separately published "Plastics Section" and "Electrical-Electronics Section" of Engineering Index. These two new bulletins were designed as a part of a plan to fragment the Index into sections according to the various engineering disciplines.

In addition to providing published bulletins for current awareness and manual search, the programs of the pilot project were designed to include a central machine search capability for selective dissemination of current information as well as retrospective search. The search capability will be the principal feature leading to the implementation of an information center for the engineering community. Engineering Index has also initiated new activities in administrative and promotional areas from which realistic forward planning can be further developed.

In its 1967 annual report, the Engineering Foundation stated that it has "given some $350,000 in support of the engineering information retrieval program. These funds have been augmented from other sources, the National Science Foundation in particular."

Unfortunately, the expected grant from the National Science Foundation to the Engineering Index was limited to $84,000 which would be allocated only on the condition that the engineering profession would also make substantial contributions to this project. As a result, the Engineering Foundation was requested by the Engineering Index, Inc. to make an additional $75,000 grant to the Engineering Information Service project. This request was granted with the provision that the Engineering Foundation's funds be utilized for the development of a computerized abstracting and indexing program which would focus on plastics, and on electrical and electronic engineering. Despite the expenditure of large sums of money by the Engineering Foundation and other funding agencies, the Engineering Information Service was never completely developed and its eventual utility proved to be quite limited. Nevertheless, the pioneering effort of the Engineering Foundation in anticipating what would eventually become the information revolution of the 1970s and 1980s must be acknowledged.

7. REGULAR RESEARCH GRANTS 1959–75

The change in direction embarked on by the Engineering Foundation in 1959 included provisions for awarding regular research grants. The principal effect of the change was to decrease the number of awards for research projects and to increase the amount of the initial award. Another change came on October 11, 1972, when the board unveiled a new grant policy which focused on the use of research to develop innovative approaches to the solution of major national problems and the development of engineering principles and techniques for the future.

Of the eighty-five new projects listed in Table 2 in the Appendix for the period from 1959 through 1975, roughly one half were classified as interdisciplinary, one quarter as civil engineering and the remaining quarter was divided between mechanical engineering (8), chemical engineering (5), mining and metallurgy (3), and electrical engineering (1). The large number of interdisciplinary projects reflected the composition of the board as well as the unwritten policy that in grant awards preference should be given to interdisciplinary work. The large number of civil engineering projects was at least partly the result of grant RC-A-61-5 awarded to ASCE in 1962 and used to establish a permanent staff position of research manager who worked closely with the Engineering Foundation's Research Committee.

Interdisciplinary Projects

One of the subjects for the Engineering Foundation Conferences came from the J. Herbert Hollomon report of the Engineers Joint Council entitled "The Nation's Engineering Research Needs 1965–1985" (RC-A-62-3). The Engineering Foundation participated in a modest way in this study by underwriting the printing of the report. The report contained additional subjects of considerable promise for Engineering Foundation programs. To make possible a second revised edition of "The Nation's Engineering Research Needs 1965–1985," a supplemental grant was given to Engineers Joint Council in 1963 to augment the Foundation's original grant made in 1962 for the first edition of the report.

Technical Projects

In the metals and materials areas the Foundation has concentrated on assistance for the development of engineering areas in need of coordination of research. Material for the *Handbook of Composite Engineering Materials and Structures* was collected. Continued effort in the area of comminution has been made. The Engineering Foundation gave $36,000 over a three-year period to the project, "Fundamental Research on Comminution as a Unit Operation," carried out at the University of California at Berkeley (RC-A-64-7) in the middle 1960s. The research was concerned with delineating factors that control breakage probabilities and product size distributions in a laboratory ball mill. Studies that were undertaken included the role of ball size, liquid-to-solid ratio, feed size, and mill speed in comminution kinetics. Because of the torque instrumentation, the university was able to obtain data for correlating input energy with the extent of comminution.

In 1973 the Franklin Institute was awarded $20,000 by the Foundation to help develop a proposed aquaculture using hot waste water from power plants (RC-A-73-3). Additional funding in excess of $100,000 was raised from other sources to provide the hard engineering and cost data necessary to create a new inland fishery for the United States and a cheap source of animal protein. Prototype facilities were established on the Susquehanna River to produce continuous crops of trout, salmon, bass, catfish, and shrimp, without polluting neighboring waters. Fish manure was used to rejuvenate ash ponds. Reports that certain aquatic plants can increase evaporation rates were also tested.

The Foundation granted $9,000 partial support for a low-energy ion-reflection spectrometer constructed in the early 1970s by Mc Master University at Hamilton, Canada (RC-A-70-3). The total cost

The Engineering Foundation Fiftieth Anniversary Dinner. Seated left to right are Ernst Weber, Eric A. Walker, Foundation Chairman Warren C. Schreiner, and Augustus B. Kinzel.

of the spectrometer was $40,000 and was mainly supported by the National Research Council of Canada. The spectrometer was used for a research program to investigate the behavior of impurities in the surface layers of solids and how these impurities react to various environmental changes, e.g., temperature annealing, thermal oxidation, etc. The spectrometer used monoenergetic ions to mass bombard the sample under investigation. Then the energy spectrum of the back scattered ions was used to identify, measure, and profile the atomic constituents of the surface layer of the solid. The versatility that this technique afforded made it a very powerful method in the study of materials and their surfaces and, therefore, had a wide application in the solid state physics and engineering fields.

During the late 60s, the Engineering Foundation sponsored three conferences on engineering in medicine relating to multiphasic health screening, which sparked a continuous dialogue among engineers and representatives of various other technical disciplines. As a result, a new organization was established, the Society for Advanced Medical Systems. These conferences were also influential in stimulating the George Washington University Medical Center at Washington, D.C. to establish a new department of Clinical Engineering. Seed money in the amount of $67,500 was supplied by the Engineering Foundation for a three-year period (RC-A-70-1) to help this new department prepare detailed plans for approaching organizations such as the Department of Health, Education and Welfare for continuing funds to support future phases. A three-year plan was envisioned to allow three engineers to carry on part-time efforts in program definition. The program resulted in a reference work in application of modern technologies to health care, a coherent description of engineers' role in the delivery of health care, assistance to government decision makers in development of a national policy, curriculum development, and better information to the medical and related professions about engineering's growth and potential importance in clinical medicine. During the second year, the study group developed a preliminary outline for content of the work reference volume, conducted and participated in workshops, meetings and professional activities to insure precise definitions of the role of the engineer in the delivery of health care. The group provided advice to the President's Advisory Committee on Management Improvement and the Department of Labor Contractors on Skill Conversion. A course,"Introduction to Clinical Engineering" was organized and conducted. The study group also consulted with a number of institutions in the reorientation of their curricula toward clinical service activities, and continued efforts to insure the widest possible consideration by the medical profession and its representative institutions.

The token financial support referred to by Chairman Peirce in his October 22, 1959 report took the form of $100 grants to each of the projects active in 1959 and in several years that followed (RC-A-62-6). The last token grants were awarded in 1965 (RC-A-64-6).

Professional Projects

In 1963 the Engineering Foundation awarded a $3,500 grant to EJC (PR-62-1) toward developing EJC publication *Engineer* as an advertising medium so that the publication could become self-supporting. The *Engineer* reached 340,000 engineers and reported on national and local news that have significance for the engineering profession.

The need for engineering participation in international activities was intensified in the early 1960s due to (1) rapid expansion of technical communication and exchange among countries, (2) the increased industrial and engineering operations abroad, and (3) U.S. technical aid programs. A grant was made to Engineers Joint Council in 1964 to develop a self-supporting international engineering relations program (IR-62-1). Clarence E. Davis, secretary emeritus of the American Society of Mechanical Engineers, was appointed as the program's director. The program was a result of a recommendation in the EJC report "The Nation's Engineering Research Needs, 1965–1985," that the engineering societies take an active leadership in the international relations area. A conference, "Engineering Research for the Developing Countries," cosponsored by the Engineering Foundation and Engineers Joint Council in August 1962, gave further attention to the problem. The EJC developed a self-supporting international engineering program, which aimed at (1) advancing international cooperation in engineering, (2) furthering international technical education, and (3) administering relations with foreign engineering societies and international organizations.

In May 1963 the Foundation made a grant that enabled Tools for Freedom to maintain their New York office (IR-63-1). Tools for Freedom was established to provide a service for United States industry in the developing nations. The program was administered solely by the United States business community. The objectives of Tools for Freedom were (1) to introduce 'Made in the USA' products to future users and buyers of machinery and equipment in growing market areas in developing countries, (2) to give United States businessmen tangible means of creating in developing nations an improved investment climate and a better understanding of the private enterprise system, (3) to raise the level of training for technicians, machinists, and mechanics essential in the efforts of these countries toward self-support, and (4) to provide a constructive, non-competitive, tax-deductible channel for the elimination of outdated manufacturing equipment and the disposal of excess company inventories. Under the Tools for Freedom program, surplus or outdated equipment was sent to schools in Africa, Latin America, and Asia. The tools were used to train students in basic skills necessary for economic growth. The equipment was donated by the participating companies and transportation was often donated by trucking and shipping companies.

A 1964 Foundation grant contributed significantly to the expansion of the cooperative effort of Volunteers for International Technical Assistance with the engineering societies (IR-63-2). VITA was an association of 1,000 technical and business people from 250 corporations and sixty-five universities throughout the United States who gave of their free time to work on problems facing the developing nations of the world. The grant enabled VITA to maintain a liaison with the engineering societies as well as a United Engineering Center address.

An apparent need existed in the early 1960s for the engineering profession to catalog, clarify, and evaluate alternate methods by which the many segments of engineering could update their technical competence and be effectively informed on new technology and concepts, particularly when such shifts result in changing occupational responsibilities. Meeting these objectives pointed toward a coordinated professional approach. Supported by a Foundation grant (ED-63-1), the Joint Advisory Committee on Continuing Engineering Studies was established to consider the roles of industry, government, academic institutions, and engineering societies. The joint committee was sponsored by the Engineers' Council for Professional Development, the Engineers Joint Council, the American Society of Engineering Education, and the National Society of Professional Engineers.

Civil Engineering Projects

The position of ASCE research manager was filled with the appointment of Donald C. Taylor in April 1962. The duties of the research manager were to work closely with the various research activities of the Society and to organize the administrative work of the position. The services of the research manager involved (1) organization of society-wide research conferences; (2) preparation of papers and techniques of civil engineering research; (3) establishment of an ASCE research fund; (4) development of research projects through research committees and research councils of the Society; and (5) to assist the ASCE Committee on Research in their program aimed at increasing civil engineering research activities.

A case study was initiated in 1971 to measure the impact of civil engineering projects on people and nature (RC-A-71-4). The objectives were to identify, evaluate, and document the socioeconomic, environmental, and ecological benefits and disadvantages of selected urban works of civil engineers constructed and implemented in metropolitan Pittsburgh and Chicago. Projects selected were representative of the array of projects that are typical of important urban works constructed and implemented in urban areas throughout the United States. These included: urban development, expressways, transit, bridges, airport facilities, skyscraper construction, office buildings, hotels, apartment complexes, civic and cultural buildings, recreation facilities, sports arenas, department stores, parking garages, exposition halls, water supply and distribution, sewage disposal, flood control, urban drainage, air and water pollution control, solid wastes disposal, land reclamation, and navigation improvements and port development. Data and information were collected and analyzed to identify the relationships of many types of urban civil engineering work to socioeconomic, environmental and ecological impacts on man and nature. The project obtained an additional grant of $50,000 from the National Science Foundation.

A six-month study was undertaken in 1969 by Harold Deutschman of Matawan, New Jersey, with a $5,500 grant from the Foundation (RC-A-69-3), to update research on transportation planning surveys. A new technique for updating transportation origin destination surveys was studied by investigating communication community interest patterns. This technique involved the correlation and calibration of travel patterns with that of telephone call distribution patterns. It was the function of this study to investigate the feasibility of obtaining a continuous set of telephone data for use in transportation planning and investigate the degree in which telephone data may be used as a proxy or indicator for the transportation demand. The study concluded that it is feasible to collect and process telephone data for use in transportation planning. It was not concluded that telephone data can be used directly for updating transportation studies, but was surmised that this area has much potential and should be explored in more detail.

Construction

To encourage new research in the field of construction engineering and management, the Foundation gave $15,000 to the ASCE Construction Division Research Council in 1967. These funds were evenly divided between three projects: "Management Control and Simulation Systems Oriented Toward Actual Field Implementation" (RC-A-67-4a) which was carried out by J. S. Arrington at Stanford University, "Mathematical Modeling of an Aggregate Processing Plant" (RC-A-67-4b) which was carried out by J. A. Havers at Purdue University, and "Equipment Selection and Evaluation for Vertical Transportation Systems" (RC-A-69-4) which was carried out by Rex Rainer at Auburn University. The last project resulted in three reports: "one related to the packaging of building materials to increase productivity of material handling elevators; the second, entitled 'Simulation of a Vertical Transport System,' presented a description of a computer simulation model of vertically transported materials for high-rise construction; and the third, 'Cost of Waiting Due to Delays in Construction,' defined the cost of waiting as it applies to the construction process." Furthermore, this project also created master data files on all appropriate rubber, track, and fixed rail erecting cranes, hoists, tower cranes, climbing cranes, and concrete pumping devices. Its Parameter File identified fifty different machines considered to be appropriate for lifting service. Each piece of equipment was described completely and evaluated as to its performance and operating characteristics. Among the most important of these characteristics were weekly and monthly rental costs, operating cost per period, installation cost, transportation cost, horizontal reach, maximum working height capability of machine, ground area required for installation, maneuverability area required, ownership cost, mobility factor, and effective load capacity of machine. Finally, a computer program was written to analyze properly the demand for lifting service, to establish the appropriate level of lifting service, to estimate the various costs associated with the satisfaction of demand, and to determine the optimum machine, or combination of machines, for the system. The completed computer output was a complete description of the machine or machines that were best suited for a particular construction project.

As a result of the Foundation research conference on building construction systems that was held in 1964, a study was undertaken with Foundation support at the University of Illinois on "A System Design Procedure for the Planning of Construction Operations" (RC-A-65-1). The Construction Division Research Committee of the American Society of Civil Engineers monitored the study. The goal of the research was twofold; namely, use of network theory in forecasting the resource requirements in the construction of a project, and the development of an automated procedure for describing critical path method networks internally in computer memory where the input is to be any problem oriented language described by any contractors. The project was started with a $21,000 grant from the Engineering Foundation. Following initiation by the Foundation, over $50,000 was contributed to the project in funds and services from other organizations. This pioneering work done by the University of Illinois in construction research led to the establishment at the university of a Center for Construction Research of the U.S. Army Corps of Engineers.

The economics of high-rise apartment buildings (RC-A-70-4) was the subject of a 1970 ASCE study of the fundamental quantitative aspects of high-rise building configurations. The intent was to (1) expand the limits of current knowledge in architectural economics, (2) put practical quantitative information in the hands of practitioners, and (3) stimulate further research and development in this neglected, yet critical area. This project resulted in the accumulation of much data that was useful to the designers of tall buildings.

In 1971, the ASCE Research Council on Underground Construction embarked on a major program of research and development to create and introduce new technology and methods for underground construction (RC-A-71-3). In addition to technological improvement, the program also indicated the need for an increased amount of underground construction throughout the country. The research program was carried out in the following four phases which were designed to: (1) provide a sound basis for delineating specific research needs, defining criteria, determining social impacts, establishing communication and education requirements for implementation, and developing administrative procedures for coordinating the research program; (2) define the broad research needs to develop a sound justification for the use of resources for research on underground construction, and to provide a plan that will produce the overall objectives; (3) develop the detailed program for achieving the objective, and to conduct an in depth study of the steps required by the skeleton plan of phase (2); and (4) accomplish the objectives by completing the research and implementation program, including ongoing evaluation of the results. In addition to a $10,000 Engineering Foundation grant to support a task force of approximately five persons, the National Science Foundation provided $54,000 and industry $6,650.

Geotechnical Projects

The 1962 reprinting of the publication *Subsurface Exploration and Sampling of Soils for Civil Engineering Purposes* by M. Juul Hvorslev, authorized by the Foundation in March, proved to be most successful. About sixty percent of the printing cost was recovered through sale of the publication before the end of the same calendar year. The book was again reprinted in 1965.

The 1971 study "Identification and Classification of Expansive Earth Materials" (RC-A-71-1) was aimed at providing a simple, fundamental, non-destructive testing technique for field measurements to indicate potential expansive behavior of earth materials. Such a technique would be of considerable practical importance to the engineer for use in designing structures of and on soils. In addition the study included an investigation of the efficacy of two new approaches to the stabilization of expansive earth materials for the prevention of swell. A number of different soils obtained from different parts of California and abnormal soils supplied by the California Division of Highways, the Federal Housing Administration, and others were used to establish the relationship between the dielectric constant of the soil particles and the swell potential of soils. Environmental factors control the swell behavior of soils; the electrical measurements were used to characterize unsaturated soils for both their composition and environmental factors.

"Rapid Determination of the Quality of Rock Used in Water Structures" (RC-A-71-5) was the title of a study initiated in July 1971 with a grant of $6,692 from the Engineering Foundation and additional support from Purdue Research Foundation in the amount of $5,774. The objective of the project was to develop a convenient rapid test method to define resistance of rock materials to weathering forces. The general conclusion derived from literature was that there is no fully accredited test to quantitatively predict the performance of stones that were used to protect structures such as earth dams, coastal engineering works, and highway bridges from the action of flowing water. The existing tests that were used to evaluate properties of concrete aggregates in most cases were inadequate to provide data for the larger rock pieces for riprap and breakwater stones. The project attempted to devise a reliable and expeditious test that can establish quantitative performance criteria for such stone.

Structures

To obtain information on the actual performance of typical contemporary structures, the Building Research Advisory Board and the American Society of Civil Engineers developed and completed a study on destructive load testing during the demolition of structures at the New York World's Fair during the winter and spring of 1966 (RC-A-65-2). Nearly $250,000 was obtained from several organizations including the Engineering Foundation. Three structures were tested. The Chimes Tower, a steel framed structure in the Belgian Village was examined for response to vibration loading. The floor of the Rathskeller in the Belgian Village (of waffle-slab, reinforced-concrete construction) was tested to destruction with vertical loading. A two-story open-web steel-joist and steel-pipe column structural frame on Bourbon Street was tested to destruction with uniform vertical loads (by vacuum method), concentrated vertical loads, and lateral loads.

The tests of the three structures at the World's Fair site in New York led to the development of another opportunity for full-scale testing of structures in conjunction with a major urban renewal project in White Plains, New York. The Foundation gave $10,000 to the ASCE Research Council on the Performance of Full-Scale Structures for a feasibility study (RC-A-67-5) of full scale testing to be conducted in conjunction with an urban renewal project in White Plains, New York. As a result of this study ASCE developed individual projects with funds for testing solicited from other sources.

A grant of $25,000 was awarded in 1975 to Colorado State University to develop new design procedures for a glue-laminated timber bridge system (RC-A-74-6). Following the energy crisis of the early 1970s, wood—a renewable material with low energy of production—was considered a potential candidate for wider use in short span bridge construction. A new bridge configuration was developed by the American Institute of Timber Construction in cooperation with state highway departments. This system employed glulam stringers and novel glulam deck members to produce a practical and easily constructed bridge. Proper accounting of component interaction, commonly considered for steel and concrete, can significantly reduce required sizes, contribute to more efficient utilization of timber resources, and improve the economy of bridge construction. The investigation included both theoretical and experimental studies. The theoretical study was directed at applying a composite analysis method to the bridge problem with proper allowance for interlayer slip and gaps in the decking. Testing was aimed at studying individual T-beams constructed of glulam girders with glulam deck and verifing the analytical model.

Projects Concerned with Water

Cornell University received a 1976 grant of $20,000 for a project entitled "Turbulence Measurement in Lakes" (RC-A-75-4). The purpose of this twelve month study was to find measurement techniques which would fill a gap in the knowledge about lakes and to provide data needed for physical and biological modeling. It was hoped that the information received could be applied towards the alleviation of thermal pollution, the limitation of temperature increase, a possible solution of wastes discharge, and towards more precise locating of fresh water intakes. The information was also applied towards the control of factors affecting oxygen modeling and towards a strong physical model for ecological monitoring.

A grant of $15,000 was awarded in 1976 to the Polytechnic Institute of New York for the support of a project entitled "Emergency Water Supplies from Groundwater in Humid Regions" (RC-A-75-5). The purpose of the twelve month project was to determine the feasibility and potential of obtaining emergency water supplies from groundwater in humid regions such as the northeastern United States. It was hoped that this scheme would lessen the amount of ground subsidence by the infrequent pumping of water instead of by the use of heavy pumping in times of emergency. Other possibilities offered by this plan were the substitution of emergency groundwater developments for large surface water reservoirs which could offer possible savings in cost and greater dependability and also an alternative to the surface water reservoirs from environmental, conservational, and socioeconomic points of view.

A grant of $10,000 was awarded in 1977 to Clarkson College to assist in their investigation into the development of feasible techniques for the recovery of acidified aquatic ecosystems (RC-A-76-4). A potential serious consequence of this country's tremendous energy utilization was the acidification of land and water resources in the northeastern United States as a result of the emission, transport, oxidation and hydrolysis, and deposition of sulfur and nitrogen oxides produced in the consumption of fossil fuels. The general approach of the project was the addition of some material or combination thereof to raise the pH of water to near neutral values and buffer the system at that level, in the expectation that acid precipitation abatement will probably not be realized for some time. It was also probable that once soils and benthic sediments in a given watershed have become acid saturated they may contribute hydrogen ions to the water for long periods of time following discontinuance of actual acid precipitation.

Environmental Projects

The study "Evaluation of Carbon Dioxide Concentration as an Indicator of Air Pollution in Urban Atmosphere" (RC-A-60-1c) was started in 1962 at the University of California, Los Angeles. The changes in concentration were used in much the same way that changes in barometric pressure are used for weather prediction. The study was underwritten by the Engineering Foundation and was sponsored by the Air Pollution Research Council of the American Society of Civil Engineers.

As a part of the development of a study on the measurement of stack emissions, a two-day workshop meeting "Stack Sampling and Monitoring" was arranged in November 1963, by the Coordinating Committee on Air Pollution (RC-A-60-1a). The objective of the meeting was to arrive at specific research project recommendations. As an aid in developing the research program the committee initiated two literature searches (RC-A-60-1b), "Measurement of Stack Gas Velocity and Methods of Sampling Stack Gases and their Particulates" and "Research into Sampling, Analysis or Monitoring of Gaseous Pollutant Emissions from Stacks."

The major objective of project "Utilization of Sewage Solids to Produce Soluble Organic Carbon" (RC-A-75-1) for which the University of Washington received a 1975 grant of $16,500, was to determine whether the organic carbon energy source required for bacterial denitrification can be inexpensively produced at the sewage treatment plant from waste solids. Continuous-flow bench-scale experiments were conducted to determine the maximum amount of soluble organic carbon, primarily volatile acids, which can be obtained from an anaerobic digester when methane production has been suppressed. Full-scale process costs were estimated and compared with the cost of denitrification using methanol as the required energy source.

In 1975 Dartmouth College received a grant of $35,400 for a study, "Assessment of Environmental Standards" (RC-A-73-2). The study was extended to the following year with another grant of the same amount. The first year's grant and approximately 25% of the second year's grant were devoted to studies in the areas of thermal discharge and solid waste regulation. Water quality criteria guidelines for developing fresh water temperature standards were used to formulate improved water temperature standards and applied to the Vermont Yankee Nuclear Power Plant on the Connecticut River. A modified effluent regulation policy was developed to control effluent heat and comparative costs were determined. Increases in plant operating costs resulting from effluent head limitations based on allowable increases in river temperatures were used to determine the relative costs of alternative temperature standards. A summary paper of this work was circulated to federal environmental officials as well as New England Electric Company representatives. The regulations for the individual New England states were studied and evaluated and the different methods of solid waste disposal examined. The dollar costs of complying with regulations based in these methods were identified, within the constraints of data availability, along with the environmental benefits of compliance. Recommendations for modification of existing rules and procedures were developed and distributed to the New England states. The effort generated an unbiased analysis on the relationship between environmental standards and engineering alternatives. Major spin-offs, in addition to public awareness, were the opportunity for the results of this study to affect educational course content in the Thayer School of Engineering of Dartmouth College and the provision of information to policy makers to allow an increased awareness of the interdependence of the design and the implementation of policy.

Mechanical Engineering Projects

In October of 1969, a $47,300 grant was awarded to the American Society of Mechanical Engineers for a two-year project to attack the problems of developing and correlating the necessary theoretical information required for the combustion aspects of incinerator design (RC-A-69-2). The study was monitored by the ASME Research Committee on Industrial Wastes. The objective was to provide the practicing engineer with the technical information required to design incinerators for the disposal of industrial wastes with a greater degree of accuracy and compliance with air pollution regulations than was then current. The study was conducted at the National Bureau of Standards and at the E. I. du Pont de Nemours and Company, Inc. The results were published in a two part monograph, with part I covering the engineering applications and part II covering the scientific applications.

"The Feeding of Pulverized Coal to High Pressure Gasifiers" (RC-A-74-7) was a study started in 1975 at West Virginia University and supported by a $20,000 Engineering Foundation grant. The purpose of the investigation was to establish the technical and economic feasibility of a new device for feeding pulverized coal to high-pressure vessels for gasification purposes. A serious gap was developing between natural gas supply and demand. The use of abundant domestic coal reserves to manufacture both substitute natural gas and low-BTU gas would help to alleviate this problem. One of the main mechanical difficulties which was encountered in this connection was a practical means of introducing solid fuel into a reactor at elevated pressures. The project was devoted to a technical and economic study of a gas dynamic injector to introduce a co-flowing subsonic jet of gas and pulverized coal into the low-pressure gaseous flow in the supersonic combustion ramjets. This configuration allowed use of standard techniques to feed the coal from an unpressurized hopper to the nozzle, followed by the use of a supersonic diffuser to raise pressure to the extent required.

In the fiscal year 1967–68, the Foundation gave a total of $20,000 to ASME for support of three research projects. The Research Committee on Corrosion and Deposits from Combustion Gasses was given $5,000 for the completion of a research program on formation of alkali non sulfates and other compounds causing corrosion in boilers and gas turbines (RC-A-67-3). The Research Committee on Fluid Meters received $10,000 for a study to determine how accurately the coefficient values for concentric orifice plates and flow nozzles can be predicted and to establish the best empirical relations for determining these values (RC-A-68-1). The flow nozzle, concentric orifice plate with flange taps, and orifice plates with other types of taps were studied in an attempt to determine a single relationship fitting a large range of diameter ratios and Reynold's numbers. Finally, $5,000 was given, as seed money for raising additional funds from industry, to the Research

Committee on Boiler Feedwater Studies for an experimental investigation of some factors which influence the accuracy of steam sampling (RC-A-68-2).

During the period 1959–1975, the Foundation supported five additional projects in the area of mechanical engineering. Two of them were of highly practical nature: "Water Requirements for Buildings" (RC-A-68-3) and "Charging Airborne Particulate Matter in High Concentrations Using Radiation and Electric and Magnetic Fields" (RC-A-73-8). The remaining three received only token support (Items 142, 143 and 144 in Table 2).

Chemical Engineering Projects

The study "Non-Ferrous Reaction Rates and Mechanisms with the Hot Filament Microscope" at Carnegie-Mellon University in Pittsburgh, Pennsylvania was started with a 1970 grant of $16,100 from the Engineering Foundation (RC-A-70-2). The broad objective of this research program was to gain a more detailed understanding of the rates and mechanisms of reactions of special interest in non-ferrous pyrometallurgy by studies with the hot filament microscope. The initial stages of the study determined what particular reactions were amenable to this study and most likely to yield useful information. This exploration phase established both the feasibility and need for a more detailed investigation of the reactions of silicate and aluminate slags with the various iron oxides which served as both oxidizing reagents and refractories in commercial practice. Necessary procedures and techniques were developed so that systematic studies could follow. Less extensive observations were made on matte oxidation and reactions at the slag-matte interface, but it was not felt that these should be continued until substantial progress was made on more fundamental metal oxide-slag reactions.

The investigation "Measurement of Boiling Heat Transfer Data with a Pyroelectric Thermometer" (RC-A-71-2) was initiated in September of 1971 at McGill University in Montreal, Canada, with a $9,860 grant from the Foundation. The objective of this project was to develop and test a technique which permitted the rapid determination of boiling heat transfer data. The proposed device was useful for (1) determining rapidly the effect of surface finish on heat transfer surface, (2) evaluating quickly the effect of liquid additives which may be used to increase boiling heat transfer rates, and (3) evaluating quickly the heat transfer effects of impurities or decomposition products found in heat transfer fluids. The project included the development of a pyroelectric device and a comparison of the data obtained by the new technique with data obtained by conventional means. A conclusion of the study was that heat fluxes may be measured using the pyroelectric effect in a

Board member Walter E. Lobo outlines a proposed project at the meeting of the Research Committee. E. O. Ohsol and S. R. Beitler are in the background.

quenching experiment. Very rapid temperature changes cannot be accurately followed by the pyroelectric sensor due to finite sensor size and finite thermal diffusivity. However, the technique has potential for use with small heat fluxes where thermocouple voltages will not be sufficiently sensitive to differentiate accurately.

The American Institute of Chemical Engineers received, in 1973, a grant (RC-A-73-1) of $20,000 from the Foundation to establish the Design Institute for Multiphase Processing. The purpose was to refine technology not only to reduce production costs and to extend capabilities, but also to control pollution. The grant enabled DIMP to draw up a first year plan, and to evaluate proposals from research teams. Additional funds for this five-year program were obtained from the industry.

Two additional chemical engineering projects received financial support from the Foundation during the period 1959–75. The investigation "Asymmetric Hollow Fiber Membranes for Enzyme Immobilization" at Stanford University was awarded a grant of $26,000 (RC-A-73-5); and $1,500 was contributed to AIChE toward reestablishment of communications exchange with mainland China (RC-A-72-2).

Metallurgical Engineering Projects

A $14,440 grant was awarded in 1969 by the Foundation to Mississippi State University for theoretical research related to fatigue in fiber-reinforced composite materials (RC-A-69-1). The problem of crack nucleation in fiber-reinforced composites and solids was investigated. Primary attention was devoted to a study of how cracks nucleate near the fibers of inclusions. Dislocation arrangements that are likely near these inclusions or fibers were determined by solving the integral equations governing dislocation behavior. Data from the literature were collected and analyzed to check the results obtained from calculations.

Carnegie-Mellon University's 1971 investigation of selective nitride formation in molten iron alloys was a study in the kinetics of coupled reactions (RC-A-71-6). Coupled rate processes occur in most complex systems, including steelmaking. The theory of the kinetics of coupled chemical reactions was poorly developed because of the paucity of data from properly controlled experiments. The selective formation of nitrides in levitated drops of iron alloys was used as a nearly ideal way to study this important class of problems in metallurgical kinetics. The Engineering Foundation contributed $20,600 toward defraying the costs of this investigation. The remaining financing was contributed by the university.

The investigation "Use of Fluidized Bed Electrodes in Extractive Metallurgy" was the third study sponsored by the Engineering Foundation in the field of metallurgy during the period 1959–75. The study was carried out at the University of California at Berkeley (RC-A-74-3).

Electrical Engineering Project

"Energy Separation in an Expanding Binary Mixture and Its Application to Energy Utilization" (RC-A-74-5) was the only electrical project supported by the Engineering Foundation during the period 1959–75. The amount of its support was $20,000 and it was awarded in 1975 to the University of California at Los Angeles.

8. NEW GRANTS INITIATIVES

The Engineering Foundation's initial activity in the area of air pollution research was a grant of $10,000 (Items RC-A-60-1a, 1b and 1d) to be used in developing a critical study of the present methods of measuring smoke stack and exhaust emissions and seeking ways to improve them. Approved at the May 3, 1962 board meeting, it was hoped that this study would result in specific projects which would result in new technical recommendations for emission controls. The research work was to be conducted in cooperation with the newly established coordinating committee on air pollution. This project quickly bore fruit with its establishment of the use of carbon dioxide concentrations as the principal indicators of air pollution in urban environments. To further this research effort, the Board of the Engineering Foundation granted an additional $40,000 (RC-A-60-1c) to this project with the stipulation that every effort be made by the Coordinating Committee on Air Pollution to raise additional funding from other sources. During the next five years this project continued to successfully investigate the problems of stack and exhaust emissions and made detailed recommendations for their continued measurement and eventual reduction.

At this same board meeting, a report was delivered on the recommendation that had been developed by the Engineering

Foundation research conference on engineering in developing countries. This report emphasized the need for the American engineering community to aid in the technological advancement of the newly independent nations of Africa and Asia. This report was one aspect of the Engineering Foundation's continuing involvement in this field.

Another pioneering effort of the Engineering Foundation during this period was its financial support for the activities of the Junior Engineering Technical Service (JETS; PR-60-2 in Table 2). The Engineering Foundation had long been concerned with the development of engineering education in America and JETS, which was devoted to sponsoring engineering interest and activities on a high school level, fitted well with the Engineering Foundation's goals in this area. At its October 4, 1962 meeting, the board of the Engineering Foundation voted $5,000 to JETS in order to subsidize the expenses of moving their headquarters to the Engineering Center in New York and to provide for their future growth in that building. A year later an additional $25,000 grant was made to JETS. Due in large part to this financial support, JETS underwent a rapid growth. It reported to the Engineering Foundation that the Foundation's grant had enabled JETS to add 102 new chapters and raise an additional $13,675 from private industry. Throughout the coming years JETS continued to grow and prosper. By 1979 it had a total annual budget of $115,000 and served 350 student chapters with a combined membership of over 5,000 young people.

The Engineering Foundation also continued its traditional support for the Engineers Council for Professional Development. Although the Engineering Foundation's earlier support for this organization had centered on its activities in accrediting engineering curricula, the Foundation now focused on the role of ECPD as a developer of programs of continuing education for professional practicing engineers. At its May 9, 1963 meeting, the board of the Engineering Foundation expressed its support for ECPD's efforts in this area by approving a grant of $5,000 to this organization for a preliminary study of the feasibility of establishing an institute of continuing engineering studies with the existing organizational framework of ECPD (ED-63-1 in Table 2). Due in large measure to this funding, ECPD was able to complete this study and develop and implement a modest but successful program of continuing education for practicing engineers.

The Engineering Foundation also early identified the increasing challenges that U.S. products would be facing from foreign competition. It is illuminating to realize in this time of the early 1990s, in which the United States annually amasses a large trade deficit, that in 1964 the Engineering Foundation held a successful research conference on the theme "Technological Challenges for the U.S. in World Markets, 1964–1974" (64(x009)03 in Table 3).

Another long-term problem was also identified by the Engineering Foundation during this period when it co-sponsored a special conference on the problems of Appalachia. Held in 1964 at the University of West Virginia, it was financed by a special $4,000 grant from the Engineering Foundation. This conference focused on the application of engineering technology to the problems posed by the economic decline of the traditional mining- and lumbering-based economies of this region of the southeastern United States. Among its recommendations were the development of new technologically oriented industries such as electronics in this region and the revitalization of the traditional extractive industries through the use of greater mechanization. As a result, this conference became a model for future efforts to apply technology to the economic problems of other underdeveloped regions of the United States.

The progressive nature of the activities of the Engineering Foundation during this period can best be summarized by the following excerpt from an editorial published in 1983: "No government body, no matter how much money it had, could show the absolute objectivity of professional idealism and technical know-how that the Engineering Foundation has displayed during recent years."

Despite its new direction the Engineering Foundation continued to aid several projects that it had formerly supported with continuing grants. Among the most important of these projects were the Corrosion Research Council and the Council on Wave Research. At its May 19, 1963 meeting, the board of the Engineering Foundation approved a special grant of $1,000 to the Corrosion Research Council (RC-A-63-7 in Table 2) provided that a recommendation be made to the Council that the Engineering Foundation receive, before the end of 1963, a report on the proposed merger of that organization with the National Association of Corrosion Engineers. The merger of these two bodies was completed by 1965. At this same meeting, the board also approved a grant of $4,000 to the Council on Wave Research for the purpose of subsidizing the publication of the *Proceedings of the Eighth Conference on*

Coastal Engineering with the stipulation that this sum was to be repaid through the proceeds of the sale of this volume. The *Proceedings* enjoyed slow sales for, at the April 9, 1964 meeting of the Engineering Foundation's Research Committee, it was reported that the Council on Wave Research still owed $830 to the Foundation and that it, with the Foundation's encouragement, was negotiating a merger with the Council on Coastal Engineering of the ASCE's Waterways and Harbor Division.

Despite the many innovations that the Engineering Foundation pursued during the 1960s, its new course was not universally accepted by its board members. Many of these individuals had serious misgivings concerning the Foundation's abandonment of its traditional emphasis on the funding of continuing research for specific engineering research projects. Although they understood the value of utilizing the limited grant resources of the Engineering Foundation as seed money for the startup of new areas of resources, they also felt that the Engineering Foundation had to

Attendees of the conference "Unconventional Protein Production" held at the University of California at Santa Barbara in August 1967.

maintain its long tradition of providing long term support for major engineering research. The Engineering Foundation had attempted to continue a shadow of its former policies by making token annual grants of $100 each as a form of moral supportive endorsement to a score of projects that it had formerly supported with regular funding, but these token grants did little to placate the supporters of the Foundation's former policies. Perhaps the most cogent statement of the reservations that were held by many board members was expressed in a letter that was written to the board of the Engineering Foundation by retiring member Ernest L. Robinson on May 1, 1963.

> As I see it, the basic problem confronting the Foundation is the change in the relative importance of a million dollars during the 50 years since the Foundation was established. In those days, a million dollars was a lot of money. Now it is a puff of smoke from another rocket, a mere penny apiece to the taxpayers.
>
> This situation has worried me since the time that I came on the Board thirteen years ago and for eight years I did not get up the courage to mention it. I don't recall that anyone else talked about it any earlier. The question was what if anything could we do about it. I confess that I have never been able to think of a program better than the kind of true engineering research projects that the board used to sponsor and I still think that our best policy would be to direct the use of our modest income so as to best accomplish the purposes which were originally intended—engineering research.
>
> As you know, the board has chosen practically to discontinue sponsoring that kind of activity and the steps that it took were such as to discourage new applications for such work. Now it looks to me that we are grasping for projects and getting some less worthwhile ones in order to get rid of our accumulating income. This policy is what I don't like. I feel that too much of our money is now spent on reports and investigations intended to help us make up our mind and on meetings and conferences which I believe would be better held within the framework of the Founder Societies conventions and transactions. Meanwhile, we let go of our opportunities for direct support of worthwhile projects within our means, some small and others not so small. I think that our present setup almost requires a full-time job of our research chairman to administer what the board expects him to whereas our research used to be administered by our director. Now we don't even have a full-time director anymore and as long as the board wishes to devote its energies to arranging for meetings and getting reports printed that is what our director must do.

> My thought is that the Engineering Foundation has a mission to seek out and sponsor engineering research where industrial efforts needs cooperative endeavor. Gentlemen, I am just a retired designer of steam turbines but I would like pleasure of donating $1,000 to the endowment of the Engineering Foundation as evidence of my faith as I wish you all farewell.

The affairs of the Engineering Foundation entered a new phase on October 15, 1965, when Harold K. Work resigned as director to become the secretary of the National Academy of Engineering. Under his leadership, the Engineering Foundation had radically reshaped its activities by shifting the focus of its funding from the long term support of continuing engineering research to the identification of gaps in existing research through the sponsorship of research conferences and the granting of seed money for reports and studies. Work went on to enjoy a distinguished career at the National Academy of Engineering and other posts.

At its May 12, 1966 meeting, the board of the Engineering Foundation decided to operate on a temporary basis without securing the services of a new director. The rationale behind this decision was stated by the board as "to see what direction the activities of the Engineering Foundation would take if left to themselves." In this manner, the Foundation activities could largely be allowed to prioritize themselves. However, it was realized that the one area where this laissez faire approach could not be allowed to operate was research conferences and as has earlier been noted in this volume, Sandford S. Cole was appointed director of the Engineering Foundation's Research Conferences on May 12, 1966.

In line with its new policies of focusing its funding activities on research conferences and short range studies, the board of the Engineering Foundation began to sever all of its remaining formal connections to research councils. At its May 12, 1966 meeting, the board voted that after 1967 research councils which had been formerly funded by the Engineering Foundation would no longer be able to use the designation "Council of the Engineering Foundation" and must switch to the terms "sponsored by" or "established by" the Engineering Foundation. It was also resolved at this same meeting to establish a policy of limiting the term of an initial Engineering Foundation grant to the period during which a council may indicate direct affiliation with the Engineering Foundation and during that period the Engineering Foundation should be represented on the council's board by the Foundation chairman and secretary.

The Engineering Foundation continued to function throughout 1966 without the services of a director and the board voted to continue this policy for the foreseeable future. It felt justified in taking this action because it was utilizing the limited funds that it had available as seed money for studies and short term projects and to fund its research conferences which were under the able Director Cole. To make a further break with its past, the board at the May 11, 1967 meeting voted to discontinue its unpopular and largely irrelevant practice of granting token grants to ongoing research projects and councils that it had formerly subsidized. As a result of this action, the sum of $2,000 which had already been authorized to fund these grants during 1967 was returned to the Engineering Foundation general fund.

The forward-looking attitude of the board of the Engineering Foundation was evident at the May 9, 1968 meeting when it resolved to support ongoing efforts by both private groups and governmental agencies to secure the adoption of the metric system of measurements in the United States.

As the decade of the 1960s came to an end, the board of the Engineering Foundation began to re-evaluate the many policy changes that it had adopted during recent years. A growing awareness became evident among the board's members that a limited number of basic engineering research projects of the type that the Foundation had traditionally funded should be once again sought out. To prepare for this action the Research Committee reported at the May 15, 1969 meeting of the board that

> the vacant position of the Director of the Foundation should be filled with an engineer who would have the responsibility for checking the progress, expenditures, reports and other information concerning projects supported by the Research Committee so that the committee as a whole could operate as an advisory body while the staff engineer (Director) would carry out the necessary supervisory work.

However, no action was taken by the board and the position of director of the Engineering Foundation remained unfilled until 1984.

At its meeting of May 17, 1970, the board voted to change the title of its conferences from the Engineering Foundation Research Conferences to the simpler Engineering Foundation Conferences. It was stated that this new title would better reflect the broader nature of these gathering's subjects and activities.

The continuing efforts of the board of the Engineering Foundation to re-evaluate its policies once again became evident when at its May 13, 1971 meeting, it voted to establish a special task force "to review and recommend (a) a set of procedures for the Engineering Foundation and a system for future review and re-establishment of Engineering Foundation objectives, (b) a set of procedures through which the Engineering Foundation may establish priority areas, solicit proposals, choose and administer grants, [and] (c) a set of priority areas in which the Engineering Foundation should concentrate its support." This task force was quickly organized and requested to render its report by February of 1972. The board voted to table all grants until that report was received and acted upon.

The report of the task force was written in the form of a series of recommendations to make the funding activities of the Engineering Foundation more effective.

1) The Engineering Foundation should continually review its objectives, its program, and its procedures. The purpose of this review and updating is to assure that the Engineering Foundation's funds are expended as effectively as possible in view of major changes that are taking place in technology, industry, education, government, and in society itself.

2) The Engineering Foundation should be more aggressive in deciding how to award its funds. We should actively solicit proposals in chosen areas of prime interest. Only proposals of high caliber should be granted funds. Unique and original proposals should be given preference. Projects chosen for support should have a high ratio of output information to input efforts and funds. Thus projects entailing large expenditures for equipment should not be viewed as a source of support for every technically defensible engineering research project, but only for the exceptional project.

3) A listing of the areas of the greatest current interest should be prepared and reviewed periodically. Projects in these priority areas may be of a general nature aiming to determine what are the important problems in an area and which approaches already proposed by others merit consideration or further research. The activities of the Engineering Foundation are known to a comparatively few engineers. To attract superior proposals we must acquaint more engineers with the Engineering Foundation's grant program. Such publicity must make clear the guidelines under which we are operating at the time.

4) Funding should be concentrated in a very few areas and grants should be both larger and fewer.

5) Priority listing and concentration of support should reflect the major problems facing the country in which engineering has a place in

finding solutions. Moreover, such problems will often interface with many disciplines besides those of engineering and thus require an interdisciplinary approach.

General Objectives of the Engineering Foundation

I. To improve the quality of engineering practice both professionally and technically.
II. To improve the interaction of engineering with other professions, with business and industry, with government and with the public.

Because these objectives are broad, it appeared to us that a few guidelines should be set up for their immediate implementation. We believe that the general objectives can stand for a substantial time, but that the guidelines in common with the priority areas should be reviewed frequently and revised as found to be appropriate. Thus, in a sense, the guidelines and priority areas taken will define the immediate *modus operandi* of the Engineering Foundation's grant functions within the scope of its general objectives.

Guidelines

A. Solicit proposals in areas of prime interest.
B. Fund only those proposals of demonstrable high caliber.
C. Give preference to unique approaches especially those in which innovations are most likely.
D. Insist on a high ratio of output information to input efforts and funds.
E. Encourage interdisciplinary approaches.
F. Do not limit funding to proposals having very specific objectives. Include proposals of a more general nature aiming to determine what are the important problems in an area of which approaches already proposed merit further research.
G. Publicize the Engineering Foundation's grant program setting forth clearly both guidelines and priority areas.
H. Fund only a few projects each year.
I. Avoid funding proposals which duplicate or parallel ongoing efforts unless the approach is substantially different and innovative.

Priority Areas

1. Production, conversion, and utilization of energy.
2. Collection and disposal of solid refuse.
3. Mass transit systems for land transportation.
4. Engineering for the future.

Review of General Objectives

> The board will review general objectives not less often than once every five years.

Review of Guidelines

> The Research Committee will review the guidelines not less often than once every two years and recommend to the board at its annual meeting proposed retentions or changes in the guidelines.

Review of Priority Areas

> The Research Committee will review priority areas not less often than once per year and recommend to the board at its annual meeting proposed retentions or changes.

Review of Organization and Operating Procedures

> No major revision in organization or in current operating procedures of either the Research Committee or the board are recommended at this time. It is believed that sufficient flexibility already exists to take care of the immediate future needs. However, this recommendation should be re-examined after the initial year of operation. One minor revision is recommended. The change of the name of the Research Committee to the Grants Committee.

After being adopted at the May 10, 1972 board meeting and publicized under the title "Innovative Approaches to the Solution of Major National Problems" in the journals of all founder societies, the policies contained in the task force report continued to serve as the primary influence on the development of the Engineering Foundation for the remainder of the 1970s.

The November 7, 1974 meeting of the Engineering Foundation Board also marked the beginning of the implementation of its new policy of making a few large grants annually to projects that proposed an innovative approach to a national problem. $27,659 was granted to the University of California for the study "Use of Fluidized Bed Electrodes in Extractive Metallurgy," while $35,400 was granted to Dartmouth College for the continuation of its project on the Assessment of Environmental Standards.

References

3.1 "Work, Harold Knowlton," *Engineers of Distinction* (1970).
3.2 "Work, Harold Knowlton," *Who's Who in Science* (1968).
3.3 Harold K. Work's Resume dated June 22, 1965.
3.4 "Annual Report, October 1, 1967, to September 30, 1968" (New York: Engineering Foundation).
3.5 "1969 Annual Report" (New York: Engineering Foundation).
3.6 "1970 Annual Report" (New York: Engineering Foundation).
3.7 "1971 Annual Report" (New York: Engineering Foundation).
3.8 "1972 Annual Report" (New York: Engineering Foundation).
3.9 "1973 Annual Report" (New York: Engineering Foundation).
3.10 "Sixty years of Service 1914–1974 and Annual Report 1973–74." (New York: Engineering Foundation).
3.11 "1975 Annual Report" (New York: Engineering Foundation).
3.12 M. L. Mendelsohn, "In Memoriam: Sanford S. Cole (1900–1986)," *Cytometry* (August 1987), pp. 111–113.
3.13 H. A. Comerer, "Sandford S. Cole, An Appreciation," *Mining Engineering* (October 1986), pp. 988, 994.
3.14 "In Memoriam. Sandford S. Cole 1900–1986," from the Engineering Foundation Annual Report 1986, pp. 3 and 4.
3.15 Vita of Sandford S. Cole.
3.16 H. K. Work "The Question of Establishing a National Academy of Engineering," *Journal of Engineering Education*, Vol. 51, No. 9 (1961), pp. 698–700.
3.17 H. K. Work, "What Is Being Done About a National Academy of Engineering," *Journal of Metals*, Vol. 14, No. 5 (1962).
3.18 "National Academy of Engineering Established Under NAS Charter," *News Report NAS-NRC*, Vol. XV, No. 1 (January 1965).
3.19 "What About Research in Developing Countries?" *Chemical Engineering Progress*, Vol. 58, No. 9 (September 1962), pp. 10–11.
3.20 J. H. Hollomon, *The Nation's Engineering Research Needs 1965–1985* (Engineers Joint Council, 1962).
3.21 J. W. Johnson, ed., *Proceedings of the Eighth Conference on Coastal Engineering* (Council on Wave Research, 1963).
3.22 H. Bieber, ed., *Engineering of Unconventional Protein Production*, Symposium Series #93 (American Institute of Chemical Engineers, 1967).
3.23 S. W. Jens and D. E. Jones, eds., *Urban Hydrology Research—Water and Metropolitan Man* (American Society of Civil Engineers, 1968).

3.24 L. B. Wingard, Jr., ed., *Enzyme Engineering* (John Wiley & Sons, 1971).
3.25 O. Salembier and A. C. Ingersoll, eds., *Women in Engineering and Management* (New York: Engineering Foundation, 1972).
3.26 J. W. Hill, ed., *Economical Construction of Concrete Dams* (American Society of Civil Engineers, 1972).
3.27 E. L. Armstrong, ed., *Inspection, Maintenance and Rehabilitation of Old Dams* (American Society of Civil Engineers, 1974).
3.28 E. K. Pye and L. B. Wingard, Jr., eds, *Enzyme Engineering* (New York: Plenum Publishing Corporation, 1973).
3.29 E. F. Casey, ed., *Need for a National Policy for the Use of Underground Space* (American Society of Civil Engineers, 1973).
3.30 S. Dubin and H. Shelton, eds., *Maintaining Professional and Technical Competence of the Older Engineer* (Washington, D.C.: American Society for Engineering Education, 1974).
3.31 J. R. Graham, ed., *Use of Shotcrete for Underground Structural Support* (American Society of Civil Engineers, 1974).
3.32 M. L. Mendelsohn, ed., "Automatic Cytology," *Journal of Histochemistry and Cytochemistry*, Vol. 22, No. 7 (1974).
3.33 M. M. Singh, ed., *Subsurface Exploration for Underground Excavation and Heavy Construction* (American Society of Civil Engineers, 1974).
3.34 K. W. Heathington, *Issues in Public Transportation*, Transportation Research Board Special Report 144 (1972).
3.35 L. Edson, *The Making of the NAE: The First 25 Years*, ed. by M. R. B. Beaudin (Washington, D.C.: National Academy of Engineering, 1989).
3.36 E. A. Walker, "Prospectives of the Engineering Societies," presented at the 25th Anniversary meeting of the National Academy of Engineering, October 1989.

CHAPTER 4

In Search for Engineering Techniques for Tomorrow's Needs 1975–89

1. COMMITTEES OF THE FOUNDATION

During the late 1970s and the 1980s the Engineering Foundation developed many innovations which greatly enhanced its services to the engineering community. Among the most important of these innovations was the development of the Research Initiation Grants, the eventual rejuvenation of its regular research grants program and the stabilization of its conferences after a period of decline. All of these innovations were made possible through the cumulative efforts of the Engineering Foundation committees.

The activities of the Engineering Foundation were generally developed and carried out in several committees established by the board and supported by the staff. The annual reports of the board to the United Engineering Trustees list thirty-three committees, twenty-one of which may be classified as administrative while the remaining twelve served either as project advisory committees or were responsible for a broader area of technical activities. All of these committees and their members are listed in chronological order of their establishment in Table 5 of Appendix A.

Administrative Committees

The Engineering Foundation's first committee was appointed at the second meeting of the board held on May 20, 1915, "to consider the applications now on record in the Secretary's hands and report its recommendations to the Board . . ." It was named Committee on Applications and had four members. In addition to reviewing the early applications for projects, the committee also developed a procedure for future handling of such applications. This procedure, adopted by the board at its December 21, 1915 meeting, implied the Committee on Applications would be a permanent institution, but no references to this committee are found after 1916. Its demise may be explained by the creation of the new Executive Committee which was established by the April 15, 1918 amendments of the United Engineering Society by-laws. As amendments to by-laws usually are made only after a period of orderly development of administrative practice, it seems probable that the duties of the Committee on Applications were taken over by the newly created Executive Committee.

The Executive Committee has been in continuous operation since its establishment. Headed by the chairmen of the Foundation, it has been composed of four or more additional members of the board generally so selected as to include at least one member from each founder society. Its overall function has been to conduct the Foundation's business between the board meetings and to plan, on a continuing basis, the future activities of the Foundation.

An examination of Table 5 in Appendix A reveals that prior to 1931 all administrative committees, except for the Executive Committee, were of short duration. In other words, the board and the Executive Committee conducted the entire business of the Foundation and created other committees to execute specific well-defined short term tasks. By 1931 the volume of work had grown to the point that a new permanent committee was appointed: the Research Procedure Committee. Its assignment was to manage all aspects of research projects sponsored by the Engineering Foundation. This new committee selected projects for consideration by the board and made recommendations to the board regarding funding for those that were selected. It also monitored all active research projects and advised the board concerning procedures for management of research.

The Research Procedure Committee was succeeded in 1960 by the Research Committee which, in turn, was replaced in 1971 by the Projects Committee. These changes reflected new developments in the activities of the Foundation. In the late 50s and throughout the 60s the board placed greater emphasis on activities concerned with the engineering profession. During the 1960s, about one half of all projects were concerned with subjects such as formation of the National Academy of Engineering, financial help to Junior Engineering Technical Society enabling it to move its headquarters to New York, helping to pay for the publication of various reports and for committee operations, supporting studies such as those of the future role of Engineering Index and defining policy development for access to airports, helping to pay for office maintenance for Volunteers for International Technical Assistance, and particularly funding several major studies aimed at modernizing the access to engineering information. These nontechnical projects were handled by the board; the other half of projects dealt with research and was managed by the Research Committee. Since the board activities generated a substantial number of the essentially nonresearch projects that required follow-up, eventually those projects were assigned to the Research Committee and, in due time, its name was changed to Projects Committee.

In 1961 a subcommittee of the Research Committee was assigned the task of managing the development of Engineering Foundation Research Conferences. By the end of the sixties, these conferences had become the principal Foundation activity and by 1969 this subcommittee was elevated to full committee status and it was renamed the Conferences Committee. Based on the precedent of the Projects Committee, it became a major operating committee of the Foundation.

Two of the committees formed in the 1920s were concerned with publications but both were short lived. The production and distribution of publications were handled by the staff while the decision of what to publish remained the prerogative of the Board. Once the publication policies were developed, no further committee activities were needed. Similarly, several policy and planning committees were established from time to time and were dissolved on the completion of their specific tasks.

The need for public relations activities was a frequent subject of Board discussions. During the 1920s as well as after World War II, outside consultants were retained for that purpose. In addition to the publicity created by these consultants, the staff conducted almost continuously various in-house efforts principally through news releases, mailers, and publications. The Committee on Relations with Founder Societies was active in 1924 and another on public relations was appointed in 1960 to provide ideas for Foundation's promotional activities.

Over the years, special committees were established from time to time to increase the financial resources of the Foundation. The first such committee became active in 1920 under the name Committee on Endowment Increase, which functioned through 1924. It appears that these committees met with some success particularly through the medium of bequests since several bequests that appear to have originated during this period resulted in additions to the Foundation's endowment in later years. In 1926, the United Engineering Society, the predecessor of the United Engineering Trustees appointed the Endowment Committee with the objectives of raising (1) five million dollars to aid engineering and allied research for general benefit not likely to be undertaken by other bodies and (2) two million dollars for the Engineering Societies Library. As of 1929, cash gifts had been contributed in the amounts of $12,750 for research and $62,500 for the library as well as several gifts in the form of insurances, annuities, and legacies.

Two committees that were concerned with finances were created in the early 1960s, but both were short lived. The Development Committee was appointed in 1983 under the chairmanship of H. Bieber and was later chaired by F. F. Aplan. It set out to organize a fund raising campaign. This book was prepared as a part of that campaign.

Technical Committees

The twelve technical committees were concerned with the following research:

Mental Hygiene of Industry, 1920
Hydraulic Research, 1920–23
Industrial Education and Training, 1920
Steam Tables Research, 1921
Paint-on-Wood Research, 1921
  Wood Finishing Research, 1922–24
Arch Dam Investigation, 1922–33
Engineering Encyclopedia, 1923
Alloys of Iron, 1929–58
Education Research, 1930–34
Study of Industrial System, 1932
Air Pollution, 1961–68
Composite Materials, 1963–68

Of these twelve technical committees, six were principally concerned with program development while the other six were associated with one particular project and may be best termed as project advisory committees. Many other technical committees, particularly of the project advisory type were active at various times but their existence and activities were not recorded in the annual reports. Two of the above listed committees, Committee on Arch Dam Investigation and the Iron Alloys Committee were active for more than a decade and had major influence on technical developments in their respective fields. Both are discussed in more detail below.

The Committee on Arch Dam Investigation was established late in 1922. It was started with nine members and three alternates. Charles Derleth, Jr., professor of civil engineering and dean of the College of Engineering at the University of California was its chairman and Fred A. Noetzli, chief engineer in the firm of Bissel and Sinnicks of San Francisco served as its secretary. H. Hawgood, consulting engineer of Los Angeles; D. C. Henry, consulting engineer of Portland, Oregon; W. F. McClure, state engineer of Sacramento; M. M. O'Shaughnessy, city engineer of San Francisco; A. Hobart Porter of Sanderson & Porter of New York; F. E. Weymouth, chief engineer of U.S. Bureau of Reclamation, Denver; Silas H. Woodard, consulting engineer of New York were the other original members. Paul Bailey, deputy state engineer of Sacramento was an alternate for McClure; R. P. McIntosh, design engineer of San Francisco, was an alternate for O'Shaughnessy; and Wayne Meredith of Sanderson & Porter of San Francisco, was an alternate for Porter. The investigations were conducted through twelve subcommittees, one each on the test dam, instruments, models, instruments and program, and on investigations of eight different existing dams. H. W. Dennis, consulting engineer with Southern Edison Company of Los Angeles, was in charge of the construction of the experimental dam, and W. A. Slater, engineer-physicist of the U.S. Bureau of Standards was in charge of the tests on the experimental dam. Finally, W. A. Breckenridge, senior vice president of Southern California Edison Company of Los Angeles was trustee for the collection and disbursement of funds raised for the construction of the test dam.

In 1925, Professor Derleth relinquished the chairmanship and was succeeded by Charles D. Marx, professor emeritus of civil engineering of Stanford University; H. W. Dennis and Alfred D. Flinn, director of the Engineering Foundation, became members of the committee; Professor R. E. Davis of the University of California became an alternate for Derleth; and Julian Hinds, research engineer at the U.S. Bureau of Reclamation became an alternate for Weymouth who left the Bureau and became president of Brock & Weymouth, Inc. of Philadelphia. Committee service of McClure and McIntosh ended in 1926, and that of Hawgood in 1928. John S. Savage, design engineer with the Bureau of Reclamation became committee member and Ivan E. Houck, engineer of the Bureau, became his alternate. George S. Binckley of Los Angeles became a member around 1930. Having completed its job, the committee was discharged in 1933. Secretary Noetzli, who originated the project, died on May 24, 1933, just a few months after the project's completion.

The Iron Alloys Committee was organized in 1929 under the chairmanship of George B. Waterhouse of MIT. Waterhouse represented the American Institute of Mining and Metallurgical Engineers. This committee had the following membership:

George K. Burgess, director, National Bureau of Standards;
Louis Jordan, NBS (representative);
Scott Turner, director, U.S. Bureau of Mines;
Charles H. Herty, Jr., Pittsburgh Experiment Station, U.S. Bureau of Mines (alternate);
R. E. Kennedy, technical secretary, American Foundrymen's Association;
H. W. Gillett, director, Battelle Memorial Institute;
Bradley Stoughton, professor of metallurgy, Lehigh University, representing American Society for Steel Treating;
Jerome Strauss, chief research engineer, Vanadium Corporation of America, representing American Society for Testing Materials;
T. H. Wickenden, metallurgical engineer, International Nickel Company, representing Society of Automotive Engineers;
James H. Critchett, vice-president, Union Carbide and Carbon Research Laboratories, representing American Electrochemical Society;
J. A. Mathews, vice-president, Crucible Steel Company of America, member-at-large.

Prior to the formation of the Iron Alloys Committee, an Advisory Committee was appointed by AIME with John Johnston, director of research and technology, United States Steel Corporation, as chairman and the following members: F. M. Becket, vice-president, Union Carbide Company; H. W. Gillett, National Bureau of Standards and later Battelle Memorial Institute; James T. MacKenzie, metallurgist, American Cast Iron Pipe Company; and A. J. Wadhams, manager, Development and Research, International Nickel Company. The committee served until the end of 1929.

The chairmanship of the Iron Alloys Committee was assumed in 1945 by H. W. Gillett of Battelle Memorial Institute and in 1947 by James L. Gregg, professor of metallurgy at Cornell University. The following additional persons served on the committee at various times:

Lyman J. Briggs, director, National Bureau of Standards;
R. S. Dean, chief engineer, Metallurgical Division, U.S. Bureau of Mines;
John W. Finch, director, U.S. Bureau of Mines;
John Johnston, director of research, United States Steel Corporation;
James T. McKenzie, chief metallurgist, American Cast Iron Pipe Company;
R. R. Sayers, director, U.S. Bureau of Mines;
Wilfred Sykes, assistant to president, Inland Steel Company;
J. G. Thompson.

Financing of the project was completed under the leadership of Alvin C. Dinkley of the Midvale Steel Company with the cooperation of the American Iron and Steel Institute through its board of directors. Frank T. Sisco served as editor, consultant and director from 1932 to the end of the project and John S. Marsh served as editor in 1940–42.

2. INTERDISCIPLINARY RESEARCH

Since its beginning, the Engineering Foundation sponsored projects which cut across not only the boundaries between two or more fields of engineering, but also across the lines between engineering and various other scientific fields. An example of the latter was Project No. 6 "Mental Hygiene of Industry" and of the former Project No. 9 "Fatigue of Metals." A particularly heavy emphasis on interdisciplinary projects occurred after the change in the direction of Engineering Foundation in the late 1950s. Although many of these interdisciplinary projects were research studies, most of them dealt with other activities such as the creation of NAE, the printing of reports, the organizing of the Engineering Information Service, and the Foundation's support for improving engineering education. Of the eighty-nine projects that were started after 1974, fifteen are designated in Table 2 in Appendix A as interdisciplinary. Of these fifteen, six dealt with multidisciplinary subject matters and six were multiproject items originating from different fields of engineering. Overall, sixteen interdisciplinary research projects were undertaken after 1974. This total was nineteen percent of the total number of research projects sponsored by the Engineering Foundation after that year.

Engineering in medicine, which overall field was certainly interdisciplinary, was a frequent theme of the Engineering Foundation Conferences. The fifth Foundation Conference, held in August 1963, was devoted to this theme. Entitled "Engineering in Medicine" it was co-chaired by H. H. Zinsser of Columbia University and R. Contini of New York University. A second conference on this subject was held in 1964. Starting in 1967 one or more conferences related to the general subject of engineering in medicine were held annually. By 1970, this topic had produced a research project: a team of researchers at George Washington University undertook the development of a formal approach to the application of engineering in medicine (RC-A-70-1). After the completion of this initial project, the Engineering Foundation sponsored further studies in this field including, among others, studies of cerebral circulation (RC-A-74-1), biomedical telemetry (RC-A-75-3-D in Table 4), cancer therapy (RC-A-77-3), leg injuries (RC-A-77-6C-1 in Table 4), and walking control (RC-A-78-6C-1 in Table 4).

A notable appreciation of the long term and important role that the Engineering Foundation had played in the sponsorship and development of interdisciplinary research became evident at its November 8, 1984 board meeting. At the request of UET, the board had appointed a special committee to investigate whether the Engineering Foundation funded research projects conformed with UET's stated objective of supporting interdisciplinary research. The Committee presented its report at the November 8 board meeting and it concluded that "in conformity with published Foundation policy encouraging interdisciplinary research, a substantial portion of the projects supported over the last seven years were interdisciplinary in character. Additional emphasis on interdisciplinary research is not needed and not recommended. A number of past projects involved medical or biomedical areas and presumably cooperation. Such projects if submitted should be evaluated on their merits."

3. INTERNATIONAL CONFERENCES

By their very nature, many of the Engineering Foundation Conferences encouraged foreign participation and in that sense were international in character. Participants from abroad attended several of the conferences that were held during the 1960s but no statistical record was kept of such participation. The 1970 conference "Engineering in Medicine—Automated Multiphasic Health Testing" in Davos, Switzerland was the first to be held abroad. It was cosponsored by the National Science Foundation; the Swiss Research Institute; F. Hoffman-LaRoche & Company Ltd., Basel; Bristol-Myers International, New York; and Searle-Medidata, Inc., Waltham, Massachusetts, all of which provided travel assistance to speakers. In addition, a German foundation provided assistance for bringing in eleven medical scientists from Germany.

G. B. Devey of the National Science Foundation and S. M. Perren of the Swiss Research Institute co-chaired the conference, with the organizing committee members coming from Switzerland, England, Germany, France, Sweden, and the United States. The conference was attended by sixty-one persons from North America, ninety-seven from Europe, and one from Japan. The participants were from the engineering and medical professions. It took seven years before another Engineering Foundation Conference went abroad. In 1977 a meeting on enzyme engineering was held at Bad Neuenahr in West Germany. Since that time, with the exception of 1980, one to four conferences have been held abroad each year. Cambridge in England, Banff and Niagara in Canada, Kashikojima in Japan, Paipa in Colombia, Galloway in Ireland, Helsingor in Denmark, and Uppsala in Sweden have been the sites of one or more Engineering Foundation Conferences. The holding of some conferences abroad helped to increase the number of foreign participants.

The staff reported at the May 8, 1975 board meeting that twenty-eight conferences had been scheduled for 1975. Attendance at these meetings showed a dramatic increase in total registration when compared with the 1974 conferences. Equally as significant, each individual conference had a ten percent increase in attendance with three times the previous year's number of foreign engineers and scientists as registrants.

The importance of the conferences to the overall activities of the Engineering Foundation was also reaffirmed at the November 6, 1975 board meeting when it was decided to make further use of the conferences to identify much needed engineering research. To determine such areas of need it was suggested that a contest of proposals be held to determine which subjects would be the focus of future conferences. However, this plan proved to be impractical and was never implemented.

By the time of the May 13, 1976 board meeting, it was decided to hold twenty-one conferences that year. Director of the Conferences, S. S. Cole was aided in their organization and supervision by Harold A. Comerer, who was appointed as assistant director of conferences at this time. Due to his experience in publishing, public relations, and other related areas, Comerer was able to successfully carry out his duties and continue his work for the Engineering Foundation through the late 1980s.

During 1976 the conferences that were sponsored by the Engineering Foundation continued to increase in popularity. It was reported at the November 18, 1976 board meeting that over the previous five years an average of twenty-five conferences had been held each year which had been attended by ever larger number of registrants. To further build on this success, the board authorized the sponsorship of thirty conferences for 1977. On the other hand, the early 1980s witnessed a decline in the success of the conferences. At the May 11, 1983 board meeting, concern was expressed about the drop in attendance. This decline had been evident since 1980–81 and it caused much concern among the board members. It was suggested by several of them that the continuing decrease in attendance could be the result of the Conferences Committee's failure to do its job. It was also suggested that a too heavy reliance on continuing conferences may have been the culprit, while it was also theorized that the decline may have been the result of increasing competition from research conferences that were now being funded and organized by other scientific and engineering organizations. The Conferences Committee was charged by the board to find out the reasons for this worrisome decline in attendance and to develop proposals to reverse it. It was also suggested by the board that the Conferences Committee examine the ongoing successful development of the Gordon Research Conferences as a means of finding the needed answers.

The temporary decline in attendance that had threatened the financial viability of the Engineering Foundation's conferences began to be reversed during 1984. At the May 20, 1984 board meeting, it was reported that nineteen to twenty meetings were planned for 1985. Average attendance per conference had also increased during 1983–84. To further strengthen the conference program, the Conferences Committee began the preparation of guidelines and the description of duties for its members. It also appointed a small subcommittee which would work in close cooperation with the board to assure that the Conferences Committee members who would be appointed in the future will have a clear understanding of their responsibilities so that their actions can accomplish the Conferences Committee's goals. As the first step in this program, the Board voted to direct the Foundation director to develop, in cooperation with the UET, a means for improved monitoring of the cash flow of individual conferences.

The efforts of the rejuvenated Conferences Committee successfully reversed the decline of the engineering conferences. At the November 10, 1988 board meeting, it was reported by the Conferences Committee that the conferences had taken on a new life as evidenced by their increased number and the greater range of new conference topics that were under development. These positive results were attributed to the improved effectiveness of the Conference Committee since 1985. To continue this momentum, Chairman P. Somasundaran of the Conferences Committee asked the members of the Founder Societies to give further assistance to the Committee's efforts by monitoring the nomination of the Conferences Committee members from their respective societies. In this way, it was hoped that only active and effective individuals would be selected to serve on this body.

4. RESEARCH INITIATION GRANTS AND STATE OF THE ART SURVEYS

One of the most important programs in the history of the Engineering Foundation's support for research began at the board meeting of May 13, 1976, when substantial interest was aroused by a proposal to provide grants to support research by young faculty members in technical areas of interest to the Founder Societies. It was felt that this type of grant would help to increase awareness of the Engineering Foundation and also increase cooperation with the Founder Societies. Furthermore, it was reasoned that a new faculty member, who had just completed his or her graduate studies or who had recently joined a university after a period in industry, frequently has developed a subject for fruitful research but he or she frequently lacks the necessary financial support to pursue it. Therefore, a grant program aimed at such individuals was likely to attract a large number of innovative proposals. Moreover, such grants may lead to new research inquiries that may warrant more substantial Engineering Foundation sponsorship at some later date.

From this initial interest the Research Initiation Grants program was developed which became the primary focus of the Engineering Foundation's support for engineering research during the late 1970s and throughout the 1980s. To begin it, the board voted $50,000 to establish five $10,000 Research Initiation Grants. The response to the new program was excellent. As a large number of high quality proposals was received in response to the initial announcement of the program, the board voted at its May 5, 1977

Allen I. Laskin, left, chairman of the Award Committee, presents the 1985 Enzyme Engineering Award to Professor Klaus Mosbach, University of Lund, Sweden, during Enzyme Engineering VIII conference held at Helsingor, Denmark, in September 1985.

meeting to increase the number of research initiation grants to ten with two of these grants to be assigned to each of the Founder Societies.

The Research Initiation Grants program involved close cooperation between the Engineering Foundation and the Founder Societies. Each year, the availability of grants was announced by the Engineering Foundation through distribution of announcements to some 230 engineering colleges and publication of the announcements in the journals of the Founder Societies. The proposers were directed to send their proposals to the appropriate Founder Society which rated the proposals and forwarded five or more highest ranked to the Engineering Foundation. The Projects Committee reviewed all proposals received from the Founder Societies and developed recommendations for board action. The selection process required about six months to complete. The deadline for receipt of proposals was December 1. The awards were made at the May meeting of the board thus giving the proposer sufficient lead time to start the project by the beginning of the new academic year.

In 1981 the Projects Committee carried out an extensive review of the Research Initiation Grants program. As a part of this review a questionnaire was sent to the recipients of completed Research Initiation Grant projects and the results of the completed forms were evaluated by a task force appointed for this purpose. The task force reported that since the inception of the Research Initiation Grants program in 1975, twenty-four projects were completed and eleven of them received additional support from various governmental and industrial organizations. In twenty-three instances, grant recipients reported that research on the concepts embodied in their projects was continuing. Based on its appreciation of this success, the task force recommended only a few changes such as an increase in the amount of individual Research Initiation Grant award even if the number of grants that were to be annually awarded would have to be reduced. The task force considered that the proposed increase in the amount of individual award was absolutely necessary to maintain the existing high quality of proposals and projects. The task force also suggested that the sponsoring societies should develop a more effective means of monitoring their Research Initiation Grant projects and to aid them in this task, the Engineering Foundation should provide more details to the monitors so that they would know exactly what was expected of them.

Research Initiation Grants awarded through 1988 are listed by chronological order in Table 4 of Appendix A. Generally, ten grants were awarded annually. The individual awards started at $10,000 in 1977 and were gradually increased until they reached $20,000 in 1986. Each award went to one individual for a particular project. Through 1988, 148 grants were made aggregating to a total of $2,279,000. The recipients were generally assistant professors at either a public or a private university in the United States. The recipients were located at eighty-four schools in thirty-five states and one recipient was at McMasters University in Canada. Pennsylvania State University had the largest number of recipients on its staff: nine individuals in the fields of metallurgical, mechanical, civil and chemical engineering. Six recipients worked at the University of Illinois at Champaign-Urbana (metallurgical, electrical and mechanical) and at the University of Minnesota at Minneapolis (metallurgical, electrical, mechanical, and chemical). Among the smaller schools, Bucknell and Rice Universities had one recipient on their staffs; in both cases, the award was in the field of civil engineering. The subject matter of research conducted under the Research Initiation Grants program covered many diverse areas and can be best appreciated by scanning the contents of Table 4 in Appendix A.

Another new category of grants was established by the board at its November 10, 1977 meeting when it decided to fund State of the Art Surveys of existing research in areas of particular interest to both the Foundation and the Founder Societies. The initial series of these grants was set at $5,000 per approved proposal and each of the Founder Societies would be allotted two such grants upon request. The ASME utilized its grants to fund the surveys of "Injection and Mixing of Turbulent Flows," which was conducted at the Virginia Polytechnic Institute, and "Dynamic Plastic Behavior of Structural Materials at Cryogenic and Elevated Temperatures" which was carried out at Brown University. The ASCE used its grants to fund the surveys of the "Conservation of Urban Water Resources" at the University of Colorado and of current research in "Offshore Geothermals" that was conducted at San Diego State University. The focus of the grant that was awarded to IEEE was the "Development of New Components Required for Optical Communications" survey that was undertaken by the Stevens Institute of Technology.

The development of both the Research Initiation Grants and the State of the Art Surveys was not universally accepted by the members of the Engineering Foundation Board. Some board members remained concerned that the direction that these new types of grants represented was contrary to the wishes of the Engineering

Foundation's founder, Ambrose Swasey. Although the board took no action concerning their concerns, these board members reservations represented another chapter in Engineering Foundation's never ceasing quest to establish its priorities, purposes, and goals.

The change in emphasis in the Engineering Foundation's grant policy from the funding of traditional research to the support of the Research Initiation Grant Program was evident at the May 3, 1979 board meeting. At this meeting only fourteen research proposals were received for traditional research, a number which reflected a considerable decline from the average number of proposals that had been received during the previous five years. Despite the limited number of proposals, one important project did receive a $40,000 grant. This proposal was concerned with the "Investigation of Fireside Corrosion of Fluidized Bed Heat Exchangers" and it was headquartered at Rennsaelar Polytechnic Institute. The Engineering Foundation committed its funds to subsidize the first two years of what was to ultimately become a very successful project.

Although the Engineering Foundation had chosen to concentrate most of its available research funding on the support of the Research Initiation Grants and the State of the Art Survey programs, it still supported a limited number of regular research projects throughout the 1970s and early 1980s. However, as the number of qualified applications for research project funding decreased during the early 1980s, the number of research projects that received support from the Engineering Foundation dwindled until it was reported by the Projects Committee at the November 8, 1984 board meeting that no research projects proposals had been received since March of that year. As a result, the Projects Committee voted to recommend to the board that it cease publication of the availability of the regular research project grant program. Although this action did not signal the end of the longest lived of all of the Engineering Foundation's activities, it did signal a temporary halt to the Foundation's regular research projects program.

Finally, it is gratifying to note that after several years of inactivity, the regular research project grants were revived at the May 14, 1987 board meeting when $12,500 was appropriated to finance up to one quarter of the costs of a project devoted to the "Instrumentation of Columns in the Norwest Bank Project at Minneapolis." The long tradition of Engineering Foundation support for cooperative research organizations was also re-established when the board voted to expend $15,000 to support the organization and start-up expenses of the Ultra-Deep Coring and Drilling Association. Unfortunately, the last project failed to get off the ground because of illness of one of its proponents.

During 1986 several earlier board initiatives began to bear fruit for the Engineering Foundation. At the April 30, 1986 board meeting, the Projects Committee reported that one additional Research Initiation Grant would be awarded to the ASCE and ASME members due to their agreement to cosponsor these additional awards with $7,000 each in matching funds. These matching funds represented the Projects Committee's earliest success in creating a greater financial involvement of the Founder Societies in the Research Initiation Grants program. Additional support from the Founder Societies for an increase in the number of Research Initiation Grants was evident at the November 6, 1986 board meeting when AIME announced its cosponsorship of a grant.

5. ENGINEERING FOUNDATION PUBLICATIONS

The first publication of the Engineering Foundation was issued under the date of February 27, 1917, and was entitled "Report on the Origin, Foundation and Scope of the National Research Council." Written by Carry T. Hutchinson, secretary of the Foundation, it described the steps that had resulted in the formation of NRC, and contained the names and affiliation of its forty-seven original members. Gano Dunn, M. I. Pupin and Ambrose Swasey were on this list. Dunn and Pupin served also on NRC's first executive committee.

The second publication was "A Progress Report to United Engineering Society" covering Foundation activities from its founding to October 1919. The cover and the title page of this report displayed the designation "Publication Number 2." Publications Number 3, 4, 6, 8, 10–12, 15, and 17 were annual reports through 1928. With the exception of 1933, 1935, and 1936, publication of annual reports continued every year through 1986 although the numbering of publications was discontinued after Number 17. The annual report for 1937 (fiscal year 1936–37) contained biographies and photographs of Swasey and Flinn.

Publications Number 5, 7, 9, 13, 14, and 16 covered a variety of subjects including a directory of "Hydraulic Laboratories in the United States of America," "Engineering Foundation—Terms of the Endowment and the Official Records of Its Establishment," and an address by Arthur D. Little entitled "Impending Changes in Our Use of Fuels." Perhaps the best known of these six publications was

Number 13 entitled "Fatigue of Metals Manual" which was prepared by Professor H. F. Moore and published in 1927. The manual was based on extensive investigations of fatigue started by Moore at the University of Illinois in 1919 as Engineering Foundation Project No. 9.

Four papers that reported the results of research that was supported by the Engineering Foundation were issued in the form

The Engineering Foundation

ADMINISTERED UNDER THE AUSPICES OF

UNITED ENGINEERING SOCIETY

AND

AMERICAN SOCIETY OF CIVIL ENGINEERS
AMERICAN INSTITUTE OF MINING ENGINEERS
AMERICAN SOCIETY OF MECHANICAL ENGINEERS
AMERICAN INSTITUTE OF ELECTRICAL ENGINEERS

33 West Thirty-ninth Street
New York

CHAIRMAN
Gano Dunn

VICE CHAIRMEN
Edward Dean Adams
Michael I. Pupin

SECRETARY
Cary T. Hutchinson

TREASURER
Joseph Struthers

Edward Dean Adams
Gano Dunn
Howard Elliott
W. F. M. Goss
Charles Warren Hunt
Michael I. Pupin
Charles F. Rand
Robert M. Raymond
J. Waldo Smith
E. Gybbon Spilsbury
Benjamin B. Thayer

Report on the
Origin Foundation and Scope of the
National Research Council

Publication
Number 1

27 February, 1917.

Title page of publication outlining the steps that resulted in forming the National Research Council.

of Foundation reprints during the early 1920s. In January 1921 began the semimonthly publication of a series of leaflets, each containing the story of some research or discovery or notable achievement in science or engineering. A total of 150 of these *Research Narratives* were published. They were also reissued in the form of three books, each containing fifty narratives. Five of these Narratives are reprinted in Appendix D.

In the middle twenties, two pamphlets entitled "Mr. Swasey's Reasons" and "Pamphlet of General Information" were issued to inform the public of the purposes and activities of the Engineering Foundation. In 1932, a vocational guidance booklet "Engineering: a Career—a Culture" was published and received wide distribution. A few years later, the remaining supply of this booklet was given to ECPD which continued its distribution.

The "Arch Dam Investigation" which was conducted from 1922 to 1933 was documented in three reports published in 1928, 1933, and 1934. The first report was published in the *ASCE Proceedings* while the other two were published by the Foundation. As far as publications are concerned, the "Alloys of Iron Research" was the most productive. Eleven monographs were published prior to World War II by McGraw-Hill Book Company covering molybdenum, silicon, tungsten, copper, carbon, chromium, and nickel, as well as the iron itself, and the phase diagram. Five more followed after the war. These were published by John Wiley & Sons, and covered aluminum, nickel, titanium, boron, calcium, columbium, and zirconium.

Five other publications were the result of Engineering Foundation sponsorship:

Preliminary Draft of the Report on the Present Status of the Art of Obtaining Undisturbed Samples of Soils published in 1940

Subsurface Exploration and Sampling of Soils for Civil Engineering Purposes by M. Juul Hvorslev published in 1949

Bibliography on Cutting of Metals prepared by an ASME committee and published in 1945

Basic Open Hearth Steel Making prepared by an AIME committee and published in 1944

The Buckling Strength of Metal Structures by Friedrich Bleich published by McGraw-Hill Book Company in 1952.

During the 1950s, the Engineering Foundation helped financially toward publications of the Council of Wave Research that included proceedings of the first six conferences on coastal engineering, the first conference on ships and waves, and the first conference on coastal engineering instruments; and the following two additional publications:

Waves, Tides, Currents, and Beaches: Glossary of Terms and List of Standard Symbols, by Robert L. Wiegel, 1952
Gravity Waves: Tables of Functions, by Robert L. Wiegel, 1954

Engineering Foundation publications issued in 1985 and 1986.

Numerous publications resulted from the Engineering Foundation conferences. Many of them are listed in References 3.22 to 3.35 and 4.17 to 4.119. Many more publications resulted from the research projects supported by the Foundation. Most of those appeared in the form of papers in the regular technical press.

The annual report for 1924 contained a two-page summary of the Foundation activities during its first ten years. For the 25th anniversary, a separate report entitled "Twenty-Five Years of Service, 1914–1939" was published which included photographs of Swasey, Flinn and Hovey. The 1949, 1954, and 1974 annual reports included thirty-five-, forty- and sixty-year summaries of Engineering Foundation's activities and contributions; the thirty-five-year report included photographs of Flinn, Hovey, Colpitts, and Sisco. A separate pamphlet entitled "A Half Century of Progress—1914–1964" was issued in 1964. It summarized the Engineering Foundation's activities up to that time.

6. REGULAR RESEARCH GRANTS 1975–89

The Engineering Foundation has supported thirty-nine projects since 1974. These projects are listed in Table 2. Six of them, involving support of young faculty members, were a part of the Research Initiation Grants Program discussed in Section 4 of this chapter. A few of the remaining thirty-three projects are discussed below. Included in the thirty-three projects were nine interdisciplinary, seven chemical, eight civil, four mechanical, three electrical, and two metallurgical projects. Most of them involved engineering research and were carried out at twenty different universities. However, six projects supported diverse activities including the production of the 1975 EJC slide show "The History of Engineering in America," the organization of the 1987 ABET National Congress on Engineering Education, organization of the still-born Ultra-Deep Coring and Drilling Association, and three publications: *Urban Water Conservation*, *Subsurface Exploration and Sampling of Soils for Civil Engineering Purposes*, and an *Engineering Foundation History*.

In 1981 S. K. Saxena of the Illinois Institute of Technology undertook to revise *Subsurface Exploration and Sampling of Soils for Civil Engineering Purposes* which had originally been written by M. J. Hvorslev under the sponsorship of the Graduate School of Harvard University, the Waterways Experiment Station of the U.S. Army Corps of Engineers, and the Engineering Foundation. The book was first published by the Waterways Experiment Station on November 19, 1949, and, because of great demand, it was reprinted by the Engineering Foundation in 1962 and again in 1965. It was soon regarded as a classic in its field. By the late 1970s, while still in use, it was in need of revisions to incorporate new testing methods and technology. The work was conducted primarily by Saxena. Assisting him in preparing new chapters and revising the existing text were R. G. Campanella of the University of British Columbia and R. M. Koerner of Drexel University. O. S. Bray, and, later, W. C. Y. Teng served as monitors for the project.

Although the Engineering Foundation was gaining greater public and professional recognition as a result of the continued success of its conferences, it also continued to seek other means of promoting itself and the American engineering community. At its May 8, 1975 meeting, the board of the Engineering Foundation voted to grant $5,000 to the Engineers Joint Council to produce a slide show on the history of engineering in America. This show was quickly finished and its showings to public and professional groups served as a useful adjunct to the Foundation's overall efforts to promote a greater understanding of the role that engineering had played in America's national development.

The development of new grants and fellowships did not mean that the Engineering Foundation had abandoned its traditional support of research projects. At its May 5, 1977 meeting, the board voted $10,000 to Rennsaelar Polytechnic Institute for the study of the Stirling Cycle heat pump and it granted $10,000 to Clarkson College to conduct the study "Recovery of Acid Lakes." This latter grant demonstrated once again the remarkable awareness of contemporary problems by the Engineering Foundation. During the late 1970s, the board also granted $40,000 to H. S. Kemp, chairman of the Executive Committee of the Design Institute of Emergency Relief Systems of the AIChE to conduct research into the improvement of high pressure relief systems for flammable liquids. The field of biomechanics also received support when $20,000 was granted to the University of Cincinnati to help in the development of a means of non-invasive aneurysm detection and analysis.

At its May 11, 1978 meeting, the board of the Engineering Foundation voted $45,400 to fund a research project devoted to the investigation "Photocatalyzed Degradation of Pesticides" at Princeton University. It also granted $50,000 for an innovative project which would utilize local microwave induced hyperthermia as a cancer therapy. This research project was carried out at Dartmouth University.

The Engineering Foundation's continuing support for the improvement of engineering education was highlighted at the November 6, 1986 board meeting when $10,000 was appropriated to the Accreditation Board for Engineering and Technology to support the National Congress on Engineering Education. The conference was held on November 20–22 of 1986 and due in part to the Engineering Foundation's financial support it was a success, producing a concise summary of the questions and problems that were facing undergraduate students. It also identified for the Engineering Foundation's Conference Committee several possible topics for future conferences.

The continuing problem of attracting high quality proposals for research projects was a recurring issue of debate among board members during the early 1980s. A suggested solution to this problem was expressed in the following excerpt of a May 21, 1981 letter from long-time board member W. E. Lobo to I. M. Viest who was then serving as the chairman of the Projects Committee.

> Relative to my suggestion which was not very well defined to the effect that the Engineering Foundation should ask the research committees of the Founder Societies to propose subjects on which research would be desirable, I should like to expand this idea in more concrete terms.
>
> Research proposals which now come to the Engineering Foundation either for Research Grants or for Research Initiation Grants deal with the subjects of the researcher's choosing. Many of the proposals are good, many are poor and the projects committee chooses the most worthwhile. On the other hand, there are without question many subjects which could benefit from research which are not pet projects of anyone. Consequently, such subjects are totally neglected. I believe that the research committees of the Founder Societies should be in a position to suggest desirable areas for research. These might be interdisciplinary or they might not. They might require large grants well beyond the means of the Foundation so that industrial or governmental financial help would be necessary. On the other hand, they might be relatively small in the $12,000 to $25,000 class. They should I think require some experimental work and not be just computational research as some of the ones we now see are. The project might require a year's work or two or three.
>
> The research committee suggesting the project would prepare an estimate of the appropriate cost, the scope of the research, possibly the approach, and some idea of how long the work might take. A description of the proposed research would then be prepared and presented to the board which would then offer consideration; give it suitable publicity,

while asking for proposals (or bids) for research. Proposals received would be treated as we now treat research grants—on their merits. If the project were one which should be of special interest to a certain industry and large enough to warrant joint funding, the Foundation would make the proper approach.

I am not in agreement with the statement made by Marcel Cordovi that the societies are in a position to go to industry themselves without the Foundation's help whenever they see the need for large research projects. Such projects do not come often as far as I know. The AIChE has really had only two since the Research Committee was formed in 1950 (I was the first chairman and served as such for five years on the first major project). Actually, I have in mind mainly research projects of the order of those which we are presently funding although I would not rule out much larger projects should they turn up which I think is not so likely to occur often.

We go further and suggest to the various research committees that inform the members of their respective societies that their suggestions for research topics would be more than welcome. Such topics would then be studied and developed. If considered worthwhile, I would much prefer to see the suggested topics coming from industry rather than from academia and some indication of this preference should be made.

The reception of this idea of suggested topics for research should not be unwelcome to university researchers. It presents them with new ideas for thesis subjects. It may open up for them new areas of interest and perhaps better opportunities to do research in fields of immediate interest and value to industry.

I think that such a program might well result in giving greater publicity to the Engineering Foundation. It might bring it more to industry's attention. It might easily be a means of securing matching funds for research projects of interest to the company for whom the original proposer works.

My purpose is not to go after the largest interdisciplinary research projects nor even to supply seed money, the old Foundation idea which we decided to drop. In the past all the Foundation has supplied was money which was fine, but with the exception of the Engineering Foundation Conferences, copies from the Gordon Conferences, the Foundation has never come forth with research ideas of its own. Sure, it funded the formation of the National Academy of Engineering, the Metal Properties Council, and others, but the ideas for these and the requests for funding were generated elsewhere. Granted that my proposal is not going to lead directly to Foundation generated research, but it will be playing a more important part in determining research efforts by making use of the research committees of its member societies. Of course, project committee members may also suggest research areas on which proposals might be requested.

I should like to see the Projects Committee make a strong recommendation to the board that we implement this idea. If it comes to naught, we have lost nothing. If it works that will be fine I think but it is certainly worth trying.

Although the Projects Committee attempted to follow Lobo's suggestions, it did not succeed in establishing lasting cooperation with the research committees of the founder societies.

7. FINANCING OF ENGINEERING FOUNDATION ACTIVITIES

The development of financial resources of the Engineering Foundation through 1930 was described in Sections 3 and 4 of Chapter 1. The financial position of the Foundation was strengthened after that time by additions to its endowment obtained primarily in response to fund raising campaigns conducted in the early 1920s, late 1920s and early 1960s. Another fund raising campaign was started in the early 1980s. The 1980s campaign is discussed in more detail in the next section. Contributions received during the 1980s campaign are not included in this presentation. The Engineering Foundation funding is discussed below in four categories: 1. endowment, 2. projects, 3. conferences, and 4. publications.

The endowment, started by Ambrose Swasey in 1915, is the basic source for funding Engineering Foundation activities. The endowment was built up from contributions listed chronologically in Table 6 in Appendix A. The total of these contributions was approximately $1,600,000. The largest contributor was Swasey whose contributions amounted to $839,000. The largest single gift was $400,000 donated by E. H. McHenry's estate in 1931 to establish the Blanche H. McHenry Memorial Fund. The following tabulation lists all donors whose names were found in the annual reports and the minutes of board meetings:

Ambrose Swasey
Edwin H. McHenry
George D. Barron
Jules Breuchaud
Edward Dean Adams
Henry R. Towne
Sophie M. Gondron
W. S. Barstow
Central Hudson Gas and Electric Company
Seeley W. Mudd
Harry de Berkeley Parsons
O. W. Boston
Richard Khuen, Jr.
Karl Emil Hilgard
John C. Parker
G. L. Knight
Alfred D. Flinn
Ernest L. Robinson
James F. Fairman

An examination of the last two columns in Table 2 in Appendix A shows that Engineering Foundation projects attracted significant support from other sources. The records are quite complete through the late 1950s. Engineering Foundation funding of $1.1 million attracted outside support of $6.8 million. After 1960, only a few records of outside support are available: they continue to indicate substantial contributions for particular projects. This situation changed after 1975 with the introduction of a large number of relatively small grants that did not require cofunding from other sources.

The "Arch Dam Investigation" (Item 19 in Table 2) was financed by contributions from many sources. The financial contributors are listed in Table 7 in Appendix A together with the so called cooperators, that is those who furnished services without compensation. Seven other projects included in Table 7 were financed in this manner. Only the names of the financial contributors are listed for these other seven projects.

Except for the first few meetings, the Engineering Foundation conferences have been self-supporting. Expenses connected with their organization were defrayed from fees charged to the attendees and from contributions received from many outside funding sources. The National Science Foundation, the National Institute of Health, several departments of U.S. government, and many industrial concerns were the main contributors. The names of known contributors are listed in Table 8 in Appendix A. It should be noted that the conferences attracted considerable financial support from abroad used primarily to subsidize attendance by foreign engineers.

Engineering Foundation publications are financed either from its endowment income—the annual reports are an example of such publications—or are paid from dedicated sources: for example, conference proceedings are paid from the income of the particular conference.

The history of Engineering Foundation funding is summarized below. The figure includes four lines presenting (1) contributions to the endowment, (2) annual income, (3) annual expenditures, and (4) the market value of the endowment. Data for the first three items are available for the full seventy-five years while the market value of the endowment was reported only in the annual reports for 1959 through 1980. The direct income and expenses of the conferences are not included in these figures.

The financial resources of the Engineering Foundation continued to develop during the 1970s. It was announced at the May 8,

1975 board meeting that the total income of the Jules Breuchaud Trust was bequested to the Foundation. Jules Breuchaud had been an active supporter of the activities of the Engineering Foundation. In his career as the president of the Underpinning and Foundation Company of New York City, he had amassed a considerable fortune. The value of his bequest to the Engineering Foundation was estimated to be a $60,000 addition to the Foundation's endowment.

The issue of increasing the Engineering Foundation's financial resources once again became a focus of board activity during the early 1980s. At the board meeting of November 23, 1982, it was suggested by the leadership of UET that the two departments of UET, the Engineering Foundation and the Engineering Societies Library, develop a long range fund raising plan for increasing their endowment. It was also strongly suggested by UET's leadership

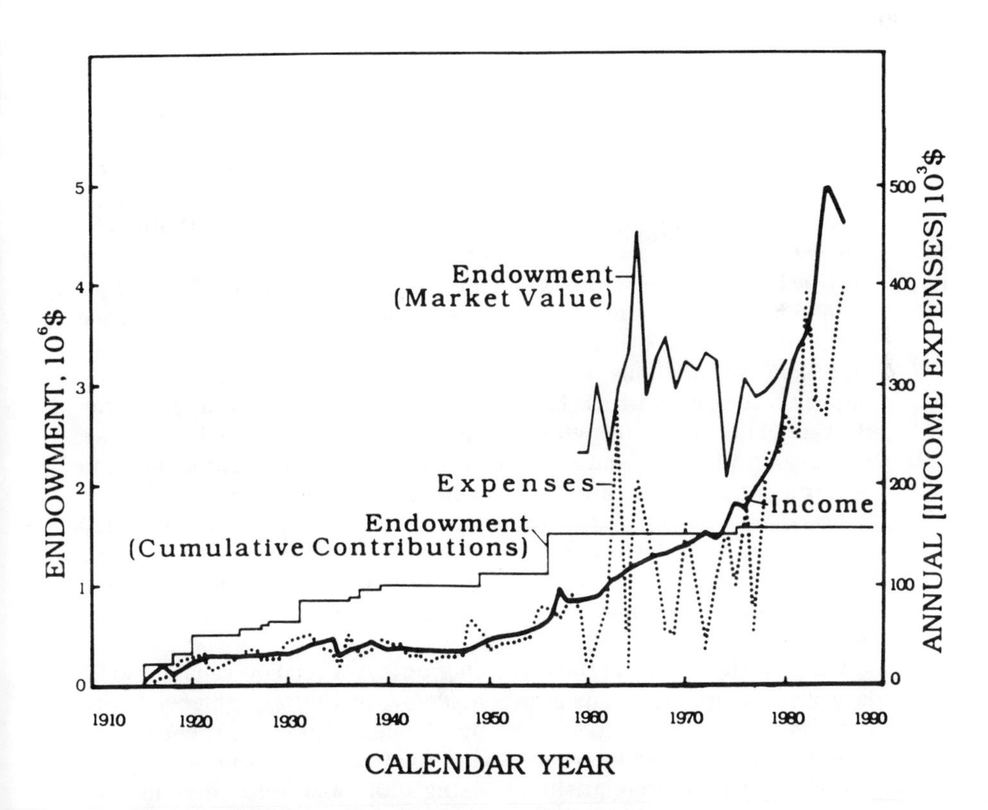

History of Endowment, Expenses and Income.

that these fund raising plans should be developed and implemented internally without employing the services of outside fund raisers. As a first step in meeting these guidelines, a development committee was formed.

8. DEVELOPMENT COMMITTEE

During 1978 the Board had appointed a special planning committee to help plot the Foundation's future goals and to also find ways to increase its endowment. The creation of this committee was symptomatic of the Engineering Foundation's almost constant preoccupation with self-evaluation during the years since the end of World War II. At the May 3, 1979 board meeting, this committee issued a report which summarized the Engineering Foundation's current thinking about itself and its role in the Engineering community.

> The goal of the planning committee is to enhance the image of the Engineering Foundation and to help in raising funds to increase the endowment. The Engineering Foundation has an impressive record of service to the engineering community—It has provided seed money for the funding of important new institutions such as the National Academy of Engineering, the ECPD, and the Corrosion Research Council. It has sponsored research projects that have produced advances in science and technology. These projects were innovative pioneering efforts in their day, but now similar formats are used by the National Science Foundation and by other, larger private foundations, such as the Ford and Sloan Foundations. It has hosted a series of important interdisciplinary conferences which have served to identify important problems that could benefit from engineering inputs to provide valuable guidance to research planners and policy makers in the private, academic, and governmental sectors and stimulated dialogue and communication between engineering and other disciplines, but its position has eroded over the years. Unfortunately, these achievements have not been adequately publicized. The Engineering Foundation is not too well-known in the engineering community to say nothing of the nation's business and political leaders.
>
> Further, the impact of the research grants has been overshadowed by larger and more numerous grants made by the National Science Foundation and other large federal agencies. Basic to this committee's task is the premise that a wider awareness of the Foundation's programs and achievements is a prerequisite to raising additional endowment monies and other gifts.

The planning committee feels some restructuring and changes in emphasis could make better use of the Foundation's limited resources and achieve higher visibility. The planning committee regards the engineering conferences as the most visible segment of the Engineering Foundation's program. By their very nature, they generate wide awareness. Extensive mailing plus reasonably good preconference publicity in society journals have fostered this awareness. The quality of the conferences has steadily improved. More attention needs to be given to publicizing the results of conferences and the occasional printings of conference proceedings are useful archival documents, but do little to stimulate current awareness of the conference operations. On the whole, the conference operation is in good shape. The planning committee therefore addressed itself primarily to a review of the grants program plus consideration of desirable new activities. The committee views the Research Initiation Grants as the most valuable and newsworthy of the projects committee operations. This relatively new program fills a real need. It has been well received by the academic community and the proposal quality has been excellent. Similar National Science Foundation programs have had a strong science bias that mitigates against proposals for applied engineering research. These awards need to be expanded in number and funding level.

On the other hand, the Research Continuation Grants had a limited utility. It is often very hard to evaluate the results of the initial grant to see whether additional funding is warranted and the incremental publicity value of such funding seems minimal. The committee questioned the value of state of the arts reviews. While these grants certainly educate the writers and are useful to people entering the field under review, it is not clear whether people working in the field need these. The consensus was that workers in the field are always several years ahead of the published literature. The committee recommends that the project committee review these programs to determine if these monies could not be used to better advantage in the research initiation area.

The old research grant program is still a viable concept, especially for programs that do not fit the National Science Foundation type of grants. However, it seems we are not getting enough top rank proposals, perhaps because of insufficient award size, perceived low probability of success, or poor awareness on the part of potential applicants. The committee feels these awards should be made only for top quality proposals even if it means none are awarded in a given year.

New programs should also be considered. The committee feels that the Engineering Foundation should consider initiating a series of prizes or sponsoring contests. This policy would enhance public awareness of the role of engineering in our society and also be a very effective way of achieving greater visibility for the Engineering Foundation. Consultations with several member societies is essential to avoid duplication or competition with existing awards. Interdisciplinary areas should be stressed, for example awards could be given for one or more of the following:

- Engineer of the Year
- Engineering Achievement in Interdisciplinary Problems
- Fostering Higher Standards in Engineering Ethics
- Contributions in the Public Policy Areas

Left to right: Ivan M. Viest, Sandford S. Cole, Curry E. Ford, and George F. Habach. Former director of conferences Sandy Cole is presented a certificate during testimonial dinner in January 1984.

- Achievements in Fostering Election of Engineering Careers by Minorities

Additional funds should be sought to enhance the scope of the impact of the Foundation's programs. It should also seek annual contributions from individuals and corporations and hire a professional public relations firm.

Proposed Action Plan

1. Appoint a member of the Board from each founder society to steward Engineering Foundation publicity in his society (through articles in each society's respective journals).
2. Keep a file of all press releases.

Conferences

1. Appoint a reporter for each conference to prepare press releases.
2. Insure good press representation.

Projects

1. Announce grant awards with appropriate fanfare at a national meeting of each recipient's professional engineering society.
2. Insist that each grant recipient prepare a press release as a part of his stewardship of his grant.
3. Enhance Project Initiation Grants by adding a large amount to each award.

Fund Raising

1. The publicity committee should solicit $11,000 from each Founder Society to support additional Research Initiation Grants.
2. The publicity committee should solicit fifteen names of likely industry prospects from each Founder Society. Such corporations will be solicited to make renewable three year commitments to support one or more Research Initiation Grants.
3. If the projects committee elects to initiate a prize program, the publicity should solicit appropriate corporate or individual donors.

Although the above report was well received by the Board of the Engineering Foundation, not all of the actions that it suggested were implemented. At the board meeting of November 15, 1979, it was decided not to establish an award program and committee.

On May 18, 1980 another major report was delivered before the board of the Engineering Foundation. This document was the report of the Projects Planning Task Force which was chaired by Viest. This report made the following recommendations:

1. The Research Initiation Grants should be continued in their present form.
2. The number of Research Initiation Grants should be expanded.
3. The amounts of individual Research Initiation Grants should be increased.
4. A new category of interdisciplinary project grants should not be created.
5. Discontinue the State of the Art Review Grants.
6. Continue unchanged the regular research project grant program.
7. No series of special achievements awards should be established by Engineering Foundation.
8. A special effort should be made to gain additional financial support from the Founder Societies for the Research Initiation Grants Program.

While no immediate actions were taken to implement the report's recommendations, all eight were eventually implemented during the next few years. The success was the slowest for item 2: while ASCE and ASME increased the number of grants to their members from two to three by providing partial support starting with the 1983 award year, and AIME joined them in 1986, it was not until 1989 that a substantial expansion of this program took place when the U.S. Air Force provided funds for twenty-five additional grants.

Shortly after its formation, the Development Committee issued its initial report. Presented at the May 11, 1983 board meeting, this report recommended that the committee sets as its goal increasing the endowment by $10,000,000. It also recommended that UET develop a combined fund raising campaign that would combine solicitations for the Engineering Foundation with the Engineering Societies Library to create a single effective appeal which would be under the direction of a professional fund raiser. In the future, the Development Committee also reported, it would consider means of providing funding for projects and restricted support for special purposes such as conferences. It would also explore the funding of additional and innovative activities for the advancement of the profession of engineering and the good of mankind.

The activities of the Development Committee entered a period of uncertainty during the early 1980s due to the fact that a decision had not yet been made concerning the proposed construction of an addition to the United Engineering Center building, the so called north tower. However, by the November 21, 1983 board meeting, the decision had been made that the north tower would not be built. This decision was an important one for it cleared the path for a more active role for the development committee, by removing the fear that its campaign would conflict with or be totally submerged by UET campaign to construct the north tower.

The Engineering Foundation continued during the 1980s to make itself better known to both the engineering community and the general public. One of the more effective means of accomplishing this goal was the development of a travelling exhibit on the achievements and activities of the Engineering Foundation. At its May 9, 1985 meeting the board authorized the expenditure of $3,000 to purchase and develop such an exhibit which has been displayed at conferences and other professional meetings.

Since the Engineering Foundation needed to bolster the efforts of its Development Committee in order to raise additional funds to support an enlargement of its Research Initiation Grants program, a special meeting of the Executive Committee was held on March 7, 1985. At this meeting, it was decided that the key role of the Foundation must be identified so that it can best utilize its present resources or those resources that it would be likely to have as a result of the development campaign. It was decided at that meeting that the strengths such as flexibility, an ability to respond quickly, and the lack of ties to any special interests, which had brought success to the Engineering Foundation in the past, should be once again utilized to aid the efforts of the Development Committee. These statements were endorsed by the entire board at its May 9, 1985 meeting.

The actions of the Development Committee continued to play a major role in the deliberations of the Engineering Foundation Board during the remainder of 1985. At the November 7, 1985 board meeting, the Development Committee announced that it had selected, after a lengthy interview process, the firm of Arthur D. Raybin Associates for the purpose of conducting a feasibility study to determine if the Engineering Foundation could successfully conduct a full-scale capital development campaign. This action was approved by the board.

The board also received at this meeting the report of the Strategic Planning Group of the Development Committee. The Strategic Planning Group had been charged by the board with the development of the essentials of a long range plan which would guide the future direction of the Engineering Foundation. This plan would also be the foundation for the Development Committee's fund raising campaign. Among the essential elements of the Strategic Planning Report were the adoption of a mission statement for the Engineering Foundation which stated that "the Engineering Foundation encourages promoting engineering leadership by promoting interdisciplinary research, technology development, and design applied for the benefit of mankind. This requires a united professional commitment to broad scientific education, technological excellence, and long term economic and societal benefits." The action plan also called for an increase in the number or Research Initiation Grants and the future expansion of the Engineering Foundation's Conference Program to include at least one workshop or conference which would be focused on finding solutions to a broad program of critical importance to engineering which affects long term economic and societal benefits, quality and reliability, and productivity. The report also advocated that the Engineering Foundation enhance its continuing efforts to assist a broad spectrum of government agencies in their missions through the Foundation Conferences.

In other areas, the report stated that the Projects Committee should develop an annual project for unsolicited research requests, exclusively for an interdisciplinary approach to the critical economic benefits gained by quality, reliability, and productivity. Engineering education should be broadened through the efforts of the Foundation by developing a program for the interchange of professionals between academia and engineering practice in industry and consulting. The report also emphasized the need for the Foundation to begin the plan for its seventy-fifth anniversary celebration as a means of generating public support for its fund raising activities. Finally, the report also called for the creation of an awards committee to build upon such recent Foundation activities as bestowing of a special award in enzyme engineering presented in September of 1985 to Professor Lemuel P. Wingard of the University of Pittsburgh. Professor Wingard was honored for his work in the development and organization of the Engineering Foundation's ongoing series of Enzyme Engineering Conferences.

The Engineering Foundation Board approved and adopted the report of the Strategic Planning Group at its November 7, 1985 meeting. This document provided not only the basis of a masterplan for the Foundation's future course, but also a critical summary of the Foundation's perception of itself as it entered the late 1980s.

As the Engineering Foundation approached its seventy-fifth anniversary, work continued on the implementation of its Development Campaign. At the May 14, 1987 board meeting, it was reported that the UET had approved the commencement of the campaign. It also authorized that creation of a working committee whose members would be drawn from the Engineering Foundation, the Engineering Societies Library, and the UET. To assist in the fund raising efforts of this campaign, UET hired a new assistant general manager to work with the development campaign.

Development also continued during 1987–88 on the preparations for the commemoration during 1989 of the Engineering Foundation's seventy-five years of achievement. Among the activities that were contemplated at the May 14, 1987 board meeting were the preparation of a comprehensive history of the Engineering Foundation, a special anniversary banquet, an historical exhibit and slide show on the Foundation's achievements, and a special public relations effort which would highlight the Foundation's ongoing activities and fund raising efforts. At the May 5, 1988 board meeting these commemorative plans began to take definite shape when it was announced that proposals had been solicited for the writing of a seventy-five-year history of the Engineering Foundation. In response to this solicitation six proposals were received of which the one submitted by L. E. Metz, the historian of the Hugh Moore Historical Park and Museums, Inc. of Easton, Pennsylvania, and I. M. Viest, a past chairman of the Engineering Foundation, was accepted.

9. HAROLD A. COMERER

A significant change in the administration of the affairs of the Engineering Foundation occurred at the beginning of 1984 when the long vacant position of director was filled. Harold A. Comerer, who had served as assistant conferences director since 1976, was appointed to this position.

Comerer was born on February 7, 1949, at Chambersburg,

Harold A. Comerer, center, director of the Engineering Foundation, Professor Brij Moudgil, left, University of Florida, and Professor P. Somasundaran, right, Columbia University at the conference "Flocculation, Sedimentation and Consolidation " at Sea Island, Georgia, in January 1985.

Pennsylvania. He received his elementary and secondary education at schools in central Pennsylvania. In 1973, he received a B.S. degree in Political Science from Shepherd's College in Shepherdstown, West Virginia. During 1973 and 1974, Comerer served as a recruiting and career counselor for the U.S. Army Reserve at Hagerstown, Maryland. In 1974 he became editor of the *Fulton Democrat*, a newspaper which was published at McConnellsburg, Pennsylvania. He held this position until 1976, when he became assistant director of the Engineering Foundation Conferences. He became the director of Engineering Foundation on January 1, 1984.

During his tenure as assistant director of conferences and as director of the Engineering Foundation, Comerer has proven to be an able administrator who has had a positive impact on the continued development of many phases of the Foundation's activities. He developed the conference publications program and brought about a greater involvement of individual conference chairmen in the planning and execution of the Engineering Foundation conferences. He has also provided staff support for the Projects Committee and played a large role in the development of the Research Initiation Grants program. Comerer negotiated the Air Force sponsorship of twenty-five additional annual research initiation grants valued at $500,000. He also streamlined and automated the operations of the Engineering Foundation by installing an in-house word processing and computer network and by retaining the Corporate Service Bureau to maintain the Engineering Foundation mailing lists. Finally, it should be noted that Comerer was instrumental in starting the Engineering Foundation's development campaign of the 1980s.

Comerer was married to Nhora Cortes-Comerer and this union produced a daughter.

References

4.1 *Report on the Present Status of the Art of Obtaining Undisturbed Samples of Soils* (New York: Engineering Foundation, 1940).

4.2 *Bibliography on Cutting Metals* (New York: ASME, 1944).

4.3 *Basic Open Hearth Steel Making* (New York: AIME, 1944).

4.4 "1975 Annual Report" (New York: Engineering Foundation).

4.5 "1976 Annual Report" (New York: Engineering Foundation).

4.6 "1977 Annual Report" (New York: Engineering Foundation).

4.7 "1978 Annual Report" (New York: Engineering Foundation).

4.8 "1979 Annual Report" (New York: Engineering Foundation).
4.9 "1980 Annual Report" (New York: Engineering Foundation).
4.10 "1981 Annual Report" (New York: Engineering Foundation).
4.11 "1982 Annual Report" (New York: Engineering Foundation).
4.12 "1983 Annual Report" (New York: Engineering Foundation).
4.13 "1984 Annual Report" (New York: Engineering Foundation).
4.14 "1985 Annual Report" (New York: Engineering Foundation).
4.15 "1986 Annual Report" (New York: Engineering Foundation).
4.16 "Conference on the Application of Engineering Technology to the Problems of Appalachia" (New York: Engineering Foundation, 1969).
4.17 "Rapid Construction of Concrete Dams," *Proceedings of Engineering Foundation Research Conference*, W. H. Price, chairman (New York: ASCE, 1970).
4.18 F. P. Huddle, ed., *Requirements for Fulfilling a National Materials Policy* (Washington, D.C.: U.S. Government Printing Office, 1975).
4.19 D. W. Hill and K. R. Wright, eds., *Urban Water Resources Management* (ASCE, 1970).
4.20 C. Berkley, ed., *Automated Multiphasic Health Testing* (New York: Engineering Foundation, 1971).
4.21 E. A. Abdun-Nur, *Control of Quality in Construction* (New York: ASCE, 1971).
4.22 K. Cheng and G. J. Kral, eds., *Technology and Social Institutions* (New York: IEEE, 1974).
4.23 A. McAdams and M. M. Henderson, eds., *Making Service Industries More Productive through Computers and Automation* (Washington, D.C.: National Technical Information Service, 1974).
4.24 *Foundations for Dams* (New York: ASCE, 1974).
4.25 *Quality System in Construction* (New York: ASCE, 1973).
4.26 W. Whipple, ed., *Urban Runoff; Quantity and Quality* (New York: ASCE, 1974).
4.27 *Safety of Small Dams* (New York: ASCE, 1975).
4.28 *Converting Existing Hydro-Electric Dams and Reservoirs into Pumped Storage Facilities* (New York: ASCE, 1975).
4.29 *Assessment of Resources and Needs in Highway Technology Education* (New York: ASCE, 1975).
4.30 L. H. Fink and L. Carlsen, eds., *Systems Engineering for Power: Status and Prospects* (Washington D.C.: National Technical Information Service, 1975).
4.31 A. S. Goodman, ed., *Civil and Environmental Engineering Aspects of Energy Complexes* (New York: ASCE, 1976).

4.32 "Conference on Automatic Cytology," *Journal of Histochemistry and Cytochemistry*, Vol. 22, No. 7; Vol. 24, No. 1; Vol. 25, No. 7; and Vol. 27, No. 1 (Baltimore: Williams & Wilkins, 1974–79).
4.33 *Engineering Manpower: A National Problem or a National Resource?* (New York: Engineering Manpower Commission, 1975).
4.34 *The Constructed Environment with Man as the Measure as Part of the Quality System in Construction* (New York: Engineering Foundation, 1975).
4.35 *Application of Signature Analysis to Machinery Reliability and Performance* (New York: Engineering Foundation, 1975).
4.36 C. S. Desai, ed., *Numerical Methods of Geomechanics*, 3 volumes (New York: ASCE, 1976).
4.37 D. L. Keairns, ed., *Fluidization Technology*, 2 volumes (Washington, D.C.: Hemisphere Publishing Company, 1976).
4.38 *Energy Storage User Needs and Technology Applications* (Oak Ridge, TN: National Technical Information Center, 1977).
4.39 J. C. Denton, S. Weber, and J. Moriarty, eds., *Energy Conservation through Effective Energy Utilization* (Washington, D.C.: U.S. Government Printing Office, 1976).
4.40 *Behavioral Demand Modeling and Valuation of Travel Time* (Washington, D.C.: Transportation Research Board, 1974).
4.41 A. S. Foss and M. M. Denn, eds., *Chemical Process Control*, Symposium Series #159 (New York: AIChE, 1976).
4.42 J. L. Holt, ed., *Engineering Implications of Chronic Materials Scarcity* (Washington, D.C.: U.S. Government Printing Office, 1977).
4.43 *Evaluation of Dam Safety* (New York: ASCE, 1977).
4.44 W. M. Alley, *Guide for Collection, Analysis, and Use of Urban Stormwater Data* (New York: ASCE, 1977).
4.45 *Land Application of Residual Materials* (New York: ASCE, 1977).
4.46 R. A. Matula, ed., *Present Status and Research Needs in Energy Recovery from Wastes* (New York: ASME, 1977).
4.47 *Responsibility and Liability of Public and Private Interests on Dams* (New York: ASCE, 1977).
4.48 *Shotcrete for Ground Support* (New York: ASCE, 1977).
4.49 *Phase Equilibria and Fluid Properties in the Chemical Industry*, Symposium Series 60 (New York: AIChE, 1978).

4.50 J. F. Davidson and D. L. Keairns, eds., *Fluidization* (Cambridge University Press, 1978).
4.51 *Enzyme Engineering*, Volume 1–7 (editor and publisher varied, 1972–84).
4.52 R. W. Bryers, ed., *Ash Deposits and Corrosion Due to Impurities in Combustion Gases* (Washington, D.C.: Hemisphere Publishing Corporation, 1978).
4.53 *Power System Reliability—Research Needs and Priorities* (Palo Alto, CA: Electric Power Research Institute, 1979).
4.54 E. B. Berman, ed., *Contingency Planning for Materials Resources* (New York: Engineering Foundation, 1979).
4.55 Y. A. Liu, ed., *Industrial Applications of Magnetic Separation* (New York: IEEE, 1979).
4.56 *Planning, Engineering, and Constructing the Superprojects* (New York: ASCE, 1979).
4.57 *Research Directions in Computer Control of Urban Traffic Systems* (New York: ASCE, 1979).
4.58 S. J. Bodilly, ed., *The Economic Impact of Energy Conservation* (Washington, D.C.: U.S. Government Printing Office, 1979).
4.59 *Biochemical Engineering* (New York: New York Academy of Sciences, 1979).
4.60 *Waste Heat Utilization* (Washington, D.C.: U.S. Department of Energy, 1979).
4.61 *Measuring and Forecasting Engineering Personnel Requirements* (New York: Engineering Manpower Commission, 1979).
4.62 *Innovative Management of Technical Information Functions* (Schenectady, NY: General Electric Company, 1977).
4.63 S. K. Saxena, ed., *Evaluation and Prediction of Subsidence* (New York: ASCE, 1979).
4.64 W. P. Henry, ed., *Improved Hydrologic Forecasting: Why and How?* (New York: ASCE, 1980).
4.65 J. Skalny, ed., *Cement Production and Use* (New York: Engineering Foundation, 1979).
4.66 *Hydropower: A National Energy Resource* (Washington, D.C.: U.S. Government Printing Office, 1979).
4.67 K. Preston, et al, eds., *Medical Imaging Techniques: A Comparison* (New York: Plenum Publishing, 1979).
4.68 *Systems Engineering for Power: Organizational Forms for Large Scale Systems*, 2 volumes (National Technical Information Service, Springfield, VA, 1980).
4.69 L. H. Fink and T. A. Trygar, eds., *Systems Engineering for*

Power: Emergency Operating State Control (Washington, D.C.: U.S. Department of Energy, 1979).

4.70 *Shotcrete for Ground Support* (New York: ASCE, 1977).

4.71 T. C. Frankiewicz, ed., *Energy from Waste* (Ann Arbor, MI: Ann Arbor Science Publishers, 1980).

4.72 *New Approaches to Nonlinear Problems in Dynamics*, ISBN 0–89871–167–3 (Philadelphia, PA: Society of Industrial and Applied Mathematics, 1980).

4.73 *Fluidization III*, ISBN 0–396–40458–3 (New York: Plenum Press, 1980).

4.74 *On-Stream Characterization and Control of Particulate Processes*, ISBN 0–939203–02–9 (New York: Engineering Foundation, 1981).

4.75 *Computerized Interpretation and the Electrocardiogram V*, ISBN 0–939204–06–1 (New York: Engineering Foundation, 1981).

4.76 *Foundations of Computer-Aided Chemical Process Design*, Vol. I, ISBN 0–8169–0189–0 & Vol. II, ISBN 0–8169–0190–9 (New York: AIChE, 1980–1981).

4.77 *Building Economics*, ISBN 0–930284–10–0 (Morgantown, WV: American Association of Cost Engineers, 1981).

4.78 *Enzyme Engineering V*, ISBN 0–306–40471–0 (New York: Plenum Publishing Company, 1981).

4.79 *Biochemical Engineering II*, Annals Vol. 369, ISBN 0–89766–127–3 (New York: New York Academy of Sciences, 1981).

4.80 *Geotechnical and Environmental Aspects of Geopressure Energy*, ISBN 0-939204–03–7 (New York: Engineering Foundation, 1981).

4.81 *Historic Preservation of Engineering Works* (New York: ASCE, 1981).

4.82 *Hazardous Waste Management*, ISBN 0–250–40459–1 (Ann Arbor, MI: Ann Arbor Science Publishers, 1981).

4.83 *Issues in Control of Urban Traffic Systems*, ISBN 0–939204–07–X (Washington, D.C.: Institute of Transportation Engineers, 1982).

4.84 *Risk/Benefit Analysis in Water Resources Planning and Management*, ISBN 0–306–40884–8 (New York: Plenum Press, 1982).

4.85 *Chemical Process Control II*, ISBN 0–8169–0203–8 (New York: AIChE, 1982).

4.86 *Modeling of Casting and Welding Processes*, ISBN

0–89520–380–4 (Warrendale, PA: Metallurgical Society of AIME, 1982).

4.87 J. E. Flack, *Urban Water Conservation*, ISBN 0–87262–297–7 (New York: ASCE, 1982).

4.88 *Engineering Education—Aims & Goals for the 80's*, ISBN 0–939204–13–4, (New York: Accreditation Board for Engineering and Technology, 1982).

4.89 *Stormwater Detention Facilities*, ISBN 0–87262–348–3 (New York: ASCE, 1983).

4.90 *Computerized Interpretation of the Electrocardiogram VI*, ISBN 0–939204–16–9 (New York: Engineering Foundation, 1983).

4.91 *Computerized Interpretation of the Electrocardiogram VII*, ISBN 0–939204–19–3 (New York: Engineering Foundation, 1983).

4.92 *Fouling and Slagging Resulting from Impurities in Combustion Gases*, ISBN 0–939204–18–5 (New York: Engineering Foundation, 1983).

4.93 *Ultrasonic Fatigue*, ISBN 0–89520–397–9 "New York: Metallurgical Society of AIME, 1983).

4.94 *Enzyme Engineering 6*, ISBN 0–306–41121–0 (New York: Plenum Press, 1983).

4.95 *Multiobjective Analysis in Water Resources*, ISBN 0–87262–406–4 (New York: ASCE, 1984).

4.96 *Fouling of Heat Exchanger Surfaces*, ISBN 0–939204–20–7 (New York: Engineering Foundation, 1984).

4.97 *Biochemical Engineering III*, ISBN 0–89766–220 (New York: New York Academy of Sciences, 1983).

4.98 *Management of Atmospheres in Tightly Enclosed Spaces*, ISBN 0–910110–40–9 (Atlanta, GA: American Society of Heating, Refrigerating and Air Conditioning Engineers, 1984).

4.99 *Fundamentals of Adsorption*, ISBN 0–8169–0265–8 (New York: AIChE, 1984).

4.100 *The Undergraduate Engineering Laboratory*, ISBN 0–939204–21–5 (New York: Engineering Foundation, 1984).

4.101 *Emerging Computer Techniques in Stormwater and Flood Management*, ISBN 0–87262–401–3 (New York: ASCE, 1984).

4.102 *Characterization and Performance Prediction of Cement and Concrete*, ISBN 0–939204–22–3 (New York: Engineering Foundation, 1984).

4.103 *Modeling of Casting and Welding Processes II*, ISBN 0–89520–477–0 (Warrendale, PA: Metallurgical Society of AIME, 1984).
4.104 *Fluidization*, ISBN 0–8169–0264–X (New York: AIChE, 1984).
4.105 *Updating Subsurface Sampling of Soils and Rocks and Their In-Situ Testing*, ISBN 0–939204–25–8 (New York: Engineering Foundation, 1985).
4.106 *Space Shuttle Environment*, ISBN 0–939204–28–2 (New York: Engineering Foundation, 1985).
4.107 *Computerized Interpretation of the Electrocardiogram VIII*, ISBN 0–939204–23–1 (New York: Engineering Foundation, 1985).
4.108 *Computerized Interpretation of the Electrocardiogram IX*, ISBN 0–939204–26–6 (New York: Engineering Foundation, 1985).
4.109 *Compressibility Phenomena in Subsidence*, ISBN 0–939204–29–0 (New York: Engineering Foundation, 1985).
4.110 *Traffic Monitoring and Control Systems*, ISBN 0–939204–24–X (Washington, D.C.: Institute of Transportation Engineers, 1985).
4.111 *Enzyme Engineering 7*, ISBN 0–89766–262–8 (New York: The New York Academy of Sciences, 1984).
4.112 *Social and Environmental Objectives in Water Resources Planning and Management*, ISBN 0–87262–575–3 (New York: ASCE, 1986).
4.113 *Risk-Based Decision Making in Water Resources*, ISBN 0–87262–575–3 (New York: ASCE, 1986).
4.114 *Computerized Interpretation of the Electrocardiogram X*, ISBN 0–939204–30–4 (New York: Engineering Foundation, 1986).
4.115 *Groundwater Contamination*, ISBN 0–939204–31–2 (New York: Engineering Foundation, 1986).
4.116 *Biochemical Engineering IV*, ISBN 0-89766-408-6 (New York: New York Academy of Sciences, 1986).
4.117 *Small Fatigue Cracks*, ISBN 0–87339–052–0 (Warrendale, PA: Metallurgical Society of AIME, 1986).
4.118 *Fluidization V*, ISBN 0–8169–0356–5 (New York: AIChE, 1986).
4.119 *Flocculation, Sedimentation and Consolidation*, ISBN 0–8169–0357–3 (New York: AIChE, 1986).
4.120 Curriculm Vitae of Harold A. Comerer, July 1989.

CHAPTER 5

Summary and Concluding Remarks

1. MAJOR ACCOMPLISHMENTS

It is difficult to summarize the many and varied achievements of the Engineering Foundation over its seventy-five year history. However, the catalytic roles that the Engineering Foundation played in the creation of both the National Research Council and the National Academy of Engineering had profound effects on the development of efforts by the federal government to support engineering research. As a result, the Engineering Foundation has had a profound effect on the course of American technological achievement during much of this century.

Equally as important was the role that the Engineering Foundation has played and continues to play as the spokesman for the American engineering community. Through its close associations with the Founder Societies and other engineering groups the Foundation has been able to perceive the needs of the engineering community and to take appropriate actions to both meet them and bring them to the notice of other elements of our society. This role has been particularly evident in the efforts of the Foundation to support the engineer as a professional person and to define his or her role in society in a broadly humanistic manner. This commitment was evidenced in the efforts that were made by the Foundation to popularize engineering and scientific achievements and investigations through the publication of its *Popular Research Narratives.* It was also the motivation for the Engineering Foundation's activities on behalf of unemployed engineers during the Great Depression. Sponsorship of the Engineers Council for Professional Development and its visiting teams did much to improve the quality of engineering education in the United States. Most importantly, the creation and continuing success of the Engineering Foundation's Research Conferences have provided a national and international forum for the exchange of information in such vital areas of scientific inquiry as biomechanics, enzyme engineering, pollution, and urban mass transit.

It has been, however, the funding and sponsorship of research projects, councils, and publications that has been the primary hallmark of its activities during the past seventy-five years. It is this activity which has probably best represented the intentions of its founder, Ambrose Swasey. Swasey provided the original funding for the Engineering Foundation at a time, 1914, when government support of scientific and engineering research was practically nonexistent. As a result, the Engineering Foundation was able to make a significant and long lasting contribution to the development of American engineering through its funding of pioneering research projects in such areas as fatigue of metals, thermal properties of steam, arch dams, lubrication, alloys of iron, impregnated paper insulation, enzyme engineering, and biomechanics. Some of these projects were truly epic in scope and resulted in monumental publications. The foremost of these projects was the "Alloys of Iron" investigation which lasted for more than two decades and resulted in the publication of a series of encyclopedic volumes on almost all aspects of iron and steel metallurgy.

The research funding and sponsorship of the Engineering Foundation has also done much during the past seventy-five years to create a climate of cooperative research among the American engineering community. The most evident example of this cooperative research were the many research councils that the Engineering Foundation has supported by providing both monetary and administrative support. Among the notable of these councils have been the Welding Research Council, the Structural Stability Research Council, the Reinforced Concrete Research Council, and the Research Council on Structural Joints. The activities of the last named council provided a graphic example of how research funded by the Engineering Foundation had provided readily apparent benefits, since it was research by this council which was largely responsible for the switch from the use of rivets to the use of high tensile strength bolts in American construction, resulting in increased structural strength, lower construction costs, and the elimination of much construction noise.

Finally, it should be noted that the Engineering Foundation's Research Initiation Grants have played a large role during the 1970s and 1980s in both opening new areas of research and launching the careers of many young and promising members of engineering faculties.

2. FOUNDATION'S ROLE IN AMERICAN TECHNOLOGY AND SOCIETY

During its seventy-five years, the Engineering Foundation has served as both a leader in the development of American engineering research and a reflection of the changing role of the American engineer in our society. As a result, its activities have often focused on new trends in the evolution of American technology and on occasion it has utilized its resources to identify and address important national problems in an attempt to utilize the resources of the engineering community to alleviate them.

The catalytic role that the Engineering Foundation played in the creation of the National Research Council has had incalculable effects on the development of American technology. The National Research Council was the first engineering organization to receive large scale research funding from the Federal Government. This funding not only set the precedent for the extensive governmental scientific and engineering research programs that would be implemented during the 1940s and 1950s, but it helped to provide a model for the organization of large-scale engineering investigations and broad subjects that took place over a prolonged period.

The activities of the Engineering Foundation also helped to define the position of the engineer in American society. It should be noted that when the Foundation was started, engineering was near the pinnacle of its popularity and social status. Many of America's largest corporations had been founded and guided to greatness by engineers, many of whom acquired the skills through pragmatic job-related experience. Ambrose Swasey came from such a background and his achievements reflected a nation in which engineering was helping to bring to maturity our industries, transportation network, and technology. However, it was also a time of transition in which the roles played by research and education in the development of engineering were rapidly increasing. Our society was becoming more complex and our technological problems were consequently greater. Trial-and-error knowledge gained from workplace experiences was no longer sufficient to meet America's needs. Ambrose Swasey recognized this change when he founded the Engineering Foundation with its primary emphasis on research and the improvement of engineering education. Under the able guidance of its first director, Alfred Flinn, the Foundation not only became internationally known for the quality of its funded projects but also for its activities to enhance the engineering profession. Under Flinn's leadership, the Foundation also became increasingly aware of the need to better integrate engineering with humanistic principles so that the role of the engineer in society would become more complete and not solely be devoted to the achievement of technological triumphs.

Despite the economic hardships that were imposed by the Great Depression, the standards that were set by Flinn and the early boards of the Engineering Foundation were maintained throughout both the 1930s and the turmoil of World War II. Both Director Hovey and Director Colpitts were brilliant engineers who had reached the pinnacle of success in their respective fields of bridge engineering and sound transmission. Although neither of them was fated to oversee the Engineering Foundation's affairs for extended periods, both Hovey and Colpitts continued the Foundation's emphasis on research and during their tenures such projects as the formation of research councils and the "Alloys of Iron" began to reach their great period of achievement. This tradition was continued and advanced under the direction of Frank T. Sisco, who maintained the high quality of the Foundation's funded projects throughout his tenure.

However, the 1940s and 1950s witnessed a monumental and permanent change in the American engineering and scientific communities. During World War II the Federal government became massively involved in the funding for scientific and engineering research. Equally as important, the rise of atomic weapons and later space rocketry heightened both the prestige and political power of the scientific community. In the age of Sputnik, research into basic engineering problems seemed too mundane in comparison to the glamour of the space race, while the thousands of dollars that the Engineering Foundation had available to fund its research projects were dwarfed by the millions of dollars that were made available to federally funded projects. In response to these changed conditions, the Engineering Foundation developed several notable and successful strategies. Through its close relationships to its founder societies and international engineering organizations, it reasserted and enhanced its role as a spokesman for the engineering community. It also helped to solidify the status of the engineering profession in the realm of government research funding, through the catalytic role that it played in the creation of the National Academy of Engineering. Finally, through the development of its research conferences, it helped to increase the relevancy of engineering research to the general society by both focusing attention on such problems as pollution and the decline of urban mass transit, and fusing the principles of engineering with such scientific pursuits as biology, chemistry, and medicine. As a result, the Engineering Foundation was able to bring much attention and support to such new and vital fields as enzyme engineering, biomechanics, and medical engineering. Finally, through its research initiation grants it was able to help provide a means of aiding young scholars whose careers would provide the Foundation with future engineering achievements, and thus assure the continued usefulness of engineering to our national development.

3. CONCLUDING REMARKS

In conclusion, the Engineering Foundation has served as a catalyst for the development of American engineering and a spokesman for the American engineering community. It has helped to define the role of the engineer in our society by focusing on the humanistic aspects of the profession. Most importantly, through its emphasis on the sponsorship of research it has enabled many projects to be undertaken which had direct and far-reaching effects on both the private industry and general public. Projects such as the research councils, the "Alloys of Iron," the "Arch Dam Investigation," and the early research conferences on such topics as pollution, biomechanics, and urban mass transit, helped to bring about significant changes in the efficiency of industry and the quality of American life. However, it is the authors' belief that these achievements have been largely misunderstood by the Engineering Foundation Board itself.

Although the deed of gift of Ambrose Swasey's original endowment and the supplemental letters that he added to it during the remainder of his life emphasized the importance of the Foundation's mission as a supporter of engineering research, the Foundation has often devoted too much of its resources and energies toward an almost continual search for its proper mission. This search is evidenced by an almost continual process of self-evaluation. Although many positive accomplishments have resulted from this process, it has also served to obscure the Foundation's true achievements. This tendency was particularly pronounced in the years during and immediately after World War

II. In this time of massive subsidies by the federal government of engineering and scientific research, the Engineering Foundation began to question the validity of its existing research grant program, due to the relative insignificance of the total volume of the Foundation's grants as compared to the federal largesse. As a result, the Foundation began to question its research mission. However, it is apparent in retrospect that in many cases the value of the Foundation's support for engineering research lay not so much in its volume, but in its flexibility, selection process and the guidance that the Foundation provided to its grantees. Much useful work was done and continues to be done with relatively modest Foundation grants, due to the fact that most of the projects were well chosen and competently managed.

It should also be noted that the Foundation's failure to attract numerous large-scale additions to its original endowment has also contributed to its penchant for self-doubt. Due to a variety of circumstances and oversights, successive boards of the Engineering Foundation were never able to launch a sustained successful endowment enhancement campaign. This failure is particularly significant since on several occasions during the seventy-five years of its existence, the Engineering Foundation has devoted much time and energy to planning such campaigns but their implementation was either incomplete or completely absent. Had such a campaign been launched, it is quite possible, based on the Foundation's excellent reputation and many contacts in the business and engineering communities, that it would have been quite successful. With an enlarged endowment, the Engineering Foundation would have been able to increase the scope of its funding for projects and thus it may have avoided the somewhat debilitating cycle of self-doubt which afflicted it during the post World War II era.

Yet, these are minor negatives compared to the many notable achievements of the Engineering Foundation. Founded by a visionary engineer-entrepreneur and guided by successive boards and directors of the highest caliber, the Engineering Foundation has been at the forefront of American engineering research for seventy-five years. By building on this legacy, it can continue to play a catalytic role in the evolution of American technology and science.

APPENDIX A

Supporting Data

TABLES 1–4

Members and officers of the Engineering Foundation who served during the first seventy-five years are listed alphabetically in Table 1. Altogether, 217 persons served as members of the board. They were appointed as representatives of the Founder Societies (ASCE, AIME, ASME, IEEE and AIChE which became the fifth Founder Society in 1958) or of the United Engineering Trustees and its predecessors, or as members at large.

Regular projects which were sponsored by the Engineering Foundation are listed in Table 2 in chronological order. The initial period through 1930 can be characterized as one of sponsorship of engineering research projects. The "Fatigue of Metals" project (Item 9 in Table 2) was carried out jointly with the National Research Council. It was the first comprehensive systematic study of the strength of metals under repeated loading undertaken in the United States. It lead to the development of the American Welding Society's welding code that became a basis for the design of machinery, rolling stock, bridges, airplanes, spacecraft, electrical equipment, and various other welded devices which were subjected to frequently repeated loads. The project "Thermal Properties of Steam" (Item 13 in Table 2) provided the first authoritative international tables relating steam pressure, temperature, and volume in a form readily usable in the design of boilers, engines, and other steam appliances. The tables covered pressures up to 3,200 psi and maximum temperature of 750 deg F thus extending the limits of the then existing data by factors of sixteen and two, respectively.

A special mention belongs to the "Arch Dam Investigation" (Item 19 in Table 2). This investigation was one of the earliest major projects that was financed by contributions from many interested individuals and small companies as well as large corporations and governmental units. It also involved many investigators studying various parts of the overall problem. Thus the "Arch Dam Investigation" may be considered a testing ground for research through cooperative efforts.

The "Alloys of Iron Research" (Item 32 in Table 2) was the major project of the 1931–45 period. It involved cooperative research and writing similar to the activities that were pioneered by the "Arch Dam Investigation." However, it went a giant step further through its truly national scope and world-wide collection of data. The second project of this period that must be highlighted is the "Welding Research Council" (Item 62 in Table 2). It was founded in 1935 as the Welding Research Committee which was sponsored jointly by AIEE and AWS. It was also an outgrowth of the earlier study "Fatigue of Metals" and the model for formation of several research councils during the post World War II period. The "Welding Research Council" successfully promoted the development and use of good welding practices and thus speeded up the acceptance of this basic connecting technique for metals.

The major characteristic of the 1946–60 period was the formation of several research councils. These independent bodies were formed on the basis of nationwide representation from academic, industrial, and governmental organizations which were interested in research in some broad field of engineering. The Structural Stability Research Council (Item 84 in Table 2), Research Council on Structural Connections (Item 90 in Table 2), and Reinforced Concrete Research Council (Item 99 in Table 2), are three very successful examples that were founded in the 1940s and which have been in continuous operation to this day. These research councils planned needed research projects, secured the necessary financing, selected the research institution, monitored or supervised the conduct of research, and channeled the research findings into practical use. The construction of record-height buildings such as the World Trade Center in New York and the Sears and John Hancock in Chicago, many structures on the Interstate Highway System, and various sophisticated structures such as the Assembly Building at Cape Canaveral, Florida, were only made possible because of the expansion of the basic knowledge through the activities of these three research councils.

The Engineering Foundation's primary focus during 1961–75 was the establishment of Engineering Foundation Conferences which are listed in chronological order in Table 3. The first such conference was held at the Tilton School, New Hampshire, in July 1962 on the topic "Engineering Research for the Developing Countries." The place and format for this meeting as well as for most other Foundation conferences was selected to provide surroundings conducive to informal meetings with emphasis on discussion by all participants. Close to 200 such conferences were held by the end of 1975 on topics ranging from "Corrosion Engineering" to "Engineering for Unconventional Protein Production" to "Engineering in Medicine." Through its conferences and research grants, the Foundation became a pioneer in focusing the attention of the engineering community on such emerging areas of concern as urban mass transportation, pollution of the environment, and the technology gap between the third world and the developed countries.

Finally, the hallmark of the Foundation's 1976–89 period was the Research Initiation Grants program listed in Table 4 in chronological order. Established in 1976 to support research by new faculty members in technical areas of interest to the Founder Societies, it was designed not only to obtain solutions to technical problems but also to help attract and keep talented individuals on the academic staffs of university faculties. These grants also promoted early transfer of research findings to the engineering profession. Since its inception, this program has been characterized consistently by large numbers of high quality research proposals in each of the five major areas of engineering. A significant extension of the program occurred in 1989 when the Air Force provided funds for twenty-five additional grants.

While, after 1960, various other activities became the principal focus of the Engineering Foundation, the engineering research projects which were so characteristic of the first fifteen-year period were not forgotten. Engineering Foundation sponsorship of research projects has been continued throughout its seventy-five years. As of 1989, almost all of the Foundation's discretionary income was channeled into engineering research, the bulk of it through the most enthusiastically received Research Initiation Grants program.

A list of principal projects subdivided into six categories follows. The list includes engineering projects and projects for the advancement of the engineering profession as well as basic research projects.

Interdisciplinary Projects

National Research Council (3)
Wood Finishing Investigation (15)
Marine Piling Investigation (18)
Fatigue of Metals (9)
Accreditation Board for Engineering and Technology (56)
Welding Research Council (62)
National Academy of Engineering (1961–65)
Fireside Corrosion of Fluidized Bed Heat Exchangers (78–1)

Chemical Engineering

Asymmetric Hollow Fiber Membranes for Enzyme Immobilization (73–5)
Design Institute for Multiphase Processing (73–1)
High Pressure Relief Systems for Flammable Liquids (76–5)
Photocatalyzed Degradation of Pesticides (77–2)

Civil Engineering

Arch Dam Investigation (19)
Plastic Flow of Concrete (41)
Earth and Foundations (38)
Hydraulic Research (67)
Structural Stability Research Council (84)
Research Council on Structural Connections (90)
Reinforced Concrete Research Council (99)
Storm Surges (118)
Study of Measurements of Stack Emissions (60–1b)
Evaluation of Carbon Dioxide Concentration as Indicator of Air Pollution in Urban Atmosphere (60–1c)
Fundamental Combustion Study for Disposal of Industrial and Municipal Wastes by Incineration (69–2)
Acid Precipitation Recovery of Acid Lakes (76–4)

Electrical Engineering

Dielectric Absorption: Electric Insulation (27)
Nitrogen in Welds (49)
Impregnated Paper Insulation (66)
Noninvasive Aneurysm Detection and Analysis (76–6)

Mechanical Engineering

Thermal Properties of Steam (13)
Lubrication (23)
Screw Threads (44)
Effect of Temperature on Properties of Metals (45)
High Temperature Steam Generation (50b)
Creep & Relaxation (61)
Plastic Flow of Metals (68)
Unsteady Heat Flow (80)
Coefficients for Concentric Orifice Plates and Flow Nozzles (68–1)

Mining, Metallurgical and Petroleum Engineering

Alloys of Iron Research (32)
Basic Open Hearth Steel Making (78)
Diffusion of Steel (97)
Comminution (104 and 64–7)
Flow of Bulk Solids (112)

TABLES 5–9

All committees of the Engineering Foundation included in the annual reports are listed in Table 5. The day-to-day business of the Foundation was conducted through three standing committees. Numerous other committees were appointed to resolve relatively short-term specific tasks. The three standing committees were the Executive Committee (No. 2 in Table 5), the Projects Committee (No. 34 and its predecessors Nos. 20 and 23), and the Conferences Committee (No. 33 and its predecessor No. 28).

The contributors to the endowment of the Engineering Foundation and the contributed amounts are listed in Table 6. The endowment was established in 1915 with the first gift from Ambrose Swasey. It was increased from time to time by funds contributed by Swasey and numerous other benefactors of the Foundation.

To finance the "Arch Dam Investigation" (19), donations were collected from a large number of companies and individuals. The total cash expenditure on the project was close to $150,000 of which only about one fifth was contributed by the Foundation. The names of all contributors as well as co-operators, that is those who furnished services without compensation, are listed in Table 7. A similar approach to financing was taken on at least seven additional projects all of which are included in Table 7.

Financial sponsors of the Engineering Foundation Conferences are listed in Table 8. The conference attendance was often subsidized from funds contributed by financial sponsors. The grants were usually awarded by the Foundation although in some instances the awards were made directly by the sponsors.

Starting in 1921, the board of the Engineering Foundation named its representatives to certain other organizations concerned with engineering research of interest to the Foundation. The names of such representatives and their periods of service are listed in Table 9.

Similarly to Table 5, all data for Tables 7, 8, and 9 were obtained from the annual reports of the Foundation.

TABLE 1 Members and Officers of the Engineering Foundation from 1914 to 1989

Name	*Office and Term*				
	Member	*Chairman*	*Vice Chairman*	*Secretary*	*Director*
*Adams, Edward Dean (+)	1915–21				
(1)	1921–26		1915–26		
Honorary Member	1926–31				
Alexander, William D. (1)	1963–75	1966–67			
Alford, Joe B. (6)	1973–77				
Allen, R. C. (3)	1954–59				
Aplan, Frank F. (2)	1976–	1985–87	1984–85		
*Arms, J. H. R.				1937–56	
Austin, J. B. (+)	1955–58				
*Bakhmeteff, B. A. (1)	1946–51	1949–51	1948–49		
*Barber, Everett M. (+)	1950–61	1955–57	1953–55		
*Barker, J. W. (4)	1947–55	1953–55	1952–53		
*Barr, John H. (3)	1922–25				
*Barrett, W. J. (6)	1955–57				
*Barron, George D. (6)	1931–34				
(2)	1934–44				
*Basford, George M. (6)	1922–23				
*Becket, Frederick M. (+)	1935–38				
*Beggs, George E. (1)	1934–39	1939	1938–39		
*Bellinger, H. C. (2)	1931–34		1933–34		
Bieber, Herman (5)	1977–				
Blecher, Franklin H. (4)	1984–				
*Borgmann, Carl W. (+)	1959–60				

TABLE 1 Members and Officers (continued)

Name	Member	Chair-man	Vice Chairman	Secretary	Director
Bowie, Robert M. (6)	1965–67				
*Boyd, James (6)	1971				
*Boylston, Herbert M. (2)	1922–28				
Brant, Austin E., Jr. (1)	1982–86				
Bray, Oscar S. (6)	1975–81				
(+)	1984–85				
Bryers, Richard W. (3)	1982–85				
*Buckley, O. E. (+)	1938–50	1939–42	1939		
*Burr, William H. (1)	1927–31				
Burtis, T. A. (6)	1965–66				
*Callow, John M. (2)	1928–31				
*Carey, W. N. (6)	1947–55				
Case, Richard P. (4)	1987–				
*Casey, J. S. (6)	1944–48				
*Chambers, Ralph H. (+)	1942–51				
*Charlesworth, H. P. (6)	1932–39	1934–36	1933–34		
*Chubb, L. W. (4)	1944–48		1945–48		
*Clevenger, Galen H. (2)	1926–32		1930–32		
*Coes, Harold V. (6)	1933–35				
*Colcord, Frank F. (2)	1936–48		1940–42		
*Cole, Sandford S. (2)	1962–66				1966–83a
*Colpitts, Edwin H.					1941–48
Comerer, Harold A.					1984–89b
*Condit, Kenneth H. (6)	1937–44		1942–44		
Cordovi, M. A. (3)	1972–84	1977–79	1975–77		
*Crawford, George G. (+)	1931–35				
*Curme, George O., Jr. (5)	1958–60		1959–60		
Damon, Richard W. (6)	1982–84				
*Davies, J. Vipond (6)	1920–28				
*DeGolyer, Everett (2)	1932–36				
*Dellinger, Brower (6)	1971–73				
Dillard, Joseph K. (6)	1979–82				
*Dinkey, Alva C. (+)	1926–31				
*Dorr, John V. N. (6)	1928–32				
(+)	1932–43				
*Dougherty, R. E. (6)	1944–47				
Doyle, Edward J. (6)	1986–				
*Dryden, Hugh L. (+)	1961–65				
Dunn, Gano (6)	1915–18	1915–17			
(4)	1925–31				
Duquette, David (+)	1987–				
*Dwight, Arthur S. (2)	1922–24				
*Elliott, Howard (+)	1915–19				
*Fairman, James F. (6)	1954–62				
*Farmer, F. M. (4)	1935–39	1936–39			
(6)	1945–47				
Fein, Richard S. (3)	1984–				
Finch, Rogers B. (+)	1984–87				

TABLE 1 Members and Officers (continued)

Name	*Member*	*Chair-man*	*Vice Chairman*	*Secretary*	*Director*
		Office and Term			
*Finlay, Walter S., Jr. (6)	1926–27				
*Fish, Edwards R. (6)	1923–34				
(+)	1935–42				
(3)	1942–44				
Fisher, Gordon P. (1)	1987–				
*Fletcher, Andrew (6)	1955–60				
(+)	1962–73				
*Flinn, Alfred D.				1917–37	1922–37
Ford, Curry E. (6)	1977–87				
Frater, Thomas J. (1)	1959–62				
Freiman, Charles V.					1990–
*Fuller, George W. (1)	1931–34	1933–34	1932–33		
*Fulweiler, Walter H. (3)	1934–42				
Gagg, R. F. (6)	1951–53				
*Gaudin, A. M. (2)	1951–61	1959–61			
Genereaux, R. P. (+)	1966–68				
(6)	1968–69				
(+)	1970–82	1980–82	1979–80		
*Gherardi, Bancroft (6)	1920–27				
Giordano, A. B. (4)	1968–80	1979–80	1977–79		
Glenn, Wayne E. (6)	1967–71		1970–71		
*Goss, W. F. M. (3)	1916–22	1919–20	1918–19		
(6)	1923–26				
Green, Estil I. (4)	1954–58				
*Greene, Arthur M., Jr. (3)	1926–28				
Guthrie, H. D. (6)	1966–67				
*Habach, G. F. (6)	1972–83		1976–77		
(+)	1983–84				
Haddad, Jerrier A. (4)	1978–87	1982–83	1980–82		
(4)			1983–84		
*Haider, Michael L. (6)	1961				
Haney, Paul D. (1)	1978–79				
Harris, William J., Jr. (2)	1968–70				
*Head, James L. (6)	1953–55				
*Heald, H. T. (+)	1966–70		1967–69		
Hellmann, Reinhard K. (4)	1980–84				
Herold, Edward W. (4)	1975–78				
*Hickernell, L. F. (+)	1961–62				
(6)	1962–63				
Hitchcock, John H. (3)	1960–68				
*Holleran, Leslie G. (1)	1950–58		1958		
*Hovey, Otis E. (1)	1930–35		1930–32		
(6)	1935–41				1937–41
*Huie, Irwing V. A. (6)	1950–51				
*Humphreys, Alex C. (6)	1915–16				
*Hunt, Charles W. (6)	1915–20				
*Hutchinson, Cary T.				1916–17	
*Hutton, Frederick R.				1915	

TABLE 1 Members and Officers (continued)

Name	*Member*	*Office and Term* *Chairman*	*Vice Chairman*	*Secretary*	*Director*
*Jacobus, D. S. (3)	1918–24		1919–20		
Jappe, K. W. (6)	1948–49				
*Jewett, Frank B. (4)	1919–25		1920–25		
Jones, C. R. (6)	1944–47				
*Justin, Joel D. (1)	1940–49	1948–49			
Kaiser, Elmer R. (2)	1948–51				
*Kidder, H. A. (6)	1932–33				
Kidder, A. H. (4)	1959–62	1961–62			
Kinckiner, R. A. (5)	1961–65				
*Kinzel, A. B. (+)	1943–55	1945–48	1944–45		
(2)	1960–64				
Kirkendall, E.				1963–65	
*Knight, George L. (6)	1927–32				
(6)	1935–36				
(6)	1947–48				
Korwek, A. D.				1982–	
Kruger, George E. (2)	1970–72				
Landis, Fred (3)	1985–				
*Lardner, Henry A. (6)	1930–41				
Latham, Henry A. (3)	1968–72				
*Ledoux, Albert R. (2)	1915–16				
Lee, E. S. (6)	1947–48				
Linder, Clarence H. (6)	1963				
*Little, Arthur D. (+)	1924–30		1927–30		
Lobo, Walter E. (6)	1958–67				
(+)	1968–82		1971–73		
Loughren, Arthur V. (4)	1963–75		1962–67 1973–75		
*Ludlow, Edwin (6)	1921–22				
MacAdam, Walter K. (6)	1967–73		1969–70		
MacDonald, B. I. (6)	1987–				
Manogue, W. H. (5)	1972–84				
Marlowe, Donald E. (6)	1981–85 1987–				
Marras, Steven W.				1957–63	
*Martin, Harold E. (6)	1955–67				
*Mason, Martin A. (1)	1958–70				
*McGeorge, R. M. (2)	1972–79				
*Meagher, E. C. (6)	1948–50				
Meier, Louis L. (6)	1982–83				
Mermel, T. W. (1)	1970–78				
Mintz, E. Forest (6)	1983–				
*Molineaux, Charles B. (6)	1955–67		1960–61		
(+)	1973–84				
*Moulton, Gail F. (6)	1949–54				
*Moultrop, Irving E. (6)	1920–22				
*Newman, W. A. (3)	1947–49				
Oishi, Satoshi (1)	1986–				

TABLE 1 Members and Officers (continued)

Name	*Office and Term*				
	Member	*Chairman*	*Vice Chairman*	*Secretary*	*Director*
*Orrok, George A. (3)	1924–30		1926–30		
*Pearse, Langdon (1)	1935–38				
*Pegram, George H. (6)	1920–26				
*Peirce, W. M. (2)	1948–59	1957–59	1955–57		
Perebinossoff, A. A. (2)	1965–68				
*Perry, J. P. H. (6)	1941–47				
*Porter, H. Hobart (+)	1918–24				
(6)	1927–32	1929–33			
(+)	1932–34				
*Potter, A. A. (3)	1939–47				
Prentis, E. A., Jr. (2)	1944–48				
*Proctor, Carlton S. (+)	1955–60				
*Pupin, Michael I. (4)	1915–19	1917–19	1916–17		
Purnell, C. S. (+)	1951–55				
*Quarles, D. A. (6)	1948–54				
*Queneau, A. L. J. (6)	1935–45	1942–45			
Queneau, Paul E. (2)	1966–76	1973–75			
*Rand, Charles F. (+)	1916–26	1920–25			
*Raymond, Robert M. (2)	1916–22				
*Rice, Calvin W.				1915	
*Rice, E. W., Jr. (4)	1918–24				
*Richards, Joseph W. (2)	1918–21				
*Ridgeway, Robert (1)	1924–30				
*Roberts, Albert (6)	1941–43				
*Roberts, George E. (+)	1927–32				
*Roberts, R. J. E. (6)	1985–89				
Robinson, Ernest L. (3)	1950–62				
*Roosevelt, R. M. (6)	1934–35				
	1945–49				
Rowley, L. N., Jr. (6)	1973–75				
*Ryder, E. M. T. (1)	1938–46				
Sangster, William M. (+)	1982–				
*Saunders, W. L. (6)	1924–27				
*Saville, Thorndike (1)	1951–54				
Schreiner, W. C. (5)	1958–72	1962–66	1961–62		
*Scribner, C. E. (6)	1915–16				
Sears, Raymond W. (6)	1971–80	1975–77			
*Seely, Warner (6)	1949–52				
Shaw, Henry (5)	1984–	1987–	1985–86		
Shaw, Milton C. (3)	1975–82				
*Sheldon, Samuel (6)	1920				
Sherrod, Gerald E. (6)	1979–83				
*Sisco, Frank T.					1949–59
*Skinner, Charles E. (4)	1930–35		1932–33		
*Slichter, W. I. (4)	1931–44				
*Smart, J. S., Jr. (6)	1962–68				
*Smith, Harold B. (4)	1931				
*Smith, H. DeWitt (6)	1958–60				

TABLE 1 Members and Officers (continued)

Name	Member	Office and Term: Chairman	Vice Chairman	Secretary	Director
*Smith, J. F. Downie (3)	1946–48				
*Smith, Jesse M. (3)	1915–16				
*Smith, J. Waldo (1)	1915–21				
*Smith, Ronald B. (6)	1967–72				
*Sperry, Elmer A. (+)	1921–27		1925–27		
*Spilsbury, E. Gybbon (6)	1918–20				
*Stephens, Charles E. (6)	1934–44				
*Stillwell, Lewis B (4)	1924–30	1925–29			
*Stuart, Francis Lee (6)	1926–31				
Suits, C. G. (4)	1948–54	1951	1949–52		
Swan, David (2)	1980–				
*SWASEY, AMBROSE - Founder	1914–37				
Tatlow, R. H., III (6)	1972–76				
Taylor, William H. (6)	1983–87				
Tebo, Julian D. (4)	1959–68				
*Thayer, B. B. (6)	1915–21				
Thompson, Willis F. (6)	1952–55				
(6)	1961				
(6)	1964				
Thornton, Stafford E. (+)	1985–				
Timby, Elmer K. (1)	1954–57				
*Townley, Calvert (6)	1919–20				
*Trinks, W. (3)	1944–46				
Tutt, Charles L., Jr. (3)	1963–75	1967–73			
*Tuttle, Arthur S. (6)	1933–35				
Viest, Ivan M. (1)	1975–87	1983–85	1982–83		
*von Wettberg, E. F., Jr. (5)	1965–77				
(+)	1982–83				
Waag, William R. (6)	1977–79				
*Walker, Arthur L. (3)	1921–26				
*Walker, Eric A. (+)	1962–66				
Walter, Richard C. (1)	1979–82				
*Weisberg, Herman (3)	1948–54	1951–53			
West, Sumner A. (6)	1985–87				
*Whittaker, M. C. (+)	1930–33				
*White, Albert E. (3)	1930–39				
Wilson, L. Edward, Jr. (6)	1987–				
*Windsor, Frank E. (6)	1930–33				
*Wisely, William H. (6)	1967–72				
Wiseman, R. J. (4)	1955–58				
*Woodard, Silas (1)	1918–24				
*Woodrow, H. R. (6)	1939–40				
*Work, Harold K.					1959–65
*Wright, Roy V. (6)	1928–30				
Yacyshyn, Rudolf J.					1989–90c
*Yarnall, D. Robert (3)	1928–34		1934–37		
(6)	1934–39				

TABLE 1 Members and Officers (continued)

Name	*Office and Term*				
	Member	*Chair-man*	*Vice Chairman*	*Secretary*	*Director*
Zecca, John A.				1965–81	

*Deceased.
1 Representing American Society of Civil Engineers.
2 Representing American Institute of Mining, Metallurgical, and Petroleum Engineers.
3 Representing The American Society of Mechanical Engineers.
4 Representing Institute of Electrical and Electronics Engineers.
5 Representing American Institute of Chemical Engineers.
6 Representing United Engineering Trustrees (including the president).
+ Member-at-Large.
a Director of Conferences.
b Served as assistant director of conferences 1976–83.
c Served as acting director from September 1989 to April 1990.

TABLE 2 Regular Projects Sponsored by the Engineering Foundation

File No.	Project Period	Project Title	Research Agency	Proponent or Principal Investigator(s)	Field	Eng. F. Grant $	Other Cntrbs $
1*	1916	Track Signal Relay Improvement	Railroad Signal Association	C. C. Rosenberg	intrd.	0	0
2	1916-20	Wear of Gear Teeth	Leland Stanford Junior University	G. H. Marx, W. F. Durand L. E. Cutter	mech.	792	0
3	1916-17	National Research Council (NRC)	National Academy of Sciences	G. E. Hale	intrd.	21,161	0
4	1917-18	Camouflage of Ships by Spray	Naval Consulting Board	H. P. Quick	intrd.	0	0
5*	1918-20	Directive Control of Wireless Communication	Columbia University	C. G. Stone, Jr. H. W. Farwell	elect.	237	0
6	1918-20	Mental Hygiene of Industry	Boston State Hospital	E. E. Southard	intrd.	2,796	0
7	1918-20	Weirs for Measuring Water	Massachusetts Institute of Technology (MIT)	C. Herschel	civil	2,089	0
8*	1919	Water Power Testing Station	Eng. Foundation (EF)	J. Alsberg	civil	0	0
9	1919-28	Fatigue of Metals	University of Illinois	H. F. Moore	intrd.	26,474	61,000
10	1919-21	Industrial Personnel Problems	NRC Committee		intrd.	0	0
11	1921-23	Advisory Board on Highway Research	NRC	A. Marston, W. K. Hatt	intrd.	3,000	0
12	1921-38	Personnel Research Federation	NRC	L. Oathwaite W. V. Bingham	intrd.	4,800	0
13	1921-34	Thermal Properties of Steam	Harvard Univ., National Bureau of Standards (NBS), MIT	H. N. Davis, N. S. Osborn F. G. Keyes, J. H. Keenan G. A. Orrok, A. Dow	mech.	2,000	156,041
14	1921-22	Graphitic Corrosion of Cast Iron	EF	J. V. Davies	intrd.	0	0
15	1922-34	Wood Finishing Investigation	U.S. Forest Products Lab.	A. H. Sabin, C. P. Winslow G. M. Hunt, F. L. Browne	intrd.	0	0
16	1922-23	Internal Stresses in Metals	Columbia University	E. P. Polushkin, A. L. Walker	intrd.	420	0
17	1922-34	Concrete and Reinforced Concrete Arches	University of Illinois, Ohio State U., et al	A. C. Janni, W. M. Wilson C. T. Morris	civil	22,500	14,000
18	1922-24	Marine Piling Investigation	NRC Div. of Engineering	R. T. Betts, Wm. G. Atwood	intrd.	0	0

TABLE 2 Regular Projects (continued)

File No.	Project Period	Project Title	Research Agency	Proponent or Principal Investigator(s)	Field	Eng. F. Grant $	Other Cntrbs $
19	1922-33	Arch Dam Investigation	Committee on Arch Dam Investigation	C. Derleth, Jr., C. D. Marx F. A. Noetzli, W. A. Slater	civil	26,551	119,215
20	1923-32	Steel Columns	ASCE Committee on Steel Columns	F. E. Turneaure	civil	14,486	7,182
21	1923-25	Mining Methods	AIME Committee on Mining Methods	R. M. Raymond	mining	847	0
22*	1923-28	Bearing Metals	ASME Committee	C. H. Bierbaum	mech.	37	0
23	1923-65	Lubrication	Harvard University, General Electric Co.	A. Kingsbury D. F. Wilcock	mech.	17,620	6,450
24	1924-39	Strength of Gear Teeth	ASME Committee	W. Lewis, R. E. Flanders	mech.	4,040	5,450
25*	1925	Airplane Surveying of Mountainous Regions	EF	F. B. Jewett R. A. Millikan	civil	0	0
26	1926-31	Blast Furnace Slags Investigation	Univ. of Wisconsin	R. S. McCaffery C. H. Herty, Jr.	met.	13,100	14,200
27	1926-31	Dielectric Absorption: Electric Insulation	Johns Hopkins Univ.	J. B. Whitehead	elect.	12,450	13,100
28	1926-34	Engineering Education Investigation	Soc. for Promotion of Engineering Education	W. E. Wickenden	intrd.	2,000	129,450
29	1927-29	Interests as Guides to Engineering as a Vocation	Stanford University	E. K. Strong, Jr.	intrd.	500	0
30a	1927-29	Fluid Flow	Univ. of Minnesota	H. E. Hartig, H. B. Wilcox	mech.	350	0
30b	1935-68	Fluid Meters	NBS, Univ. of Oklahoma Ohio State U. & others	R. J. S. Pigott S. R. Beitler	mech.	17,200	24,660
31*	1927-29	Soils	Special ASCE Committee	C. R. Grow, K. Terzaghi	civil	0	0
32	1928-65	Alloys of Iron Research	Iron Alloys Committee	G. B. Waterhouse, F. T. Sisco	met.	124,500	310,270
33	1928-37	Cutting Fluids for Working Metals	Subcommittee of ASME Special Committee on Cutting Metals	O. W. Boston, M. D. Hersey	mech.	6,250	4,210
34*	1928-30	Models of Flat Slabs	Princeton University	G. E. Beggs	civil	0	0
35	1928-34	Wire Rope Research	ASME Committee, NBS	W. H. Fulweiler	mech.	3,875	0

TABLE 2 Regular Projects (continued)

File No.	Project Period	Project Title	Research Agency	Proponent or Principal Investigator(s)	Field	Eng. F. Grant $	Other Cntrbs $
36	1929-35	Identification of Firearms and Bullets	Stanford University	C. O. Gunther	intrd.	500	0
37	1929-32	Woodworking Saws and Knives	ASME Committee	C. M. Bigelow	mech.	0	0
38	1929-48	Earth and Foundations	MIT, Techn. U.—Vienna Columbia U., Yale Univ.	L. White, C. S. Proctor K. Terzaghi and others	civil	54,737	3,257
39	1930-34	Engineering Aptitudes	University of Missouri	C. V. Mann	intrd.	1,500	0
40	1930-35	Engineering Education	EF Education Research Committee	H. N. Davis	intrd.	7,985	22
41	1930-41	Plastic Flow of Concrete	University of California	R. E. Davis	civil	14,300	28,350
42	1930-36	Pure Iron Electrodes	Lehigh University	G. E. Doan	elect.	7,722	0
43	1930-38	Barodynamic Research	Columbia University	P. B. Bucky	mining	21,000	1,300
44	1930-31	Screw Threads	ASME Comm. on Standardization and Unification of Screw Threads	R. E. Flanders	mech.	500	2,391
45	1930-65	Effect of Temperature on Properties of Metals	Battelle Memorial Inst. Crane Co. & others	N. L. Mochel, J. J. Kanter F. B. Foley, V. T. Malcolm	mech.	25,500	37,530
46	1931-33	Summer School for Engineering Teachers	Stevens Inst. of Techn. Ohio State U., et al	W. E. Wickenden H. P. Hammond	intrd.	3,000	0
47	1931	Reinforced Concrete Column Investigation	University of Illinois Lehigh University	F. E. Richart W. A. Slater, I. Lyse	civil	500	42,085
48*	1931-33	Industrial System	EF Planning Committee	H. V. Coes, J. H. Fahey	intrd.	2,472	0
49	1932-33	Nitrogen in Welds	MIT	J. W. Miller	elect.	1,200	0
50a	1931-44	Critical Pressure Steam Boiler	Purdue University	A. A. Potter, H. L. Solberg G. A. Hawkins	mech.	5,350	7,500
50b	1951-61	High Temperature Steam Generation	Purdue University	H. L. Solberg	mech.	15,350	160,000
51*	1932-33	Study of Transportation	EF	staff	intrd.	1,220	0
52	1932-41	Cottonseed Processing	Univ. of Tennessee Univ. of Texas	R. W. Morton W. R. Woolrich	mech.	3,500	130,050
53	1932-68	Boiler Feedwater Studies	Betz Laboratories, Inc.	L. D. Betz	mech.	6,475	65,880

TABLE 2 Regular Projects (continued)

File No.	*Project Period*	*Project Title*	*Research Agency*	*Proponent or Principal Investigator(s)*	*Field*	*Eng. F. Grant $*	*Other Cntrbs $*
54	1932-33	Courses for Unemployed Engineers	Engineering Societies N.Y. City Colleges	P. H. Littlefield	intrd.	0	0
55	1933	Elevator Safety	ASME	M. H. Chrisopherson	mech.	0	0
56	1933-63	Engineers' Council for Professional Development (ECPD)	Engineering Societies Council of State Boards of Examiners	C. F. Scott	intrd.	118,000	310,875
57	1934	Mental Attitude of Unemployed Engineers	Personnel Research Federation, Columbia U.	O. M. Hall	intrd.	0	0
58	1934-35	Cutting of Metals	ASME Committee	M. Marcellotti	mech.	3,750	1,400
59	1934-39	Mechanical Springs	ASME Committee	D. J. McAdam	mech.	1,500	11,670
60	1934	Riveted Joints	Research Committee		mech.	1,399	1,400
61	1934-37	Creep and Relaxation	University of Pittsburgh Westinghouse	A. Nadai	mech.	2,039	1,075
62	1935-65	Welding Research Council	Welding Research Comm. Am. Welding Society	C. A. Adams, Wm. Spraragen K. Koopman	intrd.	52,700	1,379,600
63	1935-52	Handbook on Cutting of Metals	ASME Committee	M. Marcellotti	mech.	11,250	0
64	1934	Directory of Organizations in Engineering Profession	American Engineering Council, ECPD	F. M. Feiker	intrd.	0	0
65	1935-36	Forms of Employer-Employee Cooperation	Personnel Research Federation	C. S. Slocombe	intrd.	1,400	0
66	1936-43	Impregnated Paper Insulation	Johns Hopkins University	J. B. Whitehead	elect.	9,000	15,925
67	1936-48	Hydraulic Research	Univ. of Iowa, Univ. of Minnesota, et al	L. G. Straub, F. T. Mavis and others	civil	14,300	0
68	1936-58	Plastic Flow of Metals	Westinghouse, MIT, Case Inst. of Techn., et al	A. Nadai, C. W. MacGregor G. Sachs, W. M. Baldwin	mech.	25,100	34,300
69	1936-37	Electric Shock	Johns Hopkins University	W. B. Kouwenhoven	elect.	0	0
70	1936-37	Personnel Problems and Engineering Advances	Personnel Research Federation	C. S. Slocombe	intrd.	1,200	0
71*	1937-38	Methods of Negotiating with Employees	Personnel Research Federation	C. S. Slocombe	intrd.	1,200	0

TABLE 2 Regular Projects (continued)

File No.	*Project Period*	*Project Title*	*Research Agency*	*Proponent or Principal Investigator(s)*	*Field*	*Eng. F. Grant $*	*Other Cntrbs $*
72	1938	Wind Velocity and Direction	New York University	A. F. Spilhouse	civil	500	0
73	1939	Tension Tests of Large Riveted Joints	University of California Berkeley	R. E. Davis, G. B. Woodruff H. E. Davis	civil	1,000	0
74	1940-43	Insulating Oils and Cable Saturants	MIT	J. C. Balsbaugh	elect.	6,274	43,670
75	1940-43	Engineering Index	EF, Founder Societies	H. E. Hannum	intrd.	3,000	3,925
76	1942	Calvin W. Rice Lectureship	ASME		mech.	300	
77*	1940-42	Internal Friction and Creep	The State College of Washington	C. Zener, H. H. Lester T. A. Read	intrd.	1,475	6,000
78	1942-51	Basic Open Hearth Steel Making	AIME Advisory Committee	J. Chipman, B. M. Larsen T. S. Washburn, F. T. Sisco	mining	3,000	2,000
79*	1942	Mechanics of Underground Forces	Colorado School of Mines	G. Sherman	civil	580	0
80	1942-55	Unsteady Heat Flow Investigation by Electrical Analogy	Columbia University	C. F. Kayan V. Paschkis	mech.	17,000	117,000
81	1942	Dynamic Characteristics of Materials	Pennsylvania State College	B. J. Lazan	intrd.	1,000	0
82	1942-65	Furnace Performance Factors	ASME Research Committee	A. R. Munford, G. W. Bice W. E. Somers	mech.	2,000	16,000
83	1943-45	Office of Scientific Personnel	National Academy of Sciences	H. A. Barton, H. L. Dodge M. H. Trytten	intrd.	6,500	19,000
84	1944-62	Column Research Council	Fritz Engineering and other structural labs	J. Jones, S. Hardesty B. G. Johnston, L. S. Beedle	civil	28,000	310,000
85	1944-47	Strength of Metals (Mechanical Springs)	NBS	D. J. Adams R. W. Clyne	mech.	500	2,500
87*	1944-49	Fluid Dynamic Problems in the Design and Operation of Turbines and Compressors	Purdue University	J. R. Weske	mech.	150	0
89	1946-48	Rolling Friction	MIT	M. D. Hersey	mech.	2,000	0

TABLE 2 Regular Projects (continued)

File No.	*Project Period*	*Project Title*	*Research Agency*	*Proponent or Principal Investigator(s)*	*Field*	*Eng. F. Grant $*	*Other Cntrbs $*
90	1946-59	Research Council on Riveted and Bolted Structural Joints	University of Illinois Northwestern U., et al	W. M. Wilson, C. A. Ellis G. A. Maney, T. R. Higgins	civil	24,000	433,000
91	1946-56	Properties of Gases and Gas Mixtures	MIT	F. G. Keyes	mech.	12,000	26,150
92	1946-47	Technical Services Investigation in Germany	American University U.S. Depart. of Commerce	M. S. Viteles	intrd.	500	650
93	1948	Air Survey of Soil Patterns	Cornell University	W. L. Malcolm, D. J. Belcher	civil	550	0
94	1948-54	Applied Mechanics Reviews	Illinois Institute of Technology	G. B. Pegram, M. Goland R. E. Peterson	mech.	20,000	101,700
95	1948-49	Turbulent Motion in Water	MIT	A. T. Ippen	civil	5,000	5,000
96	1947-49	True Tensile Properties of Metal in Hot Working Range	MIT	J. Chipman E. L. Batholomew, Jr.	mech.	5,500	1,500
97	1947-59	Diffusion in Steel	MIT	M. Cohen	met.	27,000	273,800
98	1948-51	Properties of Alloyed Ferrites at Low Temperatures	Univ. of Pennsylvania	R. M. Brick	met.	8,500	76,700
99	1949-59	Reinforced Concrete Research Council	University of Illinois Lehigh Univ. and others	A. J. Boase, J. M. Garrets R. F. Blanks, F. E. Richart	civil	35,000	496,200
100	1948-49	Survey of Engineering Societies Library	Engineering Societies Library (ESL) Board	J. S. Thompson R. H. Phelps, R. Wood	intrd.	5,000	0
101	1949-59	Causes and Methods of Prevention of Corrosion of Water Pipes	MIT	R. Eliassen	intrd.	12,500	53,600
102	1948-56	Building Research Advisory Board	NRC, Purdue University	F. L. Hovde, W. H. Scheick	intrd.	4,500	230,600
103**	1949-51	Simplification of Scientific Literature Searches	AIEE		elect.	0	0
104	1949-62	Comminution	Stanford University MIT	A. K. Schellinger, G. E. Agar R. J. Charles, J. H. Brown	mining	21,000	115,100
105	1950-62	Council on Wave Research	University of California	M. P. O'Brien, J. W. Johnson	civil	11,500	24,700

TABLE 2 Regular Projects (continued)

File No.	*Project Period*	*Project Title*	*Research Agency*	*Proponent or Principal Investigator(s)*	*Field*	*Eng. F. Grant $*	*Other Cntrbs $*
106	1952-55	Evaluation of Engineering Education	Am. Soc. for Engineering Education Committee on Eval. of Eng. Education	T. Saville, L. E. Grinter S. C. Hollister	intrd.	9,000	23,000
107	1953-59	Mechanical Pressure Elements	ASME Research Committee on Mech. Pressure Elem.	P. G. Exline	mech.	5,000	0
108	1953-59	Heat Conduction Charts	Columbia University	T. H. Chilton, V. Paschkis	mech.	2,000	0
109	1954-59	Wood Pole Research	U.S. Forest Products Lab.	L. J. Markwardt	elect.	12,000	256,100
110	1954-62	Corrosion Research Council	NBS, Univ. of PA, et al	W. J. Sweeney, E. H. Robie	met.	16,200	213,000
111	1954–56	High-Voltage High-Frequency Dielectric Phenomena	Stanford University	W. J. Hoover	elect.	3,000	0
112	1956-61	Flow of Bulk Solids	University of Utah	A. W. Jenike	mining	10,000	50,600
113	1955-65	Properties of Steam	Drexel Inst. of Techn.	D. W. R. Morgan	mech.	10,500	297,400
114	1954-68	Corrosion and Deposits from Combustion Gases	Battelle Memorial Inst. Babcock & Wilcox Co.	P. R. Loughin	mech.	8,000	23,200
115	1955-56	Metals Branch	AIME	R. W. Shearman	met.	5,000	46,800
116	1954-61	Thermal Resistivity Characteristics of Soils	Princeton University Con Edison Co. of N.Y.	H. F. Winterkorn R. W. Burrell	civil	8,000	53,500
117	1954-56	Forces Exerted by Oscillatory Fluid Motion on Cylinders	Univ. of California at Berkeley	A. D. K. Laird	mining	3,000	0
118	1955-58	Storm Surges	Lamont Geological Observatory	W. L. Donn	civil	5,000	0
119	1955-57	Surface Diffusion of Metals	Carnegie Inst. of Techn.	M. Simnad, G. M. Pound	met.	6,000	2,000
120	1955-61	Techniques for Forecasting Waves	Texas A&M Research Foundation	A. M. Kahan	civil	5,000	15,000
121	1955-61	Evaluation of Alloy Steels for Fatigue Resistance in Structures	University of Illinois	N. M. Newmark W. H. Munse	civil	6,000	22,250
122	1956-59	Steel Structures Painting Council	Mellon Institute	J. Bigos, J. D. Keane	civil	3,000	88,000
123	1955-59	Thickening and Thickeners	MIT	A. M. Gaudin	mining	7,500	34,300

TABLE 2 Regular Projects (continued)

File No.	Project Period	Project Title	Research Agency	Proponent or Principal Investigator(s)	Field	Eng. F. Grant $	Other Cntrbs $
124	1956-57	Heat Transfer and Pressure Drop in Fluids	Univ. of Oklahoma	O. K. Crosser	mech.	3,000	9,000
125	1956-58	Effect of Strain Waves in Blasting	Colorado School of Mines	A. W. Ruff	civil	1,500	6,000
127*	1957-58	Mining Machine Bits	University of Illinois	F. D. Wright	mining	0	0
128	1957-64	Flow of Bulk Materials	ASME Materials Handling Div. Research Committee	R. M. LaForge	mech.	3,000	0
129	1955-63	Prevention of Fracture in Metals	University of Wisconsin	J. D. Lubahn	mech.	2,000	0
131	1957-63	Nucleation and Evaporation of Metals	Carnegie Inst. of Techn.	G. M. Pound	met.	2,500	3,300
132	1955-61	Calculations in Structural Analysis	Univ. of Pennsylvania	S. Shore	civil	4,000	6,300
134	1955-61	Random Vibration	ASME Research Committee on Random Vibrations	C. E. Crede	mech.	2,000	0
135	1955-62	Documentation Research in Engineering	Western Reserve Univ.	A. Kent	intrd.	1,000	++
138	1955-63	Operating Variables in Ore Grinding	Columbia University	N. Arbiter	mining	300	+
142	1960-62	Behavior of Pressure Vessel Materials	Naval Research Laboratory	M. R. Achter	mech.	300	+
143	1960-65	Contact Fatigue of Rolling Elements	SKF Industries, Inc.	T. W. Morrison	mech.	600	+
144	1960-64	Effect of Radiation on Materials for Nuclear Installations	University of Minnesota Union Carbide Nuclear Company	R. Plunket E. C. Miller	mech.	200	+
RC-A-62-6	1959-65	Token Grants			intrd.	10,200	0
PR-60-1	1960-62	Hall of Engineering Progress	EF Public Relations Com.	J. H. Hitchcock	intrd.	1,800	0
PR-60-2	1961-63	Junior Engineering Technical Society (JETS)	JETS	R. T. Fallon	intrd.	30,000	++

TABLE 2 Regular Projects (continued)

File No.	*Project Period*	*Project Title*	*Research Agency*	*Proponent or Principal Investigator(s)*	*Field*	*Eng. F. Grant $*	*Other Cntrbs $*
none	1961-65	National Academy of Engineering Studies	Nat. Academy of Sciences Engineers Joint Council	E. A. Walker, A. B. Kinzel J. A. Stratton, H. K. Work	intrd.	25,276	
RC-A-60-6	1961	Symposium on Waterwaves	Council on Wave Research	J. W. Johnson	civil	1,000	
RC-D-60-5	1961	Assembly on Welding	American Welding Society	F. L. Plummer	intrd.	1,000	
RC-A-60-1a	1961-62	A Symposium to Advance Management and Conservation of the Air Resource	Coordinating Committee on Air Pollution	W. T. Ingram	intrd.	1,663	
RC-A-60-1b	1962-67	Study of Measurements of Stack Emissions	Coordinating Committee on Air Pollution	W. T. Ingram	intrd.	6,270	
RC-A-60-1c	1962-67	Evaluation of Carbon Dioxide Concentration as an Indicator of Air Pollution in Urban Atmosphere	University of California at Los Angeles	A. F. Bush, H. B. Nottage C. R. Scherer	intrd.	48,910	
RC-A-60-1d	1962-67	Research Council on Air Resources Engineering	Coordinating Committee on Air Pollution	W. T. Ingram	intrd.	2,000	
RC-A-61-5	1960-62	Research Stimulation and Coordination	ASCE	W. H. Wisely	civil	10,000	
none	1962-66	Reprints of M. J. Hvorslev Report	EF		civil	7,094	
RC-A-62-3	1962-63	Printing Report "The Nation's Engineering Research Needs 1965-1985"	Research Committee of the Engineers Joint Council (EJC)	J. H. Hollomon	intrd.	3,208	
PR-62-1	1963-67	"Engineer"	EJC	L. K. Wheelock	intrd.	3,473	
RC-A-63-2	1964	Analysis of Future Role of Engineering Index	Battelle Memorial Institute	F. R. Whaley, D. M. Liston	intrd.	20,000	
none	1964	Computerized Abstracting and Indexing Program in Plastics, and Electrical and Electronics Engineering	Engineering Index, Inc.	C. M. Flanagan	intrd.	75,000	
RC-A-63-3	1964-67	Engineering Information Services Program	EJC	L. K. Wheelock	intrd.	80,036	0

TABLE 2 Regular Projects (continued)

File No.	*Project Period*	*Project Title*	*Research Agency*	*Proponent or Principal Investigator(s)*	*Field*	*Eng. F. Grant $*	*Other Cntrbs $*
RC-A-63-7	1964	Corosion Research Council		F. T. Sisco, E. H. Robie	intrd.	1,000	+ +
none	1964	Coordinating Committee on Engineering Information	EJC, ESL, and EF	R. H. Phelps	intrd.	10,000	
IR-62-1	1964-65	International Program	EJC	C. E. Davies	intrd.	30,000	
IR-63-1	1964-65	Maintenance of N.Y. Office	Tools for Freedom	R. Morrow	intrd.	3,000	0
IR-63-2	1964-66	Maintenance of N.Y. Office	Volunteers for Internat. Technical Assistance	B. P. Coe	intrd.	3,000	0
ED-63-1	1964-67	Continuing Engineering Education	Joint Committee of ECPD, EJC, ASEE, and NSPE	E. Weber	intrd.	5,000	
none	1965	Handbook on Composite Materials and Structures	Joint Advisory Committee on Continuing Engineering Studies	B. R. Norton	intrd.	280	0
none	1965	R&D Work in Production Operations	Engineering Index, Inc.		intrd.	7,000	0
RC-A-64-2	1965	Metal Properties Council	ASTM & ASME Joint Comm. on Effect of Temperature on Properties of Metals	G. V. Smith	intrd.	25,000	0
RC-A-64-3	1965	Urban Transportation Research	Northwestern University	J. Snell, P. Shuldiner	civil	30,000	0
RC-A-64-4	1965	Fastener Research Council	Battelle Mem. I., et al	S. W. Marras, R. H. Ramsey	intrd.	3,300	0
RC-A-64-6	1965	Token Grants	ASCE	L. R. Shaffer, D. E. Jones C. P. Siess, J. B. Spangler	intrd.	400	0
RC-A-64-7	1965-66	Fundamental Research on Comminution as a Unit	University of California at Berkeley	D. W. Fuerstenau	intrd.	36,000	0
RC-A-64-8	1965	Development of Modern Means of Information Handling	Engineering Societies Library	R. H. Phelps	intrd.	20,250	
RC-A-63-6	1966	Proceedings of 7th and 8th Conference	Council on Wave Research	J. W. Johnson	civil	0	0
RC-A-65-1	1966	A System Design Procedure for the Planning of Construction Operations	University of Illinois	L. R. Shaffer	civil	21,000	50,000

TABLE 2 Regular Projects (continued)

File No.	*Project Period*	*Project Title*	*Research Agency*	*Proponent or Principal Investigator(s)*	*Field*	*Eng. F. Grant $*	*Other Cntrbs $*
RC-A-65-2	1966-67	Load Testing of World's Fair Structures	Building Research Advisory Board	J. R. Janney, J. W. Wiss R. C. Elstner	intrd.	15,000	231,000
RC-A-65-3	1966	Market Survey	Tripartite Committee	J. S. Sayer	intrd.	40,000	
RC-A-65-4	1967	Further Expenses of Operations	Tripartite Committee	W. E. Lobo	intrd.	6,192	
none	1967	Engineering Information and Data System	Tripartite Committee	W. E. Lobo	intrd.	35,000	
RC-A-66-2*	1967	Prestressed Concrete Piling Technology—Monograph	South Dakota School of Mines & Technology	Shu-t'ien Li	civil	0	
P-65-2	1967	Proceedings of Symposium on Technical and Social Problems of Air Pollution	Metropolitan Engineers Council on Air Resources (MECAR)	W. T. Ingram	intrd.	1,500	
RC-A-67-3	1967-68	Corrosion and Deposits from Combustion Gases	Battelle Memorial Inst.	W. T. Read	mech.	5,000	
RC-A-67-4a	1967-68	Management Control and Simulation Systems Oriented Toward Actual Field Implementation	Stanford University	J. S. Arrington	civil	5,000	
RC-A-67-4b	1967-68	Mathematical Modeling of an Aggregate Processing Plant	Purdue University	J. A. Havers	civil	5,000	
RC-A-67-5	1967-68	White Plains Full Scale Testing—Feasibility Study	ASCE Comm. on Performance of Full Scale Structures	N. FitzSimons	civil	10,000	
RC-A-68-1	1967-68	Coefficients for Concentric Orifice Plates and Flow Nozzles	University of Rhode Island	R. S. Dowdell Yu-Lin Chen	mech.	10,000	
RC-A-68-2	1967-68	Experimental Investigation of Factors Influencing the Accuracy of Steam Sampling	ASME Research Committee on Boiler Feedwater Studies	J. K. Rice	mech.	5,000	
RC-A-68-3	1968-69	Water Requirements for Buildings	Committee for Water Requirements for Buildings	L. Guss	mech.	5,000	

TABLE 2 Regular Projects (continued)

File No.	*Project Period*	*Project Title*	*Research Agency*	*Proponent or Principal Investigator(s)*	*Field*	*Eng. F. Grant $*	*Other Cntrbs $*
RC-A-68-4	1968-70	Defining Policy Development for Access to Airports	ASCE	R. F. Baker R. M. Wilmotte	civil	20,000	
RC-A-69-1	1969-70	Theoretical Research Related to Fatigue in Fiber-Reinforced Composite Materials	Mississippi State University	E. Anderle, T. Avirah M. McNeil, R. Ravella	met.	14,400	
RC-A-69-2	1969-71	Fundamental Combustion Study for Disposal of Industrial and Municipal Wastes by Incineration	NBS, E. I. du Pont de Nemours and Co., Inc.	E. S. Domalski T. R. Keane	civil	47,300	
RC-A-69-3	1969-70	Updating Transportation Planning Surveys Research	ASCE Urban Transport. Research Council	H. Deutschman	civil	5,500	
RC-A-69-4	1969	Equipment Selection and Evaluation of Vertical Transportation Systems	Auburn University	R. Rainer	civil	10,000	
RC-A-70-1	1970-73	Development of a Formal Approach to Medical Applications	George Washington University	C. A. Caceres G. Devey, J. C. Aller	intrd.	67,500	
RC-A-70-2	1970-71	Study of Non-Ferrous Reaction Rates and Mechanisms with the Hot Filament Microscope	Carnegie-Mellon University	G. Derge	chem.	16,100	8,000
RC-A-70-3	1970-72	Construction of Low Energy Ion-Reflection Spectrometer	McMaster University	H. D. Barber, S. S. Johar D. A. Thompson	intrd.	9,000	31,000
RC-A-70-4	1970-72	Economics of High Rise Apartment Buildings	ASCE	R. D. Steyert	civil	22,850	
RC-A-71-1	1971	Identification and Classification of Expansive Earth Materials	Univ. of California at Davis	K. Arulanandan J. K. Mitchell	civil	5,650	
RC-A-71-2	1971-73	Measurement of Boiling Heat Transfer with Pyroelectric Thermometer	McGill University	S. B. Lang, M. E. Weber	chem.	9,860	

TABLE 2 Regular Projects (continued)

File No.	*Project Period*	*Project Title*	*Research Agency*	*Proponent or Principal Investigator(s)*	*Field*	*Eng. F. Grant $*	*Other Cntrbs $*
RC-A-71-3	1970-72	ASCE Underground Construction Research Program	Underground Construction Research Council	R. F. Baker	civil	10,000	60,650
RC-A-71-4	1971-73	Case Study to Measure the Impact of CE Projects on People and Nature	ASCE	H. G. Poertner	civil	10,000	50,000
RC-A-71-5	1971	Rapid Determination of Quality of Rock Used in Water Structures	Purdue University	U. Saltzman, W. R. Judd	civil	6,692	5,774
RC-A-71-6	1971	Selective Nitride Formation in Molten Iron Alloys—A Study of the Kinnetics of Coupled Reactions	Carnegie-Mellon University	C. H. P. Lupis, C. Aguirre W. O. Philbrook	met.	20,645	
RC-A-72-2	1973	Reestablishment of Communications Exchange with Mainland China	AIChE	H. P. Sheng	chem.	1,500	
RC-A-73-1	1973	Design Institute for Multiphase Processing	AIChE	J. J. Rushton H. S. Kemp	chem.	20,000	
RC-A-73-2	1973-75	Assessment of Environment Standards	Dartmouth College	A. O. Converse S. R. Stearns	civil	70,800	
RC-A-73-3	1973-76	Acquaculture in Waste Hot Water	Franklin Institute	H. J. Bowen	intrd.	20,000	100,000
RC-A-73-4	1973	Environmental Impact Analysis of Geothermal Power Installations	Princeton University	R. C. Axtmann	civil	3,500	
RC-A-73-5	1973-74	Assymetric Hollow Fiber Membranes for Enzyme Immobilization	Stanford University	C. R. Robertson	chem.	26,000	
RC-A-73-6	1973	Engineering Design of Prosthetic Heart Valves	Union College	J. R. Shanebrook	intrd.	26,000	

TABLE 2 Regular Projects (continued)

File No.	*Project Period*	*Project Title*	*Research Agency*	*Proponent or Principal Investigator(s)*	*Field*	*Eng. F. Grant $*	*Other Cntrbs $*
RC-A-73-7	1973-74	Mechanics of Surface to Air Transport of Bacteria and Other Pollutants by Bursting Bubbles	University of Kentucky	H. C. Liu, J. I. Chen J. H. Lienhard	intrd.	19,500	
RC-A-73-8	1973-74	Charging Airborne Particulate Matter in High Concentrations Using Radiation and Electric and Magnetic Fields	Pennsylvania State University	R. J. Heinsohn S. H. Levine, R. J. Fjeld G. W. Malamud	mech.	16,000	
RC-A-73-9	1973	Counselling Program for Minorities and Women to Seek a Career in Engineering	ASME	J. A. Mason	intrd.	10,000	
RC-A-74-1	1974	A Study of New Methods for Measuring Cerebral Circulation	NYU Medical Center	T. Reich	intrd.	25,000	
RC-A-74-3	1974-75	Use of Fluidized Bed Electrodes in Extractive Metallurgy	Univ. of California at Berkeley	J. W. Evans	met.	27,659	
RC-A-74-4	1975-76	The Utilization of Solar Energy for Hydrogene Production by Photosynthetic Organisms	University of Miami	A. Mitsui	intrd.	20,000	
RC-A-74-5	1975-76	Energy Separation in an Expanding Binary Mixture and its Application to Energy Utilization	Univ. of California at Los Angeles	W. S. Young	elect.	20,000	
RC-A-74-6	1975-77	Tests and Analysis of Composite Action in Glulam Bridge Systems	Colorado State University	R. M. Gutkowski	civil	25,000	0
RC-A-74-7	1975-77	The Feeding of Pulverized Coal to High Pressure Gasifiers	West Virginia University	R. J. Krane	mech.	20,000	

TABLE 2 Regular Projects (continued)

File No.	*Project Period*	*Project Title*	*Research Agency*	*Proponent or Principal Investigator(s)*	*Field*	*Eng. F. Grant $*	*Other Cntrbs $*
RC-A-75-1	1975-77	Utilization of Sewage Solids to Produce Soluble Organic Carbon	University of Washington	J. F. Furguson	civil	16,500	
RC-A-75-2	1975	Production of Slide Show "The History of Engineering in America"	EJC	P. Rae	intrd.	5,000	
RC-A-75-3	1976-78	Support of Young Faculty Members in the Areas of Founder Societies	see Table 4	D. Athanasiou-Grivas J. J. K. Daemen, D. C. Jeutter N. Dagalakis, A. Atar	intrd.	50,000	0
RC-A-75-4	1976-78	Turbulence Measurement in Lakes	Cornell University	P. J. Murphy	civil	20,000	
RC-A-75-5	1976-77	Emergency Water Supplies for Groundwater in Humid Regions	Polytechnic Institute of New York	A. S. Goodman	civil	15,000	
RC-A-76-1	1976-82	Conservation of Urban Water Resources	University of Colorado	J. E. Flack	intrd.	5,000	0
RC-A-76-2	1976-79	Support of Young Faculty Members in the Areas of Interest Represented by the Founder Societies	see Table 4	H. T. Shen, L. H. Burck C. K. Law, B. U. Keane R. L. Shambaugh	intrd.	50,000	0
RC-A-76-3	1977-78	The Stirling Cycle Heat Pump	Rensselaer Polytechnic Institute	R. P. Scringe	mech.	10,000	8,500
RC-A-76-4	1977-78	Acid Precipitation: Recovery of Acid Lakes	Clarkson College	J. K. Edzwald J. V. DePinto	civil	10,000	
RC-A-76-5*	1977-78	High Pressure Relief Systems for Flammable Liquids	AIChE	H. S. Kemp. I. Swift	chem.	40,000	
RC-A-76-6	1977-78	Non-Invasive Aneurysm Detection and Analysis	University of Cincinnati	F. J. Taylor	elect.	20,000	
RC-A-77-1	1977-80	Support of Young Faculty Members in the Areas of Interest Represented by the Founder Societies	see Table 4	T. E. Higgins, R. L. Dooley G. A. Pope, R. J. Hannemann J. M. M. Rendu, R. L. Levin C. R. Johnson, J. L. LoCicero D. T. Hayhurst, A. S. Mayerson	intrd.	100,000	0

TABLE 2 Regular Projects (continued)

File No.	*Project Period*	*Project Title*	*Research Agency*	*Proponent or Principal Investigator(s)*	*Field*	*Eng. F. Grant $*	*Other Cntrbs $*
RC-A-77-2	1978-82	Photocatalyzed Degradation of Pesticides	Princeton University	F. Ollis	chem.	45,400	
RC-A-77-3	1978-81	Cancer Therapy with Local Microwave-Induced Hyperthermia	Dartmouth College	J. W. Strobelm E. B. Douple	intrd.	50,000	+ +
RC-A-77-5a	1978-81	Fellowship: Boundary Integral Equation Method Applied to Groundwater and Flow Through Porous Media	Cornell University	P. L-F. Liu, J. A. Liggett	civil	5,500	0
RC-A-77-5b	1978-81	Fellowship: Inclusion Removal Techniques for Liquid-Metal Systems	Drexel University	D. Apelian	met.	5,500	0
RC-A-77-5d	1978	Fellowship: Solar Cells of Sn02 on Si	Carnegie-Mellon University	G. N. Advani	elect.	5,500	0
RC-A-77-5e*	1978	Fellowship Transport and Interactions of Hydrogenic Gases in Metals	Princeton University	R. C. Axtmann	chem.	0	0
RC-A-77-6	1978-81	Support of Young Faculty Members in the Areas of Interest Represented by the Founder Societies	see Table 4	A. C. Singhai, C. L. Jensen L. F. Geschwindner, Jr. R. H. Yoon, W. E. King, Jr. M. Hull, F. D. Flack J. L. Mertz, P. I. Cohen D. R. Lloyd	intrd.	110,000	0
RC-A-78-1	1979-81	Fireside Corrosion of Fluidized Bed Heat Exchangers	Rensselaer Polytechnic Institute	S. R. Shatynski	intrd.	42,600	0
RC-A-78-3a	1979-84	Fellowship: Theories and Applications of Systems Analysis Techniques to the Optimal Management and Operation of a Reservoir System	Univ. of California at Los Angeles	W. W-G. Yeh	intrd.	6,000	0
RC-A-78-3b*	1979-84	Fellowship: Hot Corrosion of Metal Alloys	Univ. of Pittsburgh	F. S. Petit	met.	0	0

TABLE 2 Regular Projects (continued)

File No.	*Project Period*	*Project Title*	*Research Agency*	*Proponent or Principal Investigator(s)*	*Field*	*Eng. F. Grant $*	*Other Cntrbs $*
RC-A-78-3c	1979-81	Fellowship: Probabilistic Design of Pressure Vessels	Hartford Graduate Center	H. Kraus	mech.	6,000	0
RC-A-78-3e1	1979-83	Fellowship: Rheology and Microstructure of Multiphase Materials	Purdue University	J. M. Caruthers E. I. Franses	chem.	6,000	0
RC-A-78-3e	1979-82	Fellowship: Mixing in Continuous Flow Systems—Theory and Application of Residence Time Distributions	Rensselaer Polytechnic Institute	E. B. Nauman B. A. Buffham	chem.	6,000	0
RC-A-78-6	1979-83	Support of Young Faculty Members in the Areas of Interest Represented by the Founder Societies	see Table 4	D. Y. Tan, T. F. Anderson P. R. Taylor, R. J. Livak J. H-W. Lee, M. Donath R. Nagarajan, S. K. Case R. L. Magin, J. Mijovic J. G. Pinto	intrd.	132,000	0
RC-A-79-1	1980-84	Separation of Closely Boiling Mixture by Reactive Distillation	Univ. of Massachusetts	M. F. Doherty	chem.	31,400	
RC-A-79-1A to 1E	1980-84	Support of Young Faculty Members in the Areas of Interest Represented by the Founder Societies	see Table 4	W. C. McCarthy, H. P. Jorgl M. Y. Corapcioglu, T. D. Roy G. W. Jaworski, H. E. Hager M. A. Mahtab, R. M. German W. J. W. Wepfer, J. J. Casey J. Cheung, M. El-Sharkawi C. Houstis, G. F. Fuller S. Kovenklioglu	intrd.	210,000	0
RC-A-82-1	1982	Urban Water Conservation	ASCE Publication	S. Menaker	civil	3,500	
RC-A-82-2	1981-	Subsurface Exploration and Sampling of Soils for Civil Engineering Purposes	Illinois Institute of Technology	S. K. Saxena	civil	83,000	0

TABLE 2 Regular Projects (continued)

File No.	*Project Period*	*Project Title*	*Research Agency*	*Proponent or Principal Investigator(s)*	*Field*	*Eng. F. Grant $*	*Other Cntrbs $*
RC-A-82-3	1983-87	Kinetics and Extraction Equilibrium Models for Nickel (II)-2-Hydroxy 5-Nonylbenzo-phenon Oxime Complexation System	Syracuse University	L. L. Tavlirides	chem.	25,000	0
RC-A-83-1	1982-84	Computer Aided Design for Economic Manufacture	Univ. of Massachusetts	C. Poli, W. Knight	mech.	20,000	+ +
RC-T-86-1	1986-87	National Congress on Engineering Education	Accreditation Board for Engineering and Technology	D. R. Reyes-Guerra	intrd.	10,000	+ +
RC-T-87-1	1987	Instrumentation of Columns at the Norwest Bank Project at Minneapolis	University of Minnesota	R. T. Leon	civil	12,500	12,500
Z.20	1988-	Engineering Foundation History	Joint Venture	L. E. Metz, I. M. Viest	intrd.	25,000	0
RC-T-87-3*	1988	Organization and Formation of UDCDA	Ultra-Deep Coring and Drilling Association	J. C. Rowley W. J. Winters	intrd.	0	0

* Project discontinued or abandoned or the results are unknown
** No funds requested
\+ No information available on contribution from other sources
\+ + Substantial funding received from other sources

TABLE 3 Engineering Foundation Conferences

File Number	*Meeting Date*	*Title*	*Location*	*Chairman*	*Chairman's Affiliation*	*Att.*
1962	7/23-27/62	Engineering Research for the Developing Countries	Tilton School Tilton, NH	T. O. Paine	General Electric Co.	87
1962	8/1-3/62	Composite Materials	ditto	W. J. Harris, Jr.	Battelle Memorial Inst.	52
1963	8/5-9/63	Technology and the Civilian Economy	Proctor Academy Andover, NH	C. N. Kimball	Midwest Research Institute	44
1963	8/12-16/63	Comminution	ditto	D. W. Fuerstenau	U. of Calif.—Berkley	62
1963	8/19-23/63	Engineering in Medicine	ditto	H. H. Zinsser & R. Contini	Columbia University New York University	61
1963	8/26-30/63	Urban Transportation Research	ditto	J. A. Logan	Rose Polytech. Inst.	71
64(x007)01	7/27-31/64	Technology and the Civilian Economy II	ditto	C. N. Kimball	Midwest Research Institute	62
64(x008)02	8/3-7/64	Engineering in Medicine II	ditto	R. Contini	New York University	58
64(x009)03	8/9-14/64	Technological Challenges for the U.S. in World Markets, 1964–1974	ditto	H. G. Busignies	International Telephone & Telegraph Co.	64
64(x010)04	8/17-21/64	The Building Construction System—A Challenge to Innovation	ditto	G. P. Fisher	Cornell University	84
65(x011)01	7/26-30/65	Technology and Its Social Consequences	ditto	C. N. Kimball	Midwest Research Institute	61
65(x012)02	8/3-7/65	Urban Hydrology Research	ditto	S. W. Jens & C. F. Izzard	Reitz & Jens U.S. Bur. of Public Roads	83
65(x013)03	8/16-20/65	Engineering Design to Fit the Environment	ditto	W. S. Pollard, Jr.	Harland Batholomew & Associates	63
66(x014)01	7/25-29/66	The Role of Technological Change	ditto	J. Alcott	Midwest Research Inst.	42
66(x015)02	8/1-5/66	Industry and the Young Engineer	ditto	A. C. Ingersoll	Univ. of Southern Calif.	60
66(x016)03	8/8-12/66	Team Approaches to Complex Problem Solving	ditto	W. P. Kimball	Dartmouth College	40
66(x017)04	8/22-26/66	Particulate Matter Systems—Their Simulation and Optimization	University School Milwaukee, WI	T. P. Meloy	Allis-Chalmers	102

TABLE 3 Engineering Foundation Conferences (continued)

File Number	*Meeting Date*	*Title*	*Location*	*Chairman*	*Chairman's Affiliation*	*Att.*
66(x018)05	8/22-26/66	Technology and the City Matrix	Univ. of Calif. Santa Barbara, CA	J. P. Eberhard & W. Hausz	Nat. Bureau of Standards General Electric Co.	67
67(019)01	7/10-14/67	Engineering in Medicine—Multiphasic Health Screening	University School Milwaukee, WI	G. B. Devey	National Science Foundation	87
67(x020)02	7/17-21/67	Construction Engineering and Management Research	ditto	B. P. Bellport	Bureau of Reclamation	86
67(x021)03	7/24-28/67	Solid Waste Research and Development	ditto	P. H. McGauhey	Univ. of California—Richmond	104
67(x022)04	7/31-8/4/67	Particulate Matter Systems—Fine and Ultra-Fine Particle Technology	ditto	F. F. Aplan	Union Carbide Corp.	108
67(x023)05	7/31-8/4/67	Engineering and Behavioral Sciences	Miramar Hotel Santa Barbara, CA	H. David	National Research Council	22
67(x024)06	8/7-11/67	Engineering of Unconventional Protein Production	Univ. of Calif. Santa Barbara, CA	H. Bieber	Esso Research and Engineering Co.	85
67(x025)07	8/7-11/67	Quality Control Practices in the 1970's	Proctor Academy Andover, NH	D. Shainin	Rath & Strong, Inc.	28
67(x026)08	8/14-17/67	Informational Needs for Urban Transportation	ditto	L. J. Cusick	U.S. Dept. of Housing and Urban Development	84
67(x027)09	8/21-25/67	Complex Problem Solving	ditto	G. Bugliarello	Carnegie Inst. of Techn.	38
67(x028)10	8/28-9/1/67	Mixing Research and Applications	ditto	V. W. Uhl	University of Virginia	66
68(x029)01	7/8-12/68	Diagnosis of Corrosion Problems	Wyland Academy Beaver Dam, WI	F. L. LaQue	International Nickel Co.	29
68(x030)02	7/15-19/68	Expansive Soils	ditto	D. E. Jones	Federal Housing Administration	50
68(x031)03	7/22-26/68	Solid Waste Research and Development II	ditto	P. H. McGaughy	University of California—Richmond	94
68(x032)04	7/22-26/68	Complex Problem Solving	University School Milwaukee, WI	J. M. Ide	National Science Foundation	22
68(x033)05	7/29 - 8/2/68	Access to Airports	ditto	R. F. Baker	Consultant	42

TABLE 3 Engineering Foundation Conferences (continued)

File Number	*Meeting Date*	*Title*	*Location*	*Chairman*	*Chairman's Affiliation*	*Att.*
68(x034)06	8/5-9/68	Particulate Matter Systems—Comminution	ditto	J. M. Karpinski	Bethlehem Steel Corp.	66
68(x035)07	8/12-16/68	Behavioral and Engineering Science Applied to Social Problems	ditto	D. B. Hertz & W. J. Harris, Jr.	McKinsey & Company Battelle Memorial Institute	20
68(x038)08	8/5-9/68	Engineering in Medicine—Physical Parameters in Multiphasic Screening	Proctor Academy Andover, NH	C. A. Caceres G. B. Dewey	U.S. Public Health Service National Academy of Engineering	148
68(x037)09	8/12-16/68	Urban Hydrology Research—Water and Metropolitan Man	ditto	S. W. Jens & D. E. Jones, Jr.	Reitz & Jens Federal Housing Admin.	84
68(x038)10	8/18-23/68	Performance of Full Scale Structures	ditto	M. N. Salgo	CBS, Inc.	52
68(x039)11	8/26-30/68	Man and His Environment—Air Pollution	ditto	A. F. Bush	Univ. of California—Los Angeles	22
68(x040)12	8/26-30/68	Quality Requirements from the Systems Viewpoint	Univ. of Calif. Santa Barbara, CA	J. de S. Coutinho	Grumman Aircraft Engineering Corp.	39
69-01	2/12-16/69	Construction Engineering and Management Research—A Profile of the Future Construction System	Asilomar Conference Grounds Pacific Grove, CA	J. C. Kellogg	Al Johnson Construction Co.	114
69-02	7/14-18/69	The Persistence of Food Habits: Problem in Combatting Malnutrition	University School Milwaukee, WI	M. J. Forman	U.S. Department of State	61
69-04	7/28 - 8/1/69	Quality Assurance	ditto	H. D. Greiner	RCA	46
69-05	8/4-8/69	Industrialization in Developing Nations	ditto	B. S. Old	Arthur D. Little, Inc.	32
69-06	8/11-15/69	Fracture Prevention	ditto	H. Liebowitz	George Washington Univ.	28
69-07	8/4-8/69	Technology Assessment	Proctor Academy Andover, NH	R. A. Carpenter	Library of Congress	85
69-08	8/11-15/69	Optimizing the Economics of Urban Land Development and Management	ditto	D. E. Jones, Jr. B. R. Hanke	Federal Housing Admin. Urban Development	20
69-09	8/18-22/69	Mixing Operations	ditto	J. E. Gray	E. I. duPont de N. & Co.	77

TABLE 3 Engineering Foundation Conferences (continued)

File Number	*Meeting Date*	*Title*	*Location*	*Chairman*	*Chairman's Affiliation*	*Att.*
69-10	8/25-29/69	The Engineering Approach to Corrosion	ditto	W. H. Burton	Allied Chemical Corp.	49
69-11	7/21-25/69	Rapid Excavation	Deerfield Academy Deerfield, MA	T. P. Meloy	Melpar, Inc.	74
69-12	7/28 - 8/1/69	Joint Utilization of Right-of-Way for Utilities and Municipal Services	ditto	T. W. Mermel	U.S. Department of Interior	102
69-13	8/4-8/69	Particulate Matter Systems	ditto	L. G. Austin	Pennsylvania State U.	83
69-14	8/11-15/69	Systems Analysis in the Urban Sector	ditto	G. P. Fisher	Cornell University	59
69-15	8/18-22/69	Engineering in Medicine—Multiphasic Health Screening	ditto	C. D. Flagle	Johns Hopkins Univ.	153
69-16	8/25-29/69	Technology and the City Matrix: Requirements	ditto	G. R. Hilst	Travelers Research Corp.	38
69-17	8/18-22/69	National Goals in Water Pollution Control	Univ. of Calif. Santa Barbara, CA	F. A. Butrico	Battelle Memorial Institute	58
70-01	1/25-30/70	Waste Water Engineering in the Food Industry	Asilomar Conf. G. Pacific Grove, CA	R. I. Lachman	Cornell University	50
70-02	2/22-27/70	Building Systems—The Building Process of the 70's	ditto	E. O. Pfrang	National Bureau of Standards	97
70-03	3/1-5/70	Rapid Construction of Concrete Dams	ditto	W. H. Price	American Cement Corp.	93
70-04	7/6-10/70	Engineering in Medicine—Biomedical Engineering Optimization in the Health Sciences	New England College Henniker, NH	O. H. Schmidtt	University of Minnesota	54
70-05	7/13-17/70	National Policy for Materials	ditto	F. P. Huddle	Library of Congress	95
70-06	7/20-24/70	Urban Systems Engineering	ditto	M. Wachs & J. Schopfer	University of Illinois at Chicago Circle	46
70-08	8/3-7/70	Engineering in Medicine—Bioceramics	ditto	C. W. Hall & S. F. Hulbert	Southwest Research Ins. Clemson University	75
70-11	7/13-17/70	Tunnel Supports in Rapid Excavation	Deerfield Academy Deerfield, MA	R. H. Cornish & M. M. Singh	IIT Research Institute ditto	42

TABLE 3 Engineering Foundation Conferences (continued)

File Number	*Meeting Date*	*Title*	*Location*	*Chairman*	*Chairman's Affiliation*	*Att.*
70-12	7/20-24/70	A New Look at Research on Research in Large Organizations	ditto	J. G. Welles	University of Denver	59
70-13	7/27-31/70	Urban Water Resources Management	ditto	K. Wright & D. W. Hill	Wright & McLaughlin Engrs. Colorado State Univ.	61
70-15	8/10-14/70	Particulate Matter Systems	ditto	D. W. Fuerstenau & T. P. Meloy	Univ. of California Meloy Laboratories Inc.	33
70-16	8/17-21/70	National Metric Study Conference	ditto	R. P. Trowbridge W. K. Burton J. V. Odom	General Motors Co. Ford Motor Co. Nat. Bureau of Stand.	99
70-17	8/24-28/70	Solid Waste Research & Development —Land Disposal of Solid Wastes	ditto	E. A. Glysson	University of Michigan	113
70-19	8/17-21/70	Introduction to Prosthetic and Sensory Devices	Proctor Academy Andover, NH	E. F. Murphy & A. B. Wilson, Jr.	Veterans Administration Nat. Research Council	48
70-20	8/24-28/70	Quantitative Decision Making for the Delivery of Health Care	ditto	A. R. Jacobs	Univ. of Rochester School of Medicine	60
1970	9/14-18/70	Engineering in Medicine—Automated Multiphasic Health Testing	Swiss Research Institute Davos, Switzerland	G. B. Devey & S. M. Perren	Nat. Science Foundation Swiss Research Inst.	159
71-01	2/13-18/71	Engineering Problem Areas Interfacing with Non-Newtonian Technology	Asilomar Conf. G. Pacific Grove, CA	J. G. Savins & G. Bugliarello	Mobil Res. & Dev. Corp. University of Illinois at Chicago Circle	58
71-02	2/21-26/71	Environmental Engineering in the Food Industry	ditto	R. Lachman	Cornell University	43
71-03	2/22-26/71	Stack Sampling and Analysis	ditto	W. T. Ingram & P. M. Giever	New York University Walden Research Labs	112
71-04	4/25-30/71	Control of Quality in Construction	ditto	E. A., Abdun-Nur	Consulting Engineer	88
71-05	7/12-16/71	Owner-Engineer-Contractor Relations in Tunneling	Deerfield Academy Deerfield, MA	M. M. Singh & V. L. Stevens	Illinois Inst. of Tech. Consulting Engineer	42
71-06	7/19-23/71	Clinical Engineering	ditto	C. A. Caceres & G. B. Devey	George Washington U. Nat. Science Foundation	41
71-08	8/2-6/71	Engineering and Social Costs in Environmental Control	ditto	S. V. Margolin	Arthur D. Little, Inc.	34

TABLE 3 Engineering Foundation Conferences (continued)

File Number	*Meeting Date*	*Title*	*Location*	*Chairman*	*Chairman's Affiliation*	*Att.*
71-10	8/16-20/71	Research on Coal Mines Safety and Survival	ditto	T. P. Meloy	Meloy Laboratories Inc.	65
71-11	8/23-27/71	Solid Waste Disposal—Incineration Research	ditto	R. B. Engdahl	Battelle Memorial Institute	71
71-13	7/12-16/71	Women in Engineering	New England Col. Henniker, NH	G. Bugliarello & O. Selembier	Univ. of Ill. Chic. Circle Soc. of Women Engineers	47
71-14	7/19-23/71	Quantitative Decision Making for Delivery of Ambulatory Care	ditto	A. Jacobs	Dartmouth Medical School	92
71-15	7/26-30/71	Engineering in Medicine—Automatic Cytology	ditto	K. Preston, Jr.	Perkin-Elmer Corp.	81
71-16	8/2-6/71	Engineering in Medicine—Biotelemetry	ditto	C. W. Garrett	National Academy of Engineering	32
71-17	8/9-13/71	Enzyme Engineering	ditto	L. B. Wingard, Jr.	SUNY at Buffalo	153
71-18	8/16-20/71	Engineering Utility Tunnels in Urban Areas	ditto	M. D. Calkins	City of Kansas City	79
71-19	8/23-27/71	Research to Reduce Cost of High Voltage Underground Transmission	ditto	L. H. Fink & T. W. Mermel	Philadelphia Elect. Co. U.S. Dept. of Interior	122
71-20	8/9-13/71	Mixing Research	Proctor Academy Andover, NH	J. Y. Oldshue	Mixing Equipment Co., Inc.	60
71-21	8/16-20/71	Future Power Systems—Research, Reliability and Regulation	ditto	E. Schutzman	National Science Foundation	69
71-22	8/23-27/71	Corrosion Engineering	ditto	W. K. Boyd	Battelle Memorial Inst.	30
71-23	8/30-9/3/71	Technology Assessment—Management, Manpower and Methodologies	ditto	B. Bartocha & J. Goldhar	Nat. Science Foundation Rensselaer Pol. Inst.	76
72-01	10/18-22/71	Air Passenger Handling	Asilomar Conf. G. Pacific Grove, CA	R. D. Worrall & R. Horonjeff	Peat, Marv., Mitvh. & Co. Univ. of Calif. Richmond	58
72-02	11/10-13/71	Urban Water Policy	Arlie House Airlie, VA	M. B. McPherson	American Society of Civil Engineers	81
72-04	2/12-17/72	Recycle Implementation	Asilomar Conf. G. Pacific Grove, CA	T. D. Bath & A. Darnay	Environmental Protection Agency	48
72-05	2/22-27/72	Pattern Information Processing	Airlie House Airlie, VA	K. Preston, Jr. K. S. Fu	Perkin-Elmer Corp.	55

TABLE 3 Engineering Foundation Conferences (continued)

File Number	*Meeting Date*	*Title*	*Location*	*Chairman*	*Chairman's Affiliation*	*Att.*
72-06	3/1-6/72	Packaging Materials and Solid Waste Management	Asilomar Conf. G. Pacific Grove, CA	J. Scher & J. W. Conrad	Turner Collie & Braden Apt, Brader, Conrad & Assoc.	38
72-07	3/3-8/72	Stack Sampling and Monitoring	ditto	W. T. Ingram	New York University	70
72-08	3/16-21/72	Professional Liability and the Public Interest	ditto	E. B. Waggoner	Woodward-Clyde Consultants	85
72-09	5/14-18/72	Economical Construction of Concrete Dams	ditto	J. W. Hilf	Bureau of Reclamation	90
72-11	7/23-28/72	Energy Technologies for the Future	Vermont Academy Saxtons River, VT	R. R. Baltzhiser	Office of Science and Technology	72
72-12	7/9-14/72	Public Transportation in Urban Areas	New England Col. Henniker, NH	K. W. Heathington & J. A. Scott	Purdue University Highway Research Board	71
72-13	7/16-21/72	Women in Engineering & Management	ditto	O. Selembier & A. C. Ingersoll	Soc. of Women Engineers New York University	62
72-14	7/23-28/72	Technology Assessment Methodologies	ditto	J. Goldhar & R. Kasper	Rensselaer Pol. Inst. George Washington Univ.	90
72-15	7/30-8/4/72	Resolving Some Selected Issues of National Materials Policy	ditto	F. P. Huddle	Library of Congress	76
72-16	8/6-11/72	Making Graduate Engineering More Relevant to Industrial Needs	ditto	M. C. Shaw	Carnegie-Mellon University	49
72-17	8/13-18/72	Engineers' Role in Today's Society	ditto	C. O. Harris	General Motors Inst.	28
72-18	8/20-25/72	Research Strategy for Electric Power Needs of the Future	ditto	E. W. Greenfield	Washington State University	116
72-19	7/2-7/72	Technological Future for Offshore Oil: Accident Prevention & Response	Berwick Academy South Berwick, ME	J. White & C. Patton	University of Oklahoma	76
72-20	7/9-14/72	Health Services in the Year 2000	ditto	R. Hsieh & F. Mark	U.S. Pub. Health Serv. Hosp. Health Services and Mental Health Admin.	36
72-21	7/16-21/72	Vulnerability of Urban Areas to Subversive Disruption	ditto	D. Rosenbaum	MITRE Corporation	52
72-22	7/23-28/72	Control Strategies for Power Systems	ditto	L. H. Fink & H. G. Kwatny	Philadelphia Elect. Co. Drexel University	47
72-23	7/30 - 8/4/72	Energy Research Priorities	ditto	D. D. Dunlop	Management Consultant	38

TABLE 3 Engineering Foundation Conferences (continued)

File Number	*Meeting Date*	*Title*	*Location*	*Chairman*	*Chairman's Affiliation*	*Att.*
72-24	8/6-11/72	Research on Coal Mine Safety and Survival	ditto	T. V. Falkie	Pennsylvania State University	93
72-25	8/13-18/72	Improving Indoor Air Quality	ditto	J. R. Swanton	Arthur D. Little, Inc.	57
72-26	7/30 - 8/4/72	Short Range Load Forecasting in Electric Power Systems	Vermont Academy Saxtons River, VT	C. F. Evert	University of Cincinnati	37
72-27	8/6-11/72	Engineering in Medicine—Automatic Cytology	ditto	K. Preston, Jr.	Perkin-Elmer Corp.	128
72-28	8/13-18/72	Evaluation of Sanitary Landfill Design and Operational Practices	ditto	R. K. Ham	University of Wisconsin	109
72-29	8/20-25/72	Particulate Matter Systems	ditto	R. Klimpel	Dow Chemical Co.	41
72-31	9/5-9/72	Earthquake Protection of Underground Utility Structures	Asilomar Conf. G. Pacific Grove, CA	J. F. Szablya	Washington State University	106
73-01	3/4-9/73	Environmental Engineering in the Food Industry	ditto	C. F. Gurnham	Gurnham & Associates	44
73-02	3/11-16/73	Industrial Developments in Stack Gas Sampling and Monitoring	ditto	W. T. Ingram	New York University	109
73-03	4/1-6/73	Environmental Needs as a Part of the Quality System in Construction	ditto	W. W. Moore & G. M. Reynolds	Dames & Moore ditto	64
73-04	5/20-25/73	Technology and Social Institutions	ditto	K. Chen	Univ. of Michigan	54
73-05	6/24-29/73	Need for National Policy for the Use of Underground Space	Berwick Academy South Berwick, ME	E. F. Casey	NYC Transit Authority	63
73-06	7/1-6/73	Maintaining Professional and Technical Competence of the Older Engineer	ditto	H. Shelton & S. Dubin	Sandia Corporation Pennsylvania State University	76
73-07	7/8-13/73	Issues in Behavioral Travel and the Valuation of Travel Time	ditto	P. R. Stopher & A. H. Meyburg	Cornell University ditto	55
73-08	7/15-20/73	Use of Shotcrete for Underground Structural Support	ditto	J. R. Graham	Bureau of Reclamation	84
73-09	7/22-27/73	Engineering in Packaging	ditto	J. Goff	Michigan State Univ.	30
73-11	8/5-10/73	Goods Transportation in Urban Areas	ditto	G. P. Fisher	Cornell University	45
73-12	8/12-17/73	Mixing Research	ditto	W. C. Brasie	Dow Chemical Co.	83

TABLE 3 Engineering Foundation Conferences (continued)

File Number	*Meeting Date*	*Title*	*Location*	*Chairman*	*Chairman's Affiliation*	*Att.*
73-13	8/19-24/73	Evaluation of Health Services Delivery	ditto	R. Yaffe & D. Zalkind	U.S. Publ. Health Serv. Hosp. Univ. of North Carolina	128
73-14	8/12-17/73	Natural Hazards—Public and Private Policy	Winchendon School Winchendon, MA	D. E. Jones	Federal Housing Administration	31
73-15	8/19-24/73	Energy Conservation at Point of Use	New England Col. Henniker, NH	J. Denton	University of Pennsylvania	70
73-16	7/1-6/73	Coherent Radiation Systems	ditto	K. Preston, Jr.	Perkin-Elmer Corp.	48
73-17	7/1-6/73	Viruses in the Water Environment	ditto	G. Berg	Environ. Prot. Agency	68
73-20	7/15-20/73	Energy Research—Alternatives for Policy and Management to Meet Regional and National Needs	ditto	W. C. Ackerman	Illinois State Water Survey	71
73-21	7/22-27/73	Performance of Full Scale Structures	ditto	J. Janney	Wiss, Janney, Elstner and Associates	37
73-22	7/29 - 8/3/73	Preparation of Environmental Impact Statements	ditto	M. Blissett & L. Canter	University of Texas Univ. of Oklahoma	93
73-23	8/5-10/73	Enzyme Engineering	ditto	L. B. Wingard, Jr. & E. K. Pye	Univ. of Pittsburgh Univ. of Pennsylvania	189
73-24	8/12-17/73	Making Service Industries More Productive through Computers and Automation	ditto	A. McAdams	Cornell University	62
73-25	8/19-24/73	Career Guidance for Women Entering Engineering	ditto	N. Fitzroy	General Electric Co.	90
73-26	9/23-28/73	Rehabilitation, Maintenance and Repair of Old Dams	Asilomar Conf. G. Pacific Grove, CA	E. L. Armstrong	Consulting Engineer	173
74-01	9/25-30/73	Mine, Mill, Plant and Smelter Waste Disposal	ditto	W. A. Wahler	W. A. Wahler & Assoc.	42
74-02	12/2-7/73	Automatic Cytology	ditto	M. L. Mendelsohn	Univ. of California	166
74-03	12/9-14/73	Community Planning for Seismic Safety	ditto	W. R. Hays	Environmental Research Corp.	84
74-04	2/9-14/74	Antipollution Legislation in the Food Industry	ditto	J. & M. Beach	N-Con Systems	40

TABLE 3 Engineering Foundation Conferences (continued)

File Number	Meeting Date	Title	Location	Chairman	Chairman's Affiliation	Att.
74-05	3/18-21/74	Foundations for Dams	ditto	R. D. Harza	Harza Engineering	66
74-06	4/9-13/74	Industrial Problems in Source Sampling and Monitoring	ditto	A. Licata	York Research Corp.	118
74-07	6/9-14/74	Geothermal Energy	ditto	J. Denton	Univ. of Pennsylvania	67
74-09	7/7-14/74	Methanol as an Alternate Fuel	New England Col. Henniker, NH	R. L. Zahradnik & S. W. Gouse	Federal Energy Office US Dept. of Interior	87
74-10	7/14-19/74	Assessment of Resources and Needs in Highway Technology Education	Franklin Pierce College Rindge, NH	R. Dean & W. Thomas	Federal Highway Admin. Florida Internat. Univ.	49
74-12	7/21-26/74	Improving Indoor Air Quality	ditto	J. Yocum	Research Corp. of New E.	46
74-13	7/21-26/74	Process Design, Operation and Control for Safety and Reliability	New England Col. Henniker, NH	L. B. Evans & B. K. Adams	Mass. Inst. of Techn. Oak Ridge National Lab.	55
74-14	7/28 - 8/2/74	Productivity Improvement in the Service Sector of the Economy	Franklin Pierce Col., Rindge, NH	A. McAdams & P. Polishuk	Cornell University Nat. Bureau of Standards	64
74-17	8/4-9/74	Safety of Small Dams	New England Col. Henniker, NH	M. E. Graf	Massachusetts Division of Waterways	106
74-18	8/11-16/74	Urban Runoff—Quantity & Quality	Franklin Pierce Col., Rindge, NH	W. Whipple	Rutgers University	91
74-19	8/11-16/74	Requirements for Fulfilling a National Materials Policy	New England Col. Henniker, NH	F. P. Huddle	Library of Congress	102
74-20	8/11-16/74	Subsurface Exploration for Underground Excavation and Heavy Construction	ditto	M. M. Singh E. L Foster	IIT Research Institute Foster, Miller Associates	66
74-21	8/18-23/74	Converting Existing Hydro-Electric Dams and Reservoirs into Pump Storage Facilities	Franklin Pierce Col., Rindge, NH	E. Armstrong & T. W. Mermel	URS Systems Corp. Consultant	64
74-22	8/18-23/74	Particulate Matter Systems	New England Col. Henniker, NH	T. Mika	Univ. of California—Berkeley	42
74-23	8/18-23/74	Women in Engineering—Updating of Skills and Preparing to Reenter the Engineering Field After Temporary Retirement	ditto	E. Stephens & C. Crenshaw	Pelham Manor, NY General Electric Co.	38

TABLE 3 Engineering Foundation Conferences (continued)

File Number	Meeting Date	Title	Location	Chairman	Chairman's Affiliation	Att.
75-01	11/8-13/74	Constructed Environment with Man as the Measure	Asilomar Conf. G. Pacific Grove, CA	C. Gwathmey	Gwathmey-Seller-Crosby Architects	35
75-03	11/10-15/74	Coal Mining and Processing Legislation and Regulation	Mount Summit Inn Uniontown, PA	W. A. Wahler	W. A. Wahle & Assoc.	51
75-04	12/8-13/74	Earthquakes and Lifelines	ditto	D. B. Slemons W. W. Hays	Mackay School of Mines U.S. Geological Survey	66
75-05	1/12-17/75	Engineering Problems Interfacing with Rheology	ditto	M. E. Morrison	American Enka Corp.	65
75-06	2/23-28/75	Coping with Environmental and Safety Regulations in the Food Industry	ditto	J. & M. Beach	N-CON Systems Co., Inc.	45
75-07	3/9-14/75	Improving Coal Mine Health and Safety	Jenny Wiley State Park Prestonsburg, KY	R. L. Marovelli & D. B. Forshey	U.S. Bureau of Mines ditto	84
75-08	3/30 - 4/4/75	Evaluation of Air Pollution Emissions from Stationary Sources	Asilomar Conf. G. Pacific Grove, CA	A. Licata & R. Ostendorf	York Research Corp. Procter and Gamble Co.	120
75-09	5/18-23/75	Behavioral Travel Demand—Theory and Estimation, II	Grove Park Inn Asheville, NC	P. R. Stopher & A. H. Meyburg	Northwestern University Cornell University	68
75-10	6/8-13/75	International Conference on Automatic Cytology	ditto	L. Wheeless	U. of Rochester Medical Center	222
75-11	6/15-20/75	International Conference on Fluidization	ditto	D. L. Keairns	Westinghouse Research Laboratories	147
75-13	7/6-11/75	Socio-Technical Systems Analysis Work in Engineering Systems	Franklin Pierce Col., Rindge, NH	L. E. Davis & D. B. Miller	U. of California at LA Intern. Business Mach.	33
75-14	7/13-16/75	Engineering Manpower: A National Problem or a National Resource?	New England Col. Henniker, NH	P. Doigan	General Electric Co.	80
75-16	7/20-25/75	Technology Transfer via Enterpreneurship Development	ditto	D. A. Bauman & J. Kokalis, Jr.	Carnegie-Mellon Univ. ditto	57
75-17	7/20-25/75	Flood Plain Control	Franklin Pierce College Rindge, NH	L. S. Tucker & T. Lee	Urban Drainage and Flood Control Distr. 69 Wis. St. Wat. Res. Dep.	85

TABLE 3 Engineering Foundation Conferences (continued)

File Number	*Meeting Date*	*Title*	*Location*	*Chairman*	*Chairman's Affiliation*	*Att.*
75-18	7/27 - 8/1/75	Application of Signature Analysis to Technical Problems	New England Col. Henniker, NH	P. G. Whitterel	E. I. du Pont De Nemours & Co.	46
75-19	7/27 - 8/1/75	Real World Problems in Computer Analysis of Electro-Cardiograms	Franklin Pierce Col., Rindge, NH	R. L. Sandberg & J. K. Hooper	Phone-A-Gram US Public Health Serv.	58
75-20	8/3-8/75	Civil and Environmental Aspects of Energy Complexes	New England Col. Henniker, NH	A. S. Goodman	Polytechnic Institute of New York	58
75-22	8/10-15/75	Improving the Effectiveness and Efficiency of Scientific Technical Information	New England College Henniker, NH	B. H. Weil D. Staiger	Exxon Research and Engineering Company AIAA	105
75-23	8/10-15/75	Coal Preparation for Coal Conversion Process	Franklin Pierce Col., Rindge, NH	S. P. Babu	Institute of Gas Technology	99
75-24	8/17-22/75	Systems Engineering for Power	New England Col. Henniker, NH	L. H. Fink & K. Carlsen	Energy Research and Development Administr.	106
75-25	8/17-22/75	Mixing	Franklin Pierce Col., Rindge, NH	W. R. Penny	Monsanto Company	68
75-27	8/24-25/75	Environmental Aspects of Hydroelectric and Pumped Storage Projects	ditto	J. G. Alesi & N. J. Wilding	EBASCO Services, Inc. ditto	55
75-28	8/3-8/75	International Conference on Enzyme Engineering	Reed College Portland, OR	E. K. Pye H. Weetall	U. of Pennsylvania Corning Glass Works	189
75-29	9/7-12/75	Goods Transportation in Urban Areas	Miramar Hotel Santa Barbara, CA	G. P. Fisher	Cornell U.	68
75-31	9/14-19/75	Information Technology Systems to Provide Services to the Public	Asilomar Conf. G. Pacific Grove, CA	M. Henderson	National Bureau of Standards	50
75-32	9/21-26-75	Risk Benefit—Methodology and Application	ditto	D. Okrent	U. of California—Los Angeles	98
75-33	9/28 - 10/3/75	Responsibility and Liability of Public & Private Interests on Dams	ditto	J. Ellam	Pennslyvania Dept. of Environmental Resources	66
76(1062)01	11/9-14/75	Microbiological Recovery of Oil	Tidewater Inn Easton, MD	D. Dunlop	Florida Power and Light	77
76(1017)02	11/16-21/75	Criteria for Full Scale Performance Evaluation of Residential Structures	Virginia Polytech. Institute Blacksburg, VA	R. Ostendorf	Procter & Gamble Co.	59

TABLE 3 Engineering Foundation Conferences (continued)

File Number	*Meeting Date*	*Title*	*Location*	*Chairman*	*Chairman's Affiliation*	*Att.*
76(1043)03	12/7-12/75	Perspectives: Product Liability and Product Safety	Asilomar Conf. G. Pacific Grove, CA	A. S. Weinstein & J. M. Amos	Carnegie-Mellon Univ. U. of Missouri at Rolla	51
76(1047)04	1/11-16/76	Integrating Water Resources and Water Quality Planning	ditto	L. Dworsky	Cornell University	80
76(1040)05	1/18-23/76	Chemical Process Control	ditto	M. M. Denn & A. S. Foss	Univ. of Delaware U. of Calif.—Berkeley	83
76(1060)06	2/8-13/76	Energy Storage User Needs and Technology Applications	ditto	J. Vanderryn & V. Cooper	Energy Res. & Dev. Corp. Elect. Power Res. Inst.	131
76(1058)07	2/15-20/76	Environmental Engineering in the Food Industry	ditto	J. M. Beach & W. Bough	N-COM Systems, Inc. University of Georgia	66
76(1058)08	3/21-26/76	Decision Making for Natural Hazards	ditto	R. R. Fox	George Washington Univ.	44
76(1036)11	6/20-25/76	Numerical Methods in Geomechanics	VA Polytechnic Instit. & State U. Blacksburg, VA	C. S. Desai	Virginia Polytechnic Instit. and State Univ.	152
76(1066)12	6/27 - 7/2/76	Improving Ambulatory Health Care Delivery	New England Col. Henniker, NH	G. S. Lasdon	National Center for Health Services Research	27
76(1074)15	7/11-16/76	Computerized Interpretation of ECG. II	Franklin Pierce Col., Rindge, NH	J. K. Hooper	US Public Health Serv. Hospital, Baltimore	58
76(1059)16	7/11-16/76	Wind Energy Conversion	New England Col. Henniker, NH	R. Smith	Southwest Research Institute	47
76(1068)17	9/26 - 10/1/76	Land Application of Residual Materials	Tidewater Inn Easton, MD	R. E. Cummings & J. Murphy	R. E. Cummings Assoc. Bergen Co. Munic. Engrs.	66
76(1061)18	7/18-23/76	Transportation of Fuels for Utility Consumption	ditto	R. A. Schmidt	Electric Power Research Institute	32
76(1085)19	7/18-23/76	Full Scale Structural Testing	Franklin Pierce Col., Rindge, NH	A. Longinow	IIT Research Institute	49
76(1079)20	7/25-30/76	Technology Assessment: Evaluation of Ten Years Experience	New England Col. Henniker, NH	V. T. Coates & J. Coates	George Washington Univ. Office of Technology Assessment	48
76(1084)21	7/25-30/76	Recycling Implementation: Engineering and Economics	Franklin Pierce Col., Rindge, NH	H. Bullis	Library of Congress	67

TABLE 3 Engineering Foundation Conferences (continued)

File Number	Meeting Date	Title	Location	Chairman	Chairman's Affiliation	Att.
76(1065)22	8/1-6/76	Welding R&D: Problems and Opportunities	New England Col. Henniker, NH	P. Ramsey & R. K. Sager	A. O. Smith Corp. Aluminum Co. of America	80
76(1023)23	8/1-6/76	Particle Science and Engineering in the Process Industries	Franklin Pierce Col., Rindge, NH	P. T. Luckie J. A. Herbst	Kennedy Van Suan Co. University of Utah	74
76(1094)24	8/8-13/76	Engineering Implications of Chronic Resource Scarcity	New England Col. Henniker, NH	J. B. Wachtman F. P. Huddle	NBS Library of Congress	84
76(1086)25	8/8-13/76	Algorithms for Image Processing	ditto	K. S. Fu & A. Rosenfeld	Purdue University University of Maryland	105
76(1092)26	8/15-20/76	Comparative Productivity of Non-Invasive Techniques for Medical Diagnosis	New England College Henniker, NH	K. Preston & S. Jonson	Carnegie-Mellon Univ. Mayo Clinic	43
76(1083)28	8/22-27/76	Power Systems Planning and Operations—Future Problems and Research Needs	ditto	P. Anderson & J. L. Davidson	Elect. Power Res. Inst. Long Island Lighting Co.	94
76(1064)29	8/22-27/76	New Directions in Textile Technology	Franklin Pierce Col., Rindge, NH	T. W. George	North Carolina State University	56
76(1093)30	8/29 - 9/3/76	The Technology, Management and Economics of Information Centers and Services	Tidewater Inn Easton, MD	J. E. Creps & R. M. Mason	Engineering Index, Inc. Metrics, Inc.	90
76(1080)31	8/29 - 9/3/76	Update in Engineering Curricula	Franklin Pierce Col., Rindge, NH	D. H. Thomas	Drexel University	59
76(1070)32	9/12-17/76	Emission Sampling for Source Evaluation	Hueston Woods State Park Oxford, OH	R. Ostendorf	Procter & Gamble Co.	119
76(1070)33	9/19-24/76	Present Status and Research Needs of Energy Recovery from Wastes	ditto	R. A. Matula	Louisiana State Univ.	97
77-01	10/3-8/76	Shotcrete for Ground Support	Tidewater Inn Easton, MD	J. A. Veltrop	Harza Engineering Co.	91
77-02	10/10-15/76	Upward Mobility and Professional Development for Women Engineers	ditto	L. Bryant	Collins Radio Group Rockwell International	71
77-03	10/17-22/76	Disposal of Flue Gas Desulfurization Solids	Hueston Woods College Corner, OH	G. Merritt	Pennsylvania Dept. of Natural Resources	70

TABLE 3 Engineering Foundation Conferences (continued)

File Number	Meeting Date	Title	Location	Chairman	Chairman's Affiliation	Att.
77-04	11/28 - 12/3/76	Evaluation of Dam Safety	Asilomar Conf. G. Pacific Grove, CA	H. Willis	US Army Corps of Engineers	168
77-05	12/5-10/76	Water for Energy Development	ditto	G. M. Karadi	Univ. of Wisconsin at Milwaukee	99
77-06	12/12-17/76	Automated Cytology V	Casino, Pensacola Beach. FL	P. Mullaney	Los Alamos Scientific Laboratory	244
77-07	11/28 - 12/3/76	Instrumentation and Analysis of Urban Stormwater Data	Tidewater Inn Easton, MD	J. Biesecker & L. S. Tucker	US Geological Survey Urban Drainage and Flood Control District	68
77-08	1/2-7/77	Building Materials	Asilomar Conf. G. Pacific Grove, CA	L. T. Eby	US Gypsum Co.	49
77-09	1/16-21/77	Estimation and Correlation of Phase Equilibria and Fluid Properties in Chemical Industry	ditto	S. I. Sandler & T. S. Storvick	Univ. of Delaware University of Missouri at Columbia	136
77-10	2/6-11//77	Energy's Role in Food Production	ditto	D. R. Price	Cornell University	40
77-11	2/13-18/77	Environmental Engineering in the Food Processing Industry	ditto	R. A. Tsugita C. Kahr	James M. Montgomery Co. ditto	65
77-12	3/20-25/77	Flood Proofing for Flood Plain Management	ditto	T. M. Lee, L. S. Tucker & D. E. Jones	Wis. Dept. of Nat. Res. Urban Drainage and FCD HUD/FHA	103
77-14	4/13-17/77	Public Policy in Ground Water Protection	VA Pol. Inst. & St. U. Blacksburg, VA	W. R. Walker	Virginia Polytechnic Institute and State U.	65
77-15	6/19-24/77	Particle Science and Engineering in the Coal Processing Industry	Franklin Pierce Col. Rindge, NH	R. Hogg	Pennsylvania State University	99
77-16	6/26 - 7/1/77	Ash Deposits and Corrosions due to Impurities in Combustion Gases	New England Col. Henniker, NH	R. W. Bryers	Foster-Wheeler Energy Corporation	96
77-17	6/26-7/1/77	Product Safety—A Product Liability Antonym	Franklin Pierce Col., Rindge, NH	A. P. Szews	Marquette University	48
77-19	7/10-15/77	Mass Transfer and Scale-up of Fermentation	New England Col., Henniker, NH	H. W. Blanch & A. E. Humphrey	Univ. of Delaware Univ. of Pennsylvania	78
77-20	7/10-15/77	Non-Conventional Siting of Power Plants	Franklin Pierce Col., Rindge, NH	H. Kornberg & H. Harty	Elect. Power Res. Inst. Battelle-Northwest	56

TABLE 3 Engineering Foundation Conferences (continued)

File Number	*Meeting Date*	*Title*	*Location*	*Chairman*	*Chairman's Affiliation*	*Att.*
77-21	7/17-22/77	An Assessment of Environmental Regulation	New England College Henniker, NH	S. R. Stearns, N. Drobny & M. Pender	Dartmouth College Battelle-Columbus Town of Hempstead, NY	27
77-23	7/24-29/77	Ventilation versus Energy Conservation in Buildings	ditto	P. McNall	National Bureau of Standards	44
77-24	7/24-29/77	Application of New Signature Analysis Technology	Franklin Pierce Col., Rindge, NH	L. Mitchell	Virginia Polytechnic Institute and State U.	90
77-25	7/31 - 8/5/77	Diagnostic Immunology	New England Col. Henniker, NH	H. H. Weetall & G. Odstrchel	Corning Glass Works ditto	153
77-26	7/31 - 8/5/77	Clean Combustion of Coal	Franklin Pierce Col., Rindge, NH	V. Engelman	Science Applications, Inc.	71
77-27	8/7-12/77	Cybernetic Model of the Human Neuromuscular System	New England Col. Henniker, NH	R. W. Mann	Massachusetts Inst. of Technology	72
77-28	8/7-12/77	New Manufacturing Techniques and Needs of Industry for Improved Productivity	Franklin Pierce College Rindge, NH	M. C. Shaw	Carnegie-Mellon University	39
77-29	8/14-19/77	Considerations for an Expanded Coal Industry	New England Col. Henniker, NH	W. G. Wilson	Energy Research and Development Administr.	59
77-30	8/14-19/77	Innovative Management of Technical Information Functions in Industry	Franklin Pierce College Rindge, NH	J. A. Price	Exxon Research and Engineering	76
77-32	8/21-26/77	Systems Engineering for Power: Emergency Operating State Control	New England Col. Henniker, NH	L. Fink & K. Carlsen	ERDA ditto	55
77-33	8/21-26/77	Mixing	Franklin Pierce Col., Rindge, NH	R. J. Adler	Case-Western Reserve U.	100
77-36	9/25-30/77	Enzyme Engineering IV	Steigenberger Kurthotel, Bad Neuenahr, W. Germ.	G. Manecke	der Freie Universitat Berlin	242
78(1113)01	10/23-28/77	Engineering and Science Research for Industrial Development	Tidewater Inn Easton, MD	F. Landis	Univ. of Wisconsin at Milwaukee	75

TABLE 3 Engineering Foundation Conferences (continued)

File Number	Meeting Date	Title	Location	Chairman	Chairman's Affiliation	Att.
78(1148)03	10/23-28/77	Energy Storage in Solar Applications and Transportation	Sea Palms Club St. Simons Is., Ga	G. F. Pezdirtz	Department of Energy	62
78(1095)04	10/30 - 11/4/77	Theory, Practice and Process Principles for Physical Separations	Asilomar Conf. G Pacific Grove, CA	M. P. Freeman	Dorr-Olivar, Inc.	99
78(1074)05	1/8-13/78	Computerized Interpretation of ECG III	ditto	J. K. Hooper	US Public Health Serv.	49
78(1131)06	12/4-9/77	Contingency Planning for Materials Resources	ditto	E. B. Berman & W. Fisher	Edward B. Berman Assoc. University of Texas	47
78(1149)07	12/4-9/77	Goods Transportation in Urban Areas	The Cloister Sea Island, Ga	G. P. Fisher & A. H. Meyburg	Cornell University ditto	62
78(1123)08	1/15-20/78	Evaluation and Prediction of Subsidence	Casino Inn Pensacola, FL	S. K. Saxena	Illinois Institute of Technology	87
78(1117)09	2/19-24/78	Algorithms for Image and Scene Analysis	Asilomar Conf. G. Pacific Grove, CA	K. S. Fu & A. Rosenfeld	Purdue University Univ. of Maryland	68
78(1141)10	2/26 - 3/3/78	Environmental Engineering in the Food Processing Industry VI	ditto	R. E. Carawan	North Carolina State University	86
78(1130)11	3/5-10/78	Stack Sampling for Source Evaluation	Mid Pines Club South. Pines, NC	R. Ostendorf & J. Steiner	Procter & Gamble Co. Acurex Corp.	99
78(1156)12	3/5-9/78	Power Systems Reliability—Research Needs and Priorities	Asilomar Conf. G. Pacific Grove, CA	M. P. Bhavaraju	Electric Power Research Institute	82
78(1078)13	4/1-6/78	Fluidization II	Cambridge U. Cambridge, England	J. L. Davidson D. L. Keairns	Univ. of Cambridge Westinghouse Research	214
78(1137)14	4/16-21/78	Shotcrete for Underground Support III	Post Hotel, St. Anton am Arlberg Austria	R. Mason & I. Wolff & O. Schauritsch	A. A. Mathews, Inc. Chemat Warenhandels GmbH ditto	96
78(1135)15	4/23-29/78	Automated Cytology VI	Schloss Elmau Klais, W. Germany	T. Jovin	Max Planck Inst.	292
78(1139)16	4/23-28/78	Systems Engineering for Power, Organizational Forms for Large Scale Systems	Asilomar Conference Grounds Pacific Grove, Ca	L. Fink K. Carlsen	Department of Energy ditto	56

TABLE 3 Engineering Foundation Conferences (continued)

File Number	Meeting Date	Title	Location	Chairman	Chairman's Affiliation	Att.
78(1144)17	4/30 - 5/5/78	Planning, Engineering and Constructing the Superprojects	ditto	J. W. Ward	Arizona State U.	87
78(1132)18	5/21-26/78	Building Materials III	ditto	G. Frohnsdorf & H. Olin	Nat. Bureau of Standards US League of Savings Associations	33
78(1127)19	6/18-23/78	Particle Science and Engineering in the Process Industries	ditto	J. Herbst & K. Sastry	U. of Utah U. of Calif.—Berkeley	79
78(1138)20	6/25-30/78	Historic Preservation of Engineering Works	Franklin Pierce Col., Rindge, NH	E. Kemp	West Virginia Univ.	39
78(1115)23	7/9-14/78	Engineering Graduate Programs—Future Directions	New England Col. Henniker, NH	A. T. Murphy	Carnegie-Mellon University	58
78(1159)24	7/9-14/78	Examination of the Scientific Basis of Government Regulations	Franklin Pierce Col., Rindge, NH	K. W. Gardiner	Univ. of California—Riverside	32
78(1167)25	7/16-21/78	Urban Runoff Pollution and its Implications for the Future	New England Col. Henniker, NH	W. Whipple, Jr.	Water Res. Reas. Inst. Rutgers U.	111
78(1128)26	7/16-21/78	Unit Operations in Resource Recovery Operations	Franklin Pierce Col., Rindge, NH	G. Pearsall & A. Vesilind	Duke Univ. ditto	66
78(1157)27	7/23-28/78	Economic Impact of Energy Conservation Measures	New England Col. Henniker, NH	A. Kaufman	Library of Congress	54
78(1087)28	7/23-28/78	Design and Performance of Structures Exposed to Fire	Franklin Pierce Col., Rindge, NH	P. DeCicco	Polytechnic Institute of New York	46
78(1146)29	7/30 - 8/4/78	Building Consensus on Legislation for National Materials Policy	New England College Henniker, NH	F. P. Huddle J. Holt	Library of Congress Office of Technology Assessment	110
78(1152)30	7/30 - 8/4/78	Industrial Applications of Magnetic Separations	Franklin Pierce Col., Rindge, NH	Y. A. Liu	Auburn Univ.	90
78(1136)31	8/6-11/78	Cooperative Welding R&D—Strategy for the 1980's	New England Col. Henniker, NH	P. Ramsey	A. O. Smith Corp.	78
78(1135)32	8/6-11/78	The New Engineering Work Force	Franklin Pierce Col., Rindge, NH	R. R. O'Neil & A. C. Ingersoll	U. of Calif.—LA ditto	29
78(1134)33	8/13-18/78	Waste Heat Utilization	New England Col. Henniker, NH	R. P. Perkins	E. I. duPont de Nemours & Co.	48

TABLE 3 Engineering Foundation Conferences (continued)

File Number	Meeting Date	Title	Location	Chairman	Chairman's Affiliation	Att.
78(1158)34	8/13-18/78	Role of Education in Non-Destructive Evaluation	Franklin Pierce Col., Rindge, NH	R. Smith	Southwest Research Institute	24
78(1165)35	8/20-25/78	Biochemical Engineering	New England Col. Henniker, NH	W. R. Vieth & A. Constantinides	Rutgers Univ. ditto	109
78(1164)36	8/20-25/78	Measuring and Forecasting Engineering Manpower and Skill Requirement	Franklin Pierce College Rindge, NH	W. Hibbard	Virginia Polytechnic & State University	53
78(1172)37	9/10-15/78	Western Coal: Problems, Progress and Promises	Keystone Lodge Keystone, CO	P. S. Jacobsen & A. Melcher	Colorado School of Mines Research Inst.	86
79(1181)01	1/7-12/79	Computerized Interpretation of the ECG IV	Asilomar Conf. G. Pacific Grove, CA	T. A. Pryor	Univ. of Utah	60
79(1181)02	2/11-16/79	Research Directions in Computer Control of Urban Traffic	ditto	W. S. Levine, E. Lieberman & J. J. Fearnsides	U. of Maryland KLD Associates US Dept. of Transportation	67
79(1173)03	2/25 - 3/2/79	Environmental Engineering in the Food Processing Industry	ditto	R. E. Carawan & J. R. Geisman	N. Carolina St. U. Ohio State U.	96
79(1188)04	3/11-16/79	Hydropower: A National Energy Resource	Tidewater Inn Easton, MD	H. G. Robinson & A. J. Fredrich	US Army Institute for Water Resources	89
79(1192)05	3/25-30/79	Improved Hydrologic Forecasting	Asilomar Conf. G. Pacific Grove, CA	W. P. Henry	CH2M Hill	58
79(1184)06	4/1-6/79	Stack Sampling and Stationary Source Emission Evaluation	ditto	R. L. Ajax	Environmental Protection Agency	118
79(1176)07	9/30 - 10/5/79	Systems Engineering for Power: Probabilistic Evaluation of Storage and Transmission	Tidewater Inn Easton, MD	L. H. Fink	SEPI	32
79(1187)08	6/24-29/79	Cement—Chemistry of Production and Use	Franklin Pierce Col., Rindge, NH	J. P. Skalny & R. E. Philleo	Martin Marietta Lab Corps of Engineers	76
79(1162)09	6/24-29/79	Biomechanics of Movement	New England Col. Henniker, NH	W. Z. Rymer & G. Bekey	Northwestern U.M.C. U. of So. California	89
79(1182)10	7/8-13/79	Urban Water Conservation	Franklin Pierce College Rindge, NH	L. A. Herr P. L. Thopson M. B. Sonnen	Fed. Highway Administr. ditto Water Resources Engnrs.	48

TABLE 3 Engineering Foundation Conferences (continued)

File Number	*Meeting Date*	*Title*	*Location*	*Chairman*	*Chairman's Affiliation*	*Att.*
79(1166)13	7/15-20/79	Offshore Industrial Port Islands	New England Col. Henniker, NH	E. Chesson, R. B. Biggs & W. S. Gaither	University of Delware	41
79(1189)15	7/22-27/79	Municipal Waste as a Resource	ditto	T. C. Frankiewicz	Occidental Res. Corp.	76
79(1202)16	7/29 - 8/3/79	Technical Information Centers: Which Way Is the Future?	Franklin Pierce Col., Rindge, NH	R. G. M. Cosgrove	Exxon Research and Engineering Co.	75
79(1163)17	7/29 - 8/3/79	Enzyme Engineering V	New England Col. Henniker, NH	H. H. Weetall & G. P. Royer	Corning Glass Works U. of Delaware	174
79(1197)18	8/5-10/79	Advanced Coal Cleaning	Franklin Pierce Col., Rindge, NH	T. P. Meloy	Univ. of West Virginia	87
79(1162)19	8/5-10/79	Diagnostic Immunology II	New England Col. Henniker, NH	G. Odstrchel & H. McDonald	Corning Glass Works ditto	116
79(1196)20	8/12-17/79	The Systems Approach to Energy Supply and Demand Controversies	ditto	E. V. Sherry W. F. Allaire	Air Products & Chemicals Allied Chemical Co.	72
79(1179)21	8/12-17/79	Mixing VI	ditto	S. J. Chen	Kenics Corp.	88
79(1180)22	8/19-24/79	Development of Mine-Machine Design Curricula	Franklin Pierce Col., Rindge, NH	C. R. Peterson & W. Hustrulid	Mass. Inst. of Techn. Colo. Sch. of Mines	17
79(1175)23	8/19-24/79	Systems Engineering for Power V: Approaches to Physically Based Load Modeling	New England College Henniker, NH	L. H. Fink & T. A. Trygar	US Dept. of Energy ditto	40
79(1199)24	9/30 - 10/5/79	System Engineering for Power: US/European Workshop	Hotel Schweizerhof and Kongresshaus Davos, Switzerland	L. H. Fink	SEPI	69
80(1178)01	10/7-10/79	Hard Facing and Ware	Asilomar Conf. G. Pacific Grove, CA	K. C. Antony & R. W. Kirschner	Satellite Division of Cabot Corp.	78
80(1174)02	11/4-9/79	Systems Engineering for Power VI: Systems Effectiveness Evaluation	ditto	L. H. Fink & R. L. Sullivan	SEPI U. of Florida	38
80(1195)03	11/25-30/79	Automated Cytology VII	ditto	M. Mendelson	Lawrence Livermore Lab	260
80(1185)05	12/9-14/79	New Approaches to Non-Linear Problems in Dynamics	ditto	P. Holmes	Cornell Univ.	55
80(1190)06	1/6-11/80	Fatigue Crack Initiation and Early Crack Growth in Materials	ditto	W. Fine & R. O. Ritchie	Northwestern U. Mass. Inst. of Techn.	84

TABLE 3 Engineering Foundation Conferences (continued)

File Number	Meeting Date	Title	Location	Chairman	Chairman's Affiliation	Att.
80(1209)07	1/6/-11/80	Building Materials III	ditto	H. Olin R. Brungraber	U.S. League of S&L Assoc. Bucknell Univ.	43
80(1193)08	1/13-18/80	Geotechnical and Environmental Aspects of Geopressure Energy	The Cloister Sea Island, GA	S. K. Saxena	Illinois Institute of Technology	79
80(1205)09	1/20-25/80	Applied Rheology: Polymer Melt Processing and Two-Phase Systems	Asilomar Conf. G. Pacific Grove, Ca	J. White	U. of Tennessee	35
80(1208)10	2/24-29/80	Environmental Engineering in the Food Processing Industry	ditto	J. Geisman & G. Nelson	Ohio State U. Remcon, Inc.	81
80(1210)11	4/20-25/80	National Technological Cooperation	ditto	J. Foster	TRW Inc.	62
80(1151)12	3/23-28/80	Erosion by Particle and Droplet Impaction	ditto	I. Finnie M. Shaw	U. of Calif.—Berkeley Arizona State Univ.	46
80(1204)13	4/27 - 5/2/80	Computerized Interpretation of the ECG V	ditto	G. Tolan	US Air Force	62
80(1226)14	5/4-9/80	Load Management in Electric Utilities	Tidewater Inn Easton, MD	H. M Long & P. Overhalt	Oak Ridge National Lab. U.S. Dept. of Energy	70
80(1203)16	7/6-11/80	Particle Science and Engineering in the Process Industries	Franklin Pierce Col., Rindge, NH	K. V. S. Sastry	U. of Calif.—Berkeley	54
80(1194)17	7/6-11/80	Computer-Aided Chemical Process Design	New England Col. Henniker, NH	R. S. H. Mah & W. Seider	Northwestern Univ. U. of Pennsylvania	144
80(1221)18	7/13-18/80	Automation in Diagnostic Virology	Franklin Pierce Col., Rindge, NH	S. Spector & G. Lancz	University of South Florida	66
80(1191)19	7/13-18/80	Biochemical Engineering II	New England Col. Henniker, NH	A. Constantinides	Rutgers Univ.	93
80(1231)20	7/20-25/80	Bridge Design and Rehabilitation	Franklin Pierce Col., Rindge, NH	C. F. Scheffey & A. Custen	Fed. Highway Administr. Ammann & Whitney	66
80(1222)21	7/20-25/80	Engineering and Energy Economics	New England Col. Henniker, NH	F. K. Manasse & M. English	U. of New Hampshire U. of Calif.—LA	26
80(1230)22	7/27 - 8/1/80	Signature Analysis III	Franklin Pierce Col., Rindge, NH	R. Leon	Franklin Research Inst.	59
80(1211)23	7/27 - 8/1/80	Innovations in Materials Industries	New England College Henniker, NH	F. P. Huddle, H. Bullis & F. Richmond	Library of Congress ditto Universal Cyclops Corp.	90

TABLE 3 Engineering Foundation Conferences (continued)

File Number	*Meeting Date*	*Title*	*Location*	*Chairman*	*Chairman's Affiliation*	*Att.*
80(1229)24	8/3-8/80	Modeling of Casting & Welding Processes	Franklin Pierce Col., Rindge, NH	D. Apelian & H. Brody	Drexel University U. of Pittsburg	90
80(1186)25	8/3-8/80	Fluidization III	New England Col. Henniker, NH	J. M. Matsen & J. R. Grace	Exxon Res. and Enginrng U. of British Columbia	176
80(1232)26	8/10-15/80	Waste Heat Utilization II	Franklin Pierce Col., Rindge, NH	R. R. Perkins	E. I. du Pont de Nemours & Co.	30
80(1224)27	8/10-15/80	Research Needs in Building Economics	New England Col. Henniker, NH	C. Coulter & D. Weinroth	Consultant Weinroth Associates	32
80(1213)28	8/17-22/80	Hazardous Waste	Franklin Pierce College Rindge, NH	C. Parker, A. Vesilind & J. Pierce	U. of Virginia Duke University ditto	35
80(1223)29	9/21-26/80	Risk/Benefit Analysis in Water Resources	Asilomar Conf. G. Pacific Grove, CA	Y. Haimes	Case-Western Reserve U.	58
80(1235)30	8/17-22/80	Systems Engineering for Power VII: Advanced Control for Fossil Fueled Power Plants	New England College Henniker, NH	T. A. Trygar	US Dept. of Energy	30
81(1239)01	10/12-17/80	Stack Sampling and Stationary Source Emission Evaluation	Tidewater Inn Easton, MD	R. R. Perkins	E.I. du Pont de Nemours & Co.	82
81(1237)02	10/19-24/80	Systems Engineering for Power IX: Research in High Voltage Direct Current System Embedded in AC Networks	Miramar Hotel Santa Barbara CA	T. A. Trygar	U.S. Dept. of Energy	26
81(1233)03	11/30 - 12/5/80	Mechanical Properties of Concrete Systems	ditto	G. M. Sabnis & B. L. Meyers	Sheladia Associates Bechtel Power Corp.	33
82(1236)05	1/11-16/81	Systems Engineering for Power VIII	Mid-Pines Resort Southern Pines, NC	T. A. Trygar	U.S. Dept. of Engery	34
81(1228)06	1/18-23/81	Chemical Process Control II	The Cloister Sea Island, GA	D. E. Seborg & T. E. Edgar	U. of Cal.—Santa Barbara U.T. Austin TX	119
81(1242)09	2/22-27/81	Environmental Engineering in the Food Processing Industry XI	Miramar Hotel Santa Barbara, CA	G. R. Nelson & J. Cooper	West. Area Power Adm. National Food Processors Assoc.	75
81(1244)12	4/26 - 5/1/81	Computerized Interpretation of the ECG VI	ditto	R. Bonner	International Business Machines	63

TABLE 3 Engineering Foundation Conferences (continued)

File Number	Meeting Date	Title	Location	Chairman	Chairman's Affiliation	Att.
81(1220)14	6/7-12/81	Advances in Fermentation Recovery Process Technology	Banff Springs Hotel Alberta, Canada	A. S. Michaels H. Blanch A. Kaufman	Stanford University Univ. of California Merck, Sharpe and Dohme	207
81(1248)15	6/14-19/81	Goods Transportation in Urban Areas	Tidewater Inn Easton, MD	G. P. Fisher A. H. Meyburg	Cornell Univ. ditto	53
81(1238)16	6/28 - 7/3/81	Issues in Control of Urban Traffic Systems	New England Col. Henniker, NH	W. S. Levine E. Lieberman K. Kobetsky	Univ. of Maryland KLD Associates WVA Dept. of Highways	46
81(1247)17	7/5-10/81	Biomechanics and Neural Control of Movement II	ditto	R. B. Stein L. R. Young	Univ. of Alberta Mass. Inst. of Techn.	83
81(1252)18	7/12-17/81	Experimental Research in Fouling and Slagging	ditto	R. M. Bryers	Foster Wheeler Development Corp.	95
81(1255)19	7/17-22/81	Small Hydroelectric Development	ditto	J. L. Warren	Southwest Hydro Corp.	66
81(1253)20	7/26-31/81	Computer-Aided Design for the Energy Industry	ditto	K. C. Lu	Bechtel National, Inc.	40
81(1183)21	8/2-7/81	Advanced Coal Preparation	ditto	A. W. Deurbrouck	Dept. of Energy	103
81(1216)22	8/9-14/81	Mixing VIII	ditto	D. E. Leng	Dow Chemical Co.	88
81(1218)23	8/16-21/81	Diagnostic Immunology III	ditto	A. Luderer	Corning Glass Works	92
81(1263)29	7/26-31/81	Engineering Education: Aims and Goals of the 1980's	Franklin Pierce Col., Rindge, NH	W. R. Corcoran	Calif. Inst. of Techn.	
81(1256)30	8/2-7/81	Interfacial Phenomena in Mineral Processing	ditto	B. Yarar D. J. Spottiswood	Colorado School of Mines	58
81(1260)31	8/9-14/81	Toxicological, Ethical and Policy Issues in Hazardous Waste Management	ditto	P. A. Vesilind	Duke University	54
81(1259)32	8/16-21/81	Manufacturing Engineering for Tomorrow's Needs	ditto	G. E. Brosseau R. Nagel	National Science Found. Internat. Harvester Co.	53
81(1217)33	9/20-26/81	Enzyme Engineering VI	Shima Kanko Hotel Kashikojima, Japan	S. Fukui	Kyoto University	209
82(1266)01	10/4-9/81	Increasing Engineer Effectiveness through Improved Information Systems	Tidewater Inn Easton, MD	D. Heckard	Armco, Inc.	55

TABLE 3 Engineering Foundation Conferences (continued)

File Number	*Meeting Date*	*Title*	*Location*	*Chairman*	*Chairman's Affiliation*	*Att.*
82(1240)02	10/25-30/82	Fatigue and Corrosion Fatigue at Ultrasonic Frequencies	Seven Springs Hotel Champion, PA	J. R. Wells G. Buck J. Tien	Westinghouse El. Corp. Rockwell Internatl. Corp. Columbia Univ.	43
82(1201)03	1/3-8/82	Minerals and Metal Extraction with Minimum Water Usage and Pollution	Miramar Hotel Santa Barbara, CA	P. Somasundaran D. A. Dahlstrom	Columbia Univ. Envirotech Proc. E. Co.	58
82(1214)04	1/3-8/82	Updating Subsurface Soil Sampling and In Situ Testing	ditto	S. Saxena	Illinois Inst. of Tech.	89
82(1225)05	1/17-22/82	Modeling Neuromuscular Systems	ditto	G. C. Agarwal G. L. Gottlieb	Univ. of Illinois Rush Presbyterian Hosp.	35
82(1265)06	2/28 - 3/4/82	Environmental & Energy Engineering in the Food Industry XII	ditto	J. L. Cooper P. S. Halberstadt	Nat. Food Process. Assoc. General Foods Corp.	59
82(1264)07	3/7-12/82	Stack Sampling and Source Evaluation X	ditto	R. P. Perkins	E.I. du Pont de Nemours & Co.	81
82(1271)08	4/25-30/82	Cytometry in the Clinical Laboratory	ditto	B. H. Mayall L. J. Marton	Lawrence Livermore Lab. U. of Calif.—San Fran.	116
82(1268)09	5/2-7/82	Computerized Interpretation of ECG VII	ditto	M. Laks	Harbor UCLA Medical Center	77
82(1245)10	5/18-21/82	Microbial Enhancement of Oil Recovery	Shangri-la Hotel Afton, OK	E. C. Donaldson B. Clark	Dep. of Energy Phillips Petroleum Co.	131
82(1269)11	6/26 - 7/2/82	Enhancement of Life Safety from Fire—A Technological Response	Franklin Pierce Col., Rindge, NH	P. R. DeCicco	Polytechnic Institute of NY	26
82(1275)12	7/11-16/82	Land Policy through Taxation	ditto	W. Rybeck P. Finklestein	Cntr for Public Dialogue Cntr for Local Tax Res.	76
82(1258)13	7/18-23/82	New Advances in Signature Analysis	ditto	J. S. Mitchell	J. S. Mitchell, Inc.	41
82(1246)14	7/18-23/82	Advances in Hard Facing Technology	New England Col. Henniker, NH	S. J. Matthews P. A. Kammer	Cabot Corp. Entectic Corp.	61
82(1254)15	7/25-30/82	Characterization and Performance Prediction of Cements & Concretes	ditto	J. F. Young G. Frohnsdorf	Univ. of Illinois Nat. Bureau of Standards	53
82(1273)16	8/1-6/82	Stormwater Detention Facilities	Franklin Pierce College Rindge, NH	L. S. Tucker B. Urbonas W. DeGroot	Urban Drainage and Flood Control District ditto	78

TABLE 3 Engineering Foundation Conferences (continued)

File Number	Meeting Date	Title	Location	Chairman	Chairman's Affiliation	Att.
82(1280)17	8/8-13/82	Industry Response to the Hazardous Waste Challenge	ditto	R. E. Cummings J. Pierce G. Hanks	Cummings Associates Duke University Union Carbide Corp.	23
82(1267)18	9/5-11/82	Shotcrete IV: Water Tunnel Construction	Los Lanceros Hotel Paipa, Colombia	E. King A. Marulanda	PBQ&D Ingetech	56
82(1251)19	9/19-24/82	Biochemical Engineering III	Miramar Hotel Santa Barbara, CA	K. Venkatsubramanian	H. J. Heinz Co. and Rutgers University	183
83(1257)01	10/31 - 11/5/82	Fouling of Heat Exchange Surfaces	Pocono Hershey White Haven, PA	R. W. Bryers	Foster Wheeler Development Co.	73
83(1262)02	11/7-12/82	Combustion of Tomorrow's Fuels	Miramar Hotel Santa Barbara, CA	B. K. Biswas	Foster Wheeler Development Co.	78
83(1276)03	11/14-19/82	Multiobjective Analysis in Water Resources Planning and Management	ditto	Y. Y. Haimes	Case-Western Reserve U.	34
83(1289)06	2/27 - 3/3/83	Environment and Energy Engineering in the Food Processing Industry	ditto	P. Halberstadt	General Foods Corp.	63
83(1281)07	3/13-18/83	Double Diffusive Convection	ditto	D. Johnson C. F. Chen	SERI Univ. of Arizona	80
83(1293)08	4/22-27/83	Computerized Interpretation of the ECG VIII	Banff Spring Hotel Alberta, Canada	R. H. Sylvester	Univ. of Southern Calif.	81
83(1278)09	5/6-11/83	Fundamentals of Adsorption	Schloss Elmau Bavaria, W. Germ.	G. Belfort A. L. Myers	Rensselaer Pol. Inst. U. of Pennsylvania	104
83(1250)10	5/29 - 6/3/83	Fluidization IV	Shima Kanko Hotel Kashikojima, Japan	D. Kunii R. Toei	University of Tokyo ditto	210
83(1270)11	6/26 - 7/1/83	Traffic Monitoring and Control Systems	New England Col. Henniker, NH	S. Yagar J. Lam	Univ. of Waterloo Delcan Consultants	37
83(1277)12	7/10-15/83	Biomechanics and Neural Control of Movement	ditto	R. Nashner D. Childress	Good Samaritan Hosp. Northwestern Univ.	100
83(1296)14	7/24-29/83	The Undergraduate Engineering Laboratory	ditto	E. W. Ernst	Univ. of Illinois	103
83(1291)15	7/31 - 8/5/83	Modeling of Casting and Welding Processes	ditto	J. A. Dantzig J. T. Berry	Olin Metals Res. Labs. Georgia Inst. of Techn.	103

TABLE 3 Engineering Foundation Conferences (continued)

File Number	*Meeting Date*	*Title*	*Location*	*Chairman*	*Chairman's Affiliation*	*Att.*
83(1287)17	8/14-19/83	Mixing IX	ditto	D. W. Hubbard	Michigan Tech. Inst.	115
83(1306)18	8/21-26/83	Implementing Resource and Energy Recovery Systems	ditto	R. M. Grunninger	C. C. Johnson & Assoc.	62
83(1283)19	9/25-30/83	Enzyme Engineering VII	Pocono Hershey White Haven, PA	A. I. Laskin	Exxon Res. & Eng. Co.	213
84(1290)01	10/2-7/84	Stack Sampling and Stationary Source Evaluation	ditto	F. Hopton	Ontario Research	72
84(1302)03	10/16-21/83	Management of Atmospheres in Tightly Enclosed Spaces	Miramar Hotel Santa Barbara, CA	J. E. Janssen	Honeywell, Inc.	46
84(1372)04	10/23-28/83	Chemical Reaction Engineering	ditto	E. B. Nauman	Rensselaer Pol. Inst.	64
84(1297)05	10/30 - 11/4/83	Emerging Computer Techniques in Stormwater and Flood Control	Pillar and Post Niagara-on-the Lake, Canada	W. James H. C. Torno	McMaster Univ. EPA	66
84(1303)06	11/6-11/83	Structural Failures—Their Cause and Prevention	Miramar Hotel Santa Barbara, CA	E. O. Pfrang	American Society of Civil Engineers	63
84(1301)07	12/7-12/83	Clinical Cytometry	The Cloister Sea Island, GA	M. Andreff M. Melamed	Memorial Sloan-Kettering Cancer Inst.	89
84(1285)08	1/8-13/84	Impact of Coal Quality on Downstream Processes—Advanced Coal Cleaning	Miramar Hotel Santa Barbara, CA	R. S. Seghal P. T. Luckie	Praxis Engineering, Inc. Pennsylvania State U.	96
84(1309)09	7/29 - 8/3/84	Compressibility Phenomena in Subsidence	New England Col. Henniker, NH	S. K. Saxena E. Donaldson	Illinois Inst. of Techn. Univ. of Oklahoma	44
84(1286)10	1/29 - 2/3/84	Recovery of Fermentation Products	The Cloister Sea Island, GA	C. L. Cooney K. G. Taksen	Mass. Inst. of Techn. Pfizer, Inc.	205
84(1315)11	1/15-20/84	Professional Liability	Miramar Hotel Santa Barbara, CA	J. R. Janney	Wiss, Janney, Elstner & Associates	45
84(1317)12	2/19-24/84	Environmental and Energy Engineering in the Food Processing Industry	ditto	N. K. Talbert	Agway, Inc.	68
84(1313)13	6/24-29/84	Irradiation Technology in the Food Industries	Tidewater Inn Easton, MD	J. S. Sivinski S. Ahlstrom	CH2M Hill ditto	61
84(1322)14	6/13-18/84	Computerized Interpretation of the Electrocardiogram IX	ditto	D. W. Mortara	Mortara Instruments	78

TABLE 3 Engineering Foundation Conferences (continued)

File Number	*Meeting Date*	*Title*	*Location*	*Chairman*	*Chairman's Affiliation*	*Att.*
84(1314)16	7/8-13/84	Signature Analysis V	New England Col. Henniker, NH	G. Muller	Exxon Research and Engineering	43
84(1294)19	7/29 - 8/3/84	Recent Developments in Fouling and Slagging from Coal Combustion	Copper Mountain Resort Copper Mountain, CO	R. E. Barrett	Battelle—Columbus	85
84(1336)20	8/5-10/84	The Space Shuttle Experiment and Environment Workshop	New England Col. Henniker, NH	M. Lauriente	NASA	114
84(1305)23	8/19-24/84	Science and Technology of Energy Mineral Processing	ditto	B. Yarar	Colorado School of Mines	54
84(1319)24	9/30 - 10/5/84	Biochemical Engineering IV	Great Southern Galoway, Ireland	H. Lim	Purdue University	125
85(1288)02	10/16-21/84	New Directions in Separations Technology	Posthotel & Congress Hall Davos, Switzerland	P. Fottrell; N. Li	Univ. of Galloway; UOP, Inc.	84
85(1312)03	10/21-26/84	Combustion of Tomorrow's Fuels II	ditto	B. K. Biswas	Foster Wheeler Dev. Corp.	59
85(1321)04	10/21-26/84	Managing the Effects of Electromagnetic Energy	Pocono Hershey Resort White Haven, PA	J. Milano	Port Authority of NY & NJ	47
85(1321)05	11/11-16/84	Groundwater Contamination	Miramar Hotel Santa Barbara, CA	S. Koslov; Y. Haimes	Johns Hopkins Univ.; Case-Western Reserve U.	35
85(1351)06	12/2-7/84	Selectively Doped Heterostructure Transistors	ditto	J. H. Snyder; R. Dingle	Univ. of Calif.—Davis; AT&T Bell Laboratories	67
85(1345)07	2/24-28/85	Funding Issues of Civil Engineering Education	ditto	G. Lee	SUNY—Buffalo	42
85(1335)08	1/27 - 2/1/85	Flocculation, Sedimentation and Consolidation	The Cloister Sea Island, Ga	R. Gallagher; B. Moudgil	Worcester Polytech. Inst.; University of Florida	76
85(1342)09	2/10-15/85	Sport Biomechanics	Miramar Hotel Santa Barbara, CA	R. C. Nelson; C. J. Dillman	Penn. State University; U.S. Olympic Committee	65
85(1339)10	2/24 - 3/1/85	Environmental & Energy Engineering in the Food Processing Industry XV	ditto	R. E. Pailthorp	CH2M Hill	64
85(1341)11	3/17-22/85	Stack Sampling and Stationary Source Emission Evaluation	ditto	B. Mullins	Mullins Environmental Testing Co., Inc.	85

TABLE 3 Engineering Foundation Conferences (continued)

File Number	*Meeting Date*	*Title*	*Location*	*Chairman*	*Chairman's Affiliation*	*Att.*
85(1354)12	4/21-26/85	Computerized Interpretation of the Electrocardiogram X	ditto	G. Wagner B. Scherlag	Duke University Univ. of Oklahoma	80
85(1347)13	7/7-12/85	Sludge Dewatering	New England Col. Henniker, NH	P. A. Vesilind	Duke University	38
85(1330)14	7/14-19/85	Biomechanics and Neural Control of Movement	ditto	J. M. Hollerbach B. W. Peterson	Mass. Inst. of Techn. Northwestern University	124
85(1311)16	7/28 - 8/2/85	Cement Manufacture and Use	ditto	G. Frohnsdorf	Nat. Bureau of Standards	50
85(1350)17	8/4-9/85	Mixing X	ditto	D. B. Todd	Baker Perkins, Inc.	120
85(1334)18	8/11-16/85	Fine Coal Processing	ditto	R. H. Yoon	Virginia P. I. & State U.	118
85(1353)19	8/18-23/85	Neural Prosthesis: Motor Control	ditto	J. T. Mortimer	Case-Western Res. Univ.	127
85(1328)25	9/22-27/85	Enzyme Engineering VIII	Marienlyst Hotel Elsinor, Denmark	K. Mosbach	University of Lund	202
86(1362)01	11/3-8/85	Risk-Based Decision Making in Water Resources Planning and Management	Miramar Hotel Santa Barbara, CA	Y. Haimes E. Stakhiv	Case-Western Reserve U. US Army Institute for Water Resources	43
86(1369)02	11/10-15/85	Irradiation Technology for the Food Industry	ditto	J. S. Sivinski	CH2M Hill	73
86(1349)03	12/8-13/85	Recent Developments in Comminution	Keauhou Beach Hot. Keauhou-Kena, HI	J. A. Herbst N. Arbiter	Univ. of Utah Columbia University	47
86(1359)04	1/5-10/86	Near Treshold Fatigue and Growth of Small Cracks	Miramar Hotel Santa Barbara, CA	R. O. Ritchie J. Lankford	U. of Calif.—Berkeley Southwest Res. Inst.	73
86(1366)05	1/12-17/86	Modeling and Control of Casting and Welding Processes	ditto	S. Kou	U. of Wisconsin-Madison	75
86(1355)06	2/23-28/86	Processing of Electronic Materials	Sheraton Hotel & Spa Santa Barbara, CA	R. Pollard C. G. Law, Jr. L. R. Dawson	U. of Houston AT&T Tech. Sandia National Labs	59
86(1367)07	3/2-7/86	Environmental and Energy Engineering in the Food Processing Industry	ditto	R. Sucher	Anheuser-Bush, Inc.	72
86(1357)08	3/16-21/86	Applications of Ceramics in the Electronics Industry	ditto	B. Schwartz L. Levinson	Internat. Business Mach. General Electric Co.	103

TABLE 3 Engineering Foundation Conferences (continued)

File Number	*Meeting Date*	*Title*	*Location*	*Chairman*	*Chairman's Affiliation*	*Att.*
86(1372)10	4/6-11/86	Computerized Interpretation of the Electrocardiogram XI	ditto	P. Kliegfield D. Mortara	Cornell Medical College Mortara Instrument Inc.	82
86(1331)11	5/5-9/86	Fundamentals of Adsorption II	ditto	A. I. Liapis D. Everett K. S. W. Sing	Univ. of Missouri—Rolla Univ. of Bristol Brunel University	101
86(1364)12	5/11-16/86	Incorporating Social and Environmental Objectives in Water Resources Planning and Management	ditto	W. Viessman, Jr. K. Schilling	Univ. of Florida U.S. Army Institute for Water Resources	51
86(1346)13	5/11-16/86	Recovery of Bioproducts	Hotel Gillet and Uppsala Univ. Uppsala, Sweden	C. L. Cooney G. Schmidt-Kastner I. Olsson	Mass. Inst. of Techn. Bayer AG. Pharmacia Biotech AG	173
86(1332)14	5/18-23/86	Fluidization V	Hotel Marienlyst Helsingor, Denmark	K. Ostergaard A. Sorensen	Technical University of Denmark	162
86(1381)15	4/20-25/86	Core Drilling for Ultradeep Scientific Targets	Sky Valley Resort Dillard, GA	M. Walton J. Rowley F. Schuh	Minn. Geological Survey Los Alamos National Lab. ARCO	85
86(1344)16	6/8-13/86	Qualitative Methods for the Analysis of Nonlinear Problems in Dynamics II	ditto	F. A. Salam M. Levi P. Holmes	Michigan State Univ. Boston University Cornell University	42
86(1379)17	6/15-20/86	Alternative Technologies for Hazardous Waste Management	ditto	H. M. Freeman R. A. Olexey	U.S. Environmental Protection Agency	61
86(1363)18	6/22-27/86	Urban Runoff Quality: Its Impacts and Quality Enhancement Technology	ditto	B. Urbonas	Urban Drainage & Flood Control District—Denver	72
86(1374)19	7/16-11/86	Coordination of Water Resources Planning in the US Federal System	New England Col. Henniker, NH	L. B. Dworsky D. J. Allee R. M. North	Cornell University ditto Univ. of Georgia	27
86(1370)20	7/27 - 8/3/86	Biochemical Engineering V	New England Col. Henniker, NH	W. A. Weigand	Illinois Institute of Technology	150
86(1382)21	7/20-25/86	University-Industry-Government Cooperation in Research and Technology Transfer	ditto	C. T. Hill	Congressional Research Services	54
86(1373)22	7/13-18/86	Signature Analysis VI	ditto	J. L. Frarey	Shaker Research Corp.	25

TABLE 3 Engineering Foundation Conferences (continued)

File Number	Meeting Date	Title	Location	Chairman	Chairman's Affiliation	Att.
86(1368)24	9/14-19/86	Stack Sampling and Stationary Source Emission Evaluation	Sky Valley Resort Dillard, GA	W. G. DeWees E. McCarley	Entropy Environmentalist U.S. Environm. Prot. Agency	96
87(1360)01	11/16-21/86	Bioceramics	Sheraton Hot. & Spa Santa Barbara, CA	P. Ducheyne J. Lemons	U. of Pennsylvania Univ. of Alabama	87
87(1400)02	12/1-6/86	Selectively Doped Heterostructure Transistors II	Keauhou Beach Hot. Keauhou-Kona, HI	J. Harris	Stanford Univ.	49
87(1395)03	1/25-30/87	Fine Coal Cleaning and Control of Fine Coal Cleaning Processes	Sheraton Hot. & Spa Santa Barbara, CA	H. Huttenhain R. Sehgal	Bechtel National Inc. Praxis Engineering	53
87(1375)04	2/22-27/87	UV and VUV Lasers	ditto	J. G. Eden	University of Illinois	69
87(1401)05	3/1-6/87	Environmental and Energy Engineering in the Food Processing Industry	ditto	A. Fehr	Fehr-Graham Associates	63
87(1365)06	3/8-13/87	Chemical Reaction Engineering II	ditto	M. P. Dudukovic F. Krambeck	Washington Univ. Mobile Research & Dev.	111
87(1388)07	3/29 - 4/3/87	Financing and Amortizing Water Resources Infrastructure	Sheraton Palm Coast Resort Palm Coast, FL	R. M. North	University of Georgia	36
87(1389)08	3/22-27/87	Ballistic Electrons for Transistors	ditto	L. Eastman M. Heimblum	Cornell University Internat. Business Mach.	76
87(1397)09	5/5-10/87	Computerized Interpretation of the Electrocardiogram XII	ditto	W. K. Haisty E. Berberi	Wake Forest Univ. Univ. of Oklahoma	78
87(1377)11	4/26 - 5/1/87	Separation Technology	Schloss Elmau Bavaria, W. Germ.	N. Li H. Strathmann	Allied Signal Res. C. Fraunhoffer Institute	93
87(1338)12	5/3-8/87	Fluid-Particle Interaction	Centr. Sporthotel & Convention Cntr. Davos, Switzerland	P. Rowe W. B. Russel	Univ. College, London Princeton University	37
87(1384)13	6/7-12/87	Composite Construction	New England Col. Henniker, NH	I. M. Viest	Consultant	96
87(1387)14	6/14-19/87	Traffic Control Systems	ditto	S. Yagar	Univ. of Waterloo	58
87(1378)16	6/28 - 7/3/87	Sludge Technology: Theory and Practice of Dewatering	ditto	P. A. Vesilind K. V. Nair	Duke University Columbia University	43
87(1385)18	7/12-17/87	Dental Prostheses	ditto	J. A. Tesk	NBS	54

TABLE 3 Engineering Foundation Conferences (continued)

File Number	*Meeting Date*	*Title*	*Location*	*Chairman*	*Chairman's Affiliation*	*Att.*
87(1392)19	7/19-24/87	Biomechanics and Neural Control of Movements	New England Col. Henniker, NH	G. Loeb M. Raibert	National Inst. of Health Mass. Inst. of Techn.	141
87(1404)20	7/26-31/87	Engineering to Minimize Generation of Hazardous Waste	ditto	H. Freeman	U.S. Environm. Prot. Agency	53
87(1386)21	8/2-7/87	Mixing XI	ditto	E. B. Nauman	Rensselaer Polyt. Inst.	107
87(1413)23	5/3-8/87	Role of the Social and Behavioral Sciences in Water Resources Management	Sheraton Hot. & Spa Santa Barbara, CA	Y. Y. Haimes D. D. Bauman	Case-Western Reserve U. Southern Illinois U.	44
88(1390)01	10/4-9/87	Enzyme Engineering IX	ditto	H. W. Blanch & A. M. Klibanov	Univ. of Calif., Berkeley Mass. Inst. of Technology	208
88(1337)02	10/18-23/87	Diagnostic Immunology	Longboat Key Hilton Sarasota, FL	H. H. Weetall & G. Odstrchel	Ciba-Corning Diagnostic Corp.	79
88(1393)03	10/25-30/87	In-Situ Recovery of Minerals	Sheraton Hot. & Spa Santa Barbara, CA	K. R. Coyne & J. B Hiskey	Bechtel, Inc. Univ. of Arizona	57
88(1407)04	11/8-13/87	Risk-Based Decision Making	ditto	Y. Y. Haimes & E. Z. Stakhiv	Univ. of Virginia U.S. Army Inst. for Water Resources	47
88(1391)05	12/6-10/87	Structural Failures II	Sheraton Palm Coast Resort, FL	I. M. Viest & C. P. Siess	Consultant Univ. of Illinois	50
88(1409)06	1/17-22/88	Chemical Engineering Education in a Changing Environment	Sheraton Hot. & Spa Santa Barbara, CA	S. I. Sandler	Univ. of Delaware	76
88(1412)07	1/24-29/88	Properties of Concrete at Early Ages	ditto	E. G. Nawy & N. J Carino	Rutgers University Nat. Bureau of Standards	22
88(1402)08	1/31 - 2/5/88	Cell Culture Engineering	Sheraton Palm Coast Resort, FL	W.-S. Hu & A. J. Sinskey	Univ. of Minnesota Mass. Inst. of Technology	160
88(1398)09	2/7-12/28	Using Knowledge Based Expert Systems in the Engineered and Constructed Environment	Sheraton Hotel & Spa Santa Barbara, CA	L. F. Cohn & R. A. Harris	Univ. of Louisville ditto	48
88(1396)10	4/17-22/88	Modeling of Casting and Welding Processes	Sheraton Palm Coast Resort, FL	A. F. Giamei & G. J. Abaschian	Un. Technologies Res. C. Univ. of Florida	114

TABLE 3 Engineering Foundation Conferences (continued)

File Number	*Meeting Date*	*Title*	*Location*	*Chairman*	*Chairman's Affiliation*	*Att.*
88(1403)11	4/17-22/88	Recovery of Biological Products IV	Keauhou Beach Hotel, Kona, HI	M. R. Ladish, I. Holzberg & W. C. McGregor	Purdue University Syntex Corp. XOMA Corp.	161
88(1417)12	11/15-20/87	Irradiation Technology for the Food Industry	Sheraton Hot. & Spa Santa Barbara, CA	H. Farrar	Rockwell International	120
88(1410)13	3/6-11/88	Goods Transportation in Urban Areas V	ditto	G. P. Fisher & R. A. Staley	Cornell University R. A. Staley Consulting	30
88(1422)14	4/10-15/88	Computerized Interpretation of the Electrocardiogram XIII	Sheraton Palm Coast Resort, FL	E. Berbari & N. Flowers	Univ. of Oklahoma Medical College of GA	98
88(1415)15	1/10-15/88	Flocculation and Dewatering	ditto	B. J. Scheiner D. M. Moudgil	U.S. Bureau of Mines Univ. of Florida	57
88(1426)16	2/21-26/88	Mineral Matter and Ash Deposits in Coal	Sheraton Hot. & Spa Santa Barbara, CA	K. Vorres & R. W. Bryers	Argonne National Lab. Foster Wheeler Corp.	114
88(1436)17	2/28 - 3/4/88	Environmental & Energy Engineering in the Food Processing Industry	ditto	D. G. Kirk & M. N. Beach	Heinz USA N-Con Systems Inc.	61
88(1425)18	3/13-18/88	Stack Sampling and Stationary Source Emission Evaluation	ditto	E. McCarley & C. Duncan	U.S. Envir. Prot. Agency Monsanto	116
88(1414)19	7/17-22/88	Neural Prosthesis-Motor System	Trout Lodge Potosi, MO	D. R. McNeal	Rancho Rehabilitation Center	122
88(1423)20	7/10-15/88	Urban Runoff	ditto	L. A. Roesner & B. Urbonas	University of Idaho Rensselaer Pol. Inst.	82
88(1383)21	7/31 - 8/5/88	Advances in Cement Manufacture and Use	ditto	E. Gartner	W. R. Grace & Co.	70
88(1418)24	3/20-25/88	Advanced Materials and Processes for High Density Interconnect	Sheraton Hot. & Spa Santa Barbara, CA	L. M. Levinson & R. W. Vest	General Electric Co. Purdue University	104
88(1428)26	2/21-26/88	Industrial Application in Synchrotron Radiation	Sheraton Palm Coast Resort, FL	B. Batterman & G. Shenoy	Cornell University Argonne National Lab.	33
88(1427)27	7/10-15/88	Multidisciplinary Approaches to Snow Engineering	Sheraton Hot. & Spa Santa Barbara, CA	R. L. Sack & M. J. O'Rourke	University of Idaho Rensselaer Pol. Inst.	50
88(1431)28	7/31 - 8/5/88	US Technology Policy	Mercersburg Academy, PA	C. T. Hill & R. L. Stern	Congressional Res. Ser. Consultant	56

TABLE 3 Engineering Foundation Conferences (continued)

File Number	*Meeting Date*	*Title*	*Location*	*Chairman*	*Chairman's Affiliation*	*Att.*
88(1440)29	8/7-12/88	Hazardous Waste Management	ditto	R. S. Magee & A. N. Lindsey	NJ Inst. of Techn. Environ. Prot. Agency	53
88(1430)30	9/18-23/88	Bulk Power System Voltage Stability & Security	Trout Lodge Potosi, MO	L. H. Fink & J. E. Koehler	Carlsen & Fink Assoc. NY State Port Authority	66
89(1406)01	10/2-7/88	Biochemical Engineering VI	Sheraton Hotel Santa Barbara, CA	W. E. Goldstein	ESCA Genetic Corp.	138
89(1435)02	10/9-14/88	Engineering for Waste Reduction II	ditto	H. A. Freeman & A. Purcell	Environ. Prot. Agency Resource Policy Inst.	84
89(1433)03	10/30 - 11/4/88	Biotechnology Applications in Hazardous Waste Treatment	Longboat Key Hilton Sarasota, FL	G. Lewandowski, E. Batzis & P. Armenante	New Jersey Institute of Technology	75
89(1420)04	12/5-10/88	Advanced Heterostructure Transistors	Keauhou Beach Hotel, Kona, HI	R. A. Kiehl	International Business Machines	66

TABLE 4 Research Initiation Grants

File No.	Project Period	Grant Recipient	Project Title	Research Agency	Founder Society	Grant $
RC-A-75-3-A	1977-78	D. Athanasiou-Grivas	Correspondence Between the Conventional Factor of Safety and the Probability of Failure of Soil Structures	Rensselaer Polytechnic Institute	ASCE	10,000
RC-A-75-3-B	1977-81	J. J. K. Daemen	Roof Support Mechanics of Fully Grouted Untensioned Bolts in Flat Bedded Deposits	U. of Arizona	AIME	10,000
RC-A-75-3-C	1977-79	N. G. Dagalakis	Design and Testing of a Special Fiber Reinforced Composite for Human Joint Prosthesis Applications	U. of Maryland	ASME	10,000
RC-A-75-3-D	1977-78	D. C. Jeutter	A Multiple Channel Biomedical Telemetry System Using CMOS Technology	Marquette U.	IEEE	10,000
RC-A-75-3-E	1977-78	A. Attar	The Kinetics of Pore Nucleation in American Coal During Coking	U. of Houston	AIChE	10,000
RC-A-76-2-A	1977-78	H. T. Shen	Effects of Uniform Currents on Wave Forces	Clarkson College	ASCE	10,000
RC-A-76-2-B	1977-78	L. H. Burck	Development of High Toughness Brazed Joints	U. of Wisconsin—Milwaukee	AIME	10,000
RC-A-76-2-C	1977-79	C. K. Law	Combustion of Oil/Water Emulsions in Diesel Environments	Northwestern U.	ASME	10,000
RC-A-76-2-D	1977-78	B. Keane	Period/Amplitude/ Pattern Analysis of the Electroencephalogram by Microcomputer	Clemson Univ.	IEEE	10,000
RC-A-76-2-E	1977-79	R. L. Shambaugh	Three-Phase Ozonation of Aqueous Organics	Syracuse U.	AIChE	10,000
RC-A-77-1A	1978-79	T. E. Higgins	Heavy Metals and Wastewater Reuse	Arizona St. U.	ASCE	10,000
RC-A-77-1B-1	1978-79	J. M. M. Rendu	Geostatical Methods for the Evaluation of Low Grade Mineral Deposits	U. of Wisconsin	AIME	10,000
RC-A-77-1B-2	1978-80	G. A. Pope	Improved Oil Recovery by Chemical Flooding	U. of Texas	AIME	10,000
RC-A-77-1C-1	1978-80	R. L. Dooley	Pattern Recognition Applied to Cardiovascular Measurements	Mississ. St. U.	ASME	10,000
RC-A-77-1C-2	1978-80	R. L. Levin	The Water Permeability of Biological Cells During Freezing	Cornell U.	ASME	10,000
RC-A-77-1D-1	1978-79	C. R. Johnson	Extended Input Matching for Adaptive Control	Virginia Polyt. Inst. & St. U.	IEEE	10,000

TABLE 4 Research Initiation Grants (continued)

File No.	*Project Period*	*Grant Recipient*	*Project Title*	*Research Agency*	*Founder Society*	*Grant $*
RC-A-77-1D-2	1978-81	J. L. LoCicero	A Variable Bit Rate Adaptive Delta Modulation System for Improved Source Encoding	Illinois Inst. of Technology	IEEE	10,000
RC-A-77-1E-1	1978-80	D. T. Hayhurst	Binary Diffusion of Small Inorganic Gases into Molecular Sieve Zeolites	Pennsylvania State Univ.	AIChE	10,000
RC-A-77-1E-2	1978-79	A. S. Myerson	Bacterial Processes for the Removal of Pyritic Sulphur from Coal	U. of Dayton	AIChE	10,000
RC-A-77-6A-1	1979-80	A. C. Singhai	Strength Characteristics of Buried Jointed Pipeline	Arizona St. U.	ASCE	11,000
RC-A-77-6A-2	1979-80	L. F. Geschwinder, Jr.	Nonlinear Dynamic Analysis of Cable Roof Networks	Pennsylvania State Univ.	ASCE	11,000
RC-A-77-6B-1	1979-82	C. L. Jensen	Effect of Hydrogene on the Tensile Properties of Niobium-Tantalum Alloys	U. of Minnesota—Minneapolis	AIME	11,000
RC-A-77-6B-2	1979-81	R. H. Yoon	Coal Flotation in Inorganic Salt Solutions	Virginia Pol. I.	AIME	11,000
RC-A-77-6C-1	1979-81	M. Hull	Biomechanics of Human Leg Injuries	U. of Cal.—Davis	ASME	11,000
RC-A-77-6C-2	1979-80	R. D. Flack	An Experimental Investigation of Multilobe Bearings	U. of Virginia	ASME	11,000
RC-A-77-6D-1	1979-80	J. L. Merz	Investigation of Integrated and Guided-Wave Optical Devices	U. of Calif.—Santa Barbara	IEEE	11,000
RC-A-77-6D-2	1979-81	P. I. Cohen	Interface Crystallography	U. of Minnesota—Minneapolis	IEEE	11,000
RC-A-77-6E-1	1979-80	W. E. King, Jr.	An Experimental Investigation of a Novel Technique for Removing Heavy-Metal Ions from Dilute Wastewater	U. of Maryland	AIChE	11,000
RC-A-77-6E-2	1979-80	D. R. Lloyd	Reverse Osmosis Separation of Aqueous Aromatic Hydrocarbon Solutions	Virginia Polyt. Inst. & St. U.	AIChE	11,000
RC-A-78-6A-1	1980-81	D. Y. Tan	Effect of Interface Geometry on the Performance of an Earth Dam	U. of Calif. Los Angeles	ASCE	12,000
RC-A-78-6A-2	1980-82	J. H.-W. Lee	Multiport Thermal Diffusers as Line Momentum Sources in Shallow Water	U. of Delaware	ASCE	12,000

TABLE 4 Research Initiation Grants (continued)

File No.	Project Period	Grant Recipient	Project Title	Research Agency	Founder Society	Grant $
RC-A-78-6B-1	1980-81	P. R. Taylor	A Theoretical and Experimental Investigation of Removal of Arsenic from Hydrometallurgical Systems	U. of Idaho	AIME	12,000
RC-A-78-6B-2	1980-81	R. J. Livak	The Effect of Copper Addition and Zn/Mg Ratio on the Quench Sensitivity of Al-Zn-Mg Alloys	Washington St. U.	AIME	12,000
RC-A-78-6C-1	1980-81	M. Donath	Synthesis of Walking Control for the Lower Extremity Disabled	U. of Minnesota Twin Cities	ASME	12,000
RC-A-78-6C-2	1980-83	J. G. Pinto	A Constitutive Law for the Active Cardiac Muscle	San Diego State University	ASME	12,000
RC-A-78-6D-1	1980-81	S. K. Case	Wavelength-Coded Image Transmission	U. of Minnesota	IEEE	12,000
RC-A-78-6D-2	1980-81	R. L. Magin	A New Strategy for Selective Drug Delivery to Microwave Heated Tumors	U. of Illinois Urbana-Champaign	IEEE	12,000
RC-A-78-6E-1	1980-83	T. F. Anderson	Measurement and Correlation of Liquid-Liquid Equilibrium Data	U. of Connecticut	AIChE	12,000
RC-A-78-6E-2	1980-81	J. Mijovic	Long Term Durability of Thermosetting Polymers	Polytech. Inst. of New York	AIChE	12,000
RC-A-78-6E-3	1980-82	R. Nagarajan	Development of Surfacant Vesicles for Chemical Separation Processes	Pennsylvania State University	AIChE	12,000
RC-A-79-1A(1)	1981-84	W. C. McCarthy	Plate Bending Elastoplastic Finite Element with Application to Bridge Structures	New Mexico State Univ.	ASCE	14,000
RC-A-79-1A(2)	1981-83	M. Y. Corapcioglu	Averaged Subsidence Equations for Saturated-Unsaturated Deformable Porous Medium	U. of Delaware	ASCE	14,000
RC-A-79-1A(3)	1981-84	G. W. Jaworski	Mechanics of Hydraulic Fracturing in Soils	U. of New Hamps.	ASCE	14,000
RC-A-79-1B(1)	1981-83	M. A. Mahtab	Mechanics of Gas Outbursts in Louisiana Salt Mines	Columbia U.	AIME	14,000
RC-A-79-1B(2)	1981-83	T. D. Roy	Laser Welding of Nickel-Chromium-Molybdenum Containing Steels	Pennsylvania State Univ.	AIME	14,000

TABLE 4 Research Initiation Grants (continued)

File No.	*Project Period*	*Grant Recipient*	*Project Title*	*Research Agency*	*Founder Society*	*Grant $*
RC-A-79-1B(3)	1981-82	R. M. German	Structural Refractory Metals Processed by Activated Sintering	Rensselaer Polytech. Inst.	AIME	14,000
RC-A-79-1C(1)	1981-82	H. P. Jorgl	Integrated Turbocompressor Surge and Process Control	Arizona St. U.	ASME	14,000
RC-A-79-1C(2)	1981-84	W. J. W. Wepfer	Thermo-Economic Analysis, Design and Optimization of Energy Systems	Georgia Inst. of Technology	ASME	14,000
RC-A-79-1C(3)	1981-82	J. J. Casey	Restrictions on the Behavior of Finitely Deforming Elastic-Plastic Materials	U. of Houston	ASME	14,000
RC-A-79-1D(1)	1981-84	J. Cheung	Implementation Issues in Parallel Algorithms	U. of Oklahoma	IEEE	14,000
RC-A-79-1D(2)	1981-82	M. El-Sharkawi	Identification of Dynamic Equivalents—Choice of Model and Topology for the Equivalent External Power System	U. of Washington at Seattle	IEEE	14,000
RC-A-79-1D(3)	1981-84	C. Houstis	Software Partitioning in a Distributed Environment	Univ. of South Carolina	IEEE	14,000
RC-A-79-1E(1)	1981-82	G. F. Fuller	Dynamics of Adsorbed Polymer Molecules to Flow	Stanford Univ.	AIChE	14,000
RC-A-79-1E(2)	1981-82	S. Kovenklioglu	Determination of Heat Transfer Coefficients in Packed Beds with Chemical Reaction Accompanied by Severe Temperature Gradients	Stevens Institute of Technology	AIChE	14,000
RC-A-79-1E(3)	1981-83	H. E. Hager	Reaction Kinetics of Heavily Doped Wide Bandgap Semiconductors	U. of Washington	AIChE	14,000
RI-A-82-1	1982-84	D. J. Goodings	Centrifugal Modeling of Slope Failures	U. of Maryland	ASCE	16,000
RI-A-82-2	1982–85	J. D. Aristizabal -Ochoa	Proportioning of Walls and Beams for Optimum Seismic Structural Performance of Slender Coupled Wall Systems	Vanderbilt U.	ASCE	16,000
RI-A-82-3	1982-85	J. R. O'Leary	Development of a Stiffened Plate Finite Element	Illinois Inst. of Technology	ASCE	16,000
RI-A-82-4	1982-84	G. A. Irons	Investigations in Injection Metallurgy	McMaster Univ.	AIME	16,000
RI-A-82-5	1982-84	A. H. King	Diffusion Induced Grain Boundary Migration	SUNY at Stony Brook	AIME	16,000

TABLE 4 Research Initiation Grants (continued)

File No.	*Project Period*	*Grant Recipient*	*Project Title*	*Research Agency*	*Founder Society*	*Grant $*
RI-A-82-6	1982-83	I. Weiss	Effect of Accumulating Strain on the Recrystallization of Austenite	Wright State U.	AIME	16,000
RI-A-82-7	1982-87	M. C. Leu	Modeling and Analysis of Flow-Induced Vibration in Circular Saws	Cornell Univ.	ASME	16,000
RI-A-82-8	1982-84	S. Ramadhyavi	Experimental and Computational Studies of Solid-Liquid Phase Change	Tufts Univ.	ASME	16,000
RI-A-82-9	1982-84	J. W. Grant	Platelet Deposition on Arterial Walls at Flow Separation Sites	Virginia Polyt. and State U.	ASME	16,000
RI-A-82-10	1982-84	J. R. Burger	Extended Data-stationary Instructions to Simplify the Microprogramming of Pipelined Cellular Array Central Processors	University of the Pacific	IEEE	16,000
RI-A-82-11	1982-84	L. M. Li	Load Balancing Protocol Design for Local Computer Networks	Michigan State Univ.	IEEE	16,000
RI-A-82-12	1982-84	P. Markenscoff	Parallel Computation Problems: Task Partitioning and Bus Scheduling in Multiple Processor Systems and Performance Evaluation of Locking Algorithms in Database Management Systems	Univ. of Houston	IEEE	16,000
RI-A-82-13	1982-84	P. K. Agrawal	Bimetallic Supported Clusters in Fischer-Tropsch Synthesis	Georgia Inst. of Technology	AIChE	16,000
RI-A-82-14	1982-84	K. F. Jensen	Experimental and Theoretical Studies of the Transient Pore Structure of Coal	U. of Minnesota	AIChE	16,000
RI-A-82-15	1982-84	V. L. Punzi	Theoretical and Experimental Investigations on the Solute Rejection Mechanism in Reverse Osmosis	Villanova U.	AIChE	16,000
RI-A-83-1	1983-85	A. J. Durani	Effect of Slab on the Behavior of External R/C Beam to Column Connections Subjected to Earthquake Type Loading	Rice Univ.	ASCE	17,000
RI-A-83-2	1983-86	H. H. Shen	The Stress-Strain Rate Relationship in a Fluid-Solid Mixture with Solids of Two Sizes	Clarkson College of Technology	ASCE	17,000

TABLE 4 Research Initiation Grants (continued)

File No.	Project Period	Grant Recipient	Project Title	Research Agency	Founder Society	Grant $
RI-A-83-3	1983-84	J. A. Van Lund III	Biaxial Bending and Lateral-Torsional Buckling of Cold-Formed Steel Z-Beams	Colorado St. U.	ASCE	17,000
RI-A-83-4	1983-85	C. J. Bise	Mathematical Modelling of Longwall Mining Systems	Pennsylvania State Univ.	AIME	17,000
RI-A-83-5	1983-85	J. A. Todd	Kinetics and Mechanisms of the Interphase Precipitation Reaction	U. of Southern California	AIME	17,000
RI-A-83-6	1983-84	J. M. Prusa	Radial Transformation Methods in Engineering Problems with Irregular Moving Boundaries	Iowa State U.	ASME	17,000
RI-A-83-7	1983-85	C. W. Somerton	Natural Convection and Heat Transfer in Multilayer Porous Media	Louisiana State Univ.	ASME	17,000
RI-A-83-8	1983-85	M. W. Trethewey	Development of an Experimental Acoustical Modal Analysis Technique for Control of Interior Vehicle Noise	Pennsylvania State Univ.	ASME	17,000
RI-A-83-9	1983-85	B. Jabbari	Protocols for Multichannel Satellite Packet Reservation Systems	Southern Illinois Univ.	IEEE	17,000
RI-A-83-10	1983-85	S. H. Leung	Fixed-Point Implementation of Digital Controller	U. of Colorado	IEEE	17,000
RI-A-83-11	1983-85	J. M. Munro	The Influence of Pyritic Materials on the Spontaneous Combustion of Coal	S. Dakota School of Mines & Tech.	AIChE	17,000
RI-A-83-12	1983-86	A. P. Zioudas	Diffuse Reflectance FT/IR Studies of Gas Solid-Catalyzed Reacting Systems	Stevens Inst. of Technology	AIChE	17,000
RI-A-84-1	1984-86	R. J. Finno	Evaluation of Creep-Induced Effective Stress Changes in Clay	Illinois Inst. of Technology	ASCE	17,000
RI-A-84-2	1984-87	J. L. Kauschinger	Extracting Multi-Yield Surface Model Parameters from Pressuremeter Data: Theoretical Considerations	Tufts University	ASCE	17,000
RI-A-84-3	1984-87	T.-K. Tsay	Numerical Modeling of Water-Wave Propagation	Syracuse Univ.	ASCE	17,000
RI-A-84-4	1984-85	J. E. Indacochea	Effect of Inclusion Morphology and Size on the Microstructure and Toughness of HSLA Steel Welds	Univ. of Illinois	AIME	17,000

TABLE 4 Research Initiation Grants (continued)

File No.	*Project Period*	*Grant Recipient*	*Project Title*	*Research Agency*	*Founder Society*	*Grant $*
RI-A-84-5	1984-86	R. H. Wagoner	Strain Localization: Effect of Temperature Gradients	Ohio State Univ.	AIME	17,000
RI-A-84-6	1984-86	R. C. Brown	An Assessment of Ionic Mechanisms in Soot Formation	Iowa State Univ.	ASME	17,000
RI-A-84-7	1984-86	J. M. Cuschieri	Active Vibration Control for Machinery Noise Control	Florida Atlantic Univ.	ASME	17,000
RI-A-84-8	1984-86	H. N. Hashemi	Nondestructive Evaluation of Composite Materials Subjected to Combined Fatigue and Impact Loading	Northeastern U.	ASME	17,000
RI-A-84-9	1984-86	T.-H. Chao	Large Capacity Optical Correlator by Wavelength-Angle Multiplexing	Univ. of Utah	IEEE	17,000
RI-A-84-10	1984-85	C.-H. Wu	A Kinematic Error Model for Improving the Accuracy of Robot Manipulators	Northwestern U.	IEEE	17,000
RI-A-84-11	1984-86	R. Bizios	A Model for Pulmonary Neutrophil Dynamics	Rensselaer P. I.	AIChE	17,000
RI-A-84-12	1984-86	A. V. Someshwar	Corona Discharge Effects in Gas-Solid Adsorption	University of New Hampshire	AIChE	17,000
RI-A-85-1	1985-88	C. J. Miller	Flow Through Clay Liners: Model Prediction and Field Observation	Wayne State U.	ASCE	17,000
RI-A-85-2	1985-87	R. Siddharthan	Dynamic Effective Stress Response of Surface and Embedded Footings in Sand	Univ. of Nevada, Reno	ASCE	17,000
RI-A-85-3	1985-87	R. O. Mines, Jr.	Evaluation of Oxygen Uptake Rate as an Activated Sludge Process Control Parameter	Univ. of South Florida	ASCE	17,000
RI-A-85-4	1985-87	D. A. Chin	A Study of the Influence of Surface Waves on Ocean Outfall Dilution	Univ. of Miami	ASCE	17,000
RI-A-85-5	1985-87	K. Okazaki	Rapid Solidification—Powder Metallurgy Materials	U. of Kentucky	AIME	17,000
RI-A-85-6	1985-88	S. A. Marinello	A Numerical Study of Oil Recovery by Steam Injection Following Water Flooding	U. of Alabama	AIME	17,000

TABLE 4 Research Initiation Grants (continued)

File No.	*Project Period*	*Grant Recipient*	*Project Title*	*Research Agency*	*Founder Society*	*Grant $*
RI-A-85-7	1985-87	P. D. Platkowski	Further Development and Application of a Carrier Fringe Technique for Elimination of the Ambiguity in Holographic Interferometry	GMI Engineering and Management Institute	ASME	17,000
RI-A-85-8	1985-87	K. P. Rajurkar	In-Process Detection of Off-line Optimization of ECM	U. of Nebraska-Lincoln	ASME	17,000
RI-A-85-9	1985-88	R. M. McKillip, Jr.	Kinematic Observers for Rotor Vibration Control	Princeton U.	ASME	17,000
RI-A-85-10	1985-88	F. Lombardi	Diagnosable Systems For Fault Tolerant Computing	U. of Colorado	IEEE	17,000
RI-A-85-11	1985-88	C. Nguyen	Decentralized Stabilization of Linear Time-Varying Large Scale Systems	Catholic U. of America	IEEE	17,000
RI-A-85-12	1985-87	B. J. Van Wie	Development and Characterization of a Continuous Centrifugal Bio-Reactor (CCBR)	Washington State Univ.	AIChE	17,000
RI-A-85-13	1985-86	A. K. Datye	A High Resolution Electron Microscope Study of Structural Changes in Heterogeneous Catalysts Caused by Metal-Support Interactions	U. of New Mexico	AIChE	17,000
RI-A-85-14	1985-86	M. J. McCready	Characterization of Turbulent Mass Transit of a Solid Boundary	U. of Notre Dame	AIChE	17,000
RI-A-86-1	1986-87	K. M. Strzepek	An Expert System for Hazardous Waste Management in the Aquatic Environment	Texas A&M Univ.	ASCE	17,000
RI-A-86-2	1986-88	M. Picornell	Design of Clay Liners to Minimize the Effect of Shrinkage Cracks	U. of Texas, El Paso	ASCE	17,000
RI-A-86-3	1986-88	D. Znidarcic	Influence of Drains on Slope Stability	U of Colorado, Boulder	ASCE	17,000
RI-A-86-4	1986-	B. G. Thomas	Application of Mathematical Models to Investigate Defect Formation During Direct-Charging of Continuously-Cast Steel	U. of Illinois, Champaign-Urbana	AIME	17,000
RI-A-86-5	1986-87	R. J. Barnes	Optimal Estimation of Mine Ventilation System Parameters	U. of Minnesota	AIME	17,000

TABLE 4 Research Initiation Grants (continued)

File No.	*Project Period*	*Grant Recipient*	*Project Title*	*Research Agency*	*Founder Society*	*Grant $*
RI-A-86-6	1986-88	J. Cimbala	An Experimental Investigation of the Turbulent Structure in Two-Dimensional Momentumless Wakes	Pennsylvania State Univ.	ASME	17,000
RI-A-86-7	1986-88	M. Gharib	Flow Velocity Measurement by Digital Image Processing of Variable Intensity Particle Traces	U. of Calif.—San Diego	ASME	17,000
RI-A-86-8	1986-87	M. Ahmadian	Kinematic Modeling and Positioning Error Analysis of Flexible Gears	Clemson U.	ASME	17,000
RI-A-86-9	1986-88	C. L. DeMarco	Security and Robustness Measures for Electric Power Systems Subject to Random Load Disturbances	U. of Wisconsin	IEEE	17,000
RI-A-86-10	1986-	C. A. Zukowski	Using Management of Bounded Uncertainty in Digital MOS Circuit Simulation	Columbia U.	IEEE	17,000
RI-A-86-11	1986-88	S. J. Parulekar	Investigation of Effect of Culture and Plasmid Size on Behavior of Recombitant Ascherichia and Bacillus Subtilis Cultures	Illinois Inst. of Technology	AIChE	17,000
RI-A-86-12	1986-87	R. M. Enick	Direct Viscosity Enhancement of High Pressure Carbon Dioxide	U. of Pittsburgh	AIChE	17,000
RI-A-86-13	1986-88	D. Elsworth	Rock Mass Hydraulic Conductivity Enhancement Resulting from Longwall Mining	Pennsylvania State Univ.	AIME	17,000
RI-A-87-1	1987-88	J. C. Evans	Hydraulic Fracturing of Soil-Bentonite Slurry Trench Cutoff Walls	Bucknell U.	ASCE	20,000
RI-A-87-2	1987-88	L. C. Bank	A Theoretical and Experimental Investigation of Thin-Walled Fiber-Reinforced Plastic Beams	Rensselaer Polyt. Inst.	ASCE	20,000
RI-A-87-4	1987-88	A. W. Cramb	Grain Refinement During the Solidification of Steel	Carnegie Melon Univ.	AIME	20,000
RI-A-87-5	1987-88	J. Ari-Gur	Iso-Response Curve for Anisotropic Plates Under Lateral Pressure Pulse	Western Michigan U.	AIME	20,000
RI-A-87-6	1987-88	J. Bentsman	Vibrational Control of Systems with Time Delay	U. of Illinois, Champaign-Urbana	ASME	20,000
RI-A-87-7	1987-	B. Javidi	Application of Optical Computing Techniques to Spread Spectrum Signal Detection	Michigan State University	ASME	20,000

TABLE 4 Research Initiation Grants (continued)

File No.	*Project Period*	*Grant Recipient*	*Project Title*	*Research Agency*	*Founder Society*	*Grant $*
RI-A-87-8	1987-	L. P. Trombetta	Oxide Charging and Interface Trap Generation in Thin MOS Oxides	U. of Houston	IEEE	20,000
RI-A-87-9	1987-88	V. A. Burrows	Real Time In-Situ Analysis of the Kinetics of GaAs Anodic Oxidation	Arizona State University	AIChE	20,000
RI-A-87-10	1987-	R. R. Reinhart	Reaction Kinetics of a CF4 Plasma Si02 Semiconductor Etch	Texas Tech. University	AIChE	20,000
RI-A-87-11	1987-88	C. P. Lion	Monitoring Pipeline Integrity by Real Time Flow Simulations	U. of Idaho	ASCE	20,000
RI-A-87-12	1987-	S. S. Adams	Development of Engineering Approaches for the Practical Application of Geologic Information to Mineral Exploration and Prospect Evaluation	Colorado School of Mines	AIME	20,000
RI-A-87-13	1987-	D. A. Streit	An Analytical and Experimental Investigation of the Effects of Parametric Excitation in Robotic Manipulators	Pennsylvania State Univ.	ASME	20,000
RI-A-88-1	1988-	K. Kasai	Design Equations for Eccentric Bracing Based on Linear and Nonlinear Modeling	Illinois Inst. of Techn.	ASCE	20,000
RI-A-88-2	1988-	J. E. Tobiason	Fundamental Aspects of Particle Deposition in Porous Media: Effect of Particle Size Distribution	U. of Mass. at Amherst	ASCE	20,000
RI-A-88-3	1988-	A. W. Blair	Optimization Techniques for Minimizing Surface Irrigation Efficiencies Using Real-Time Simulation Modeling	New Mexico State University	ASCE	20,000
RI-A-88-4	1988-	A. B. Doucet	A New Description of Yielding and Its Application to Complex Loading-Path Problems	Louisiana State Univ.	AIME	20,000
RI-A-88-5	1988-	S. D. Thompson	Development of Design Methodologies for Sizing Segregation Bunkers in Conveyor Belt Networks	U. of Illinois, Champaign-Urbana	AIME	20,000
RI-A-88-6	1988-	M. A. Adenremi	Compositional Hydrodynamic Modeling of Gas Condensate Flow in Natural Gas Pipelines	Pennsylvania State Univ.	AIME	20,000

TABLE 4 Research Initiation Grants (continued)

File No.	*Project Period*	*Grant Recipient*	*Project Title*	*Research Agency*	*Founder Society*	*Grant $*
RI-A-88-7	1988-	C. L. Chan	New Method of Numerical Modeling for Laser Materials Processing	U. of Arizona	ASME	20,000
RI-A-88-8	1988-	K. D. Cole	Transient Surface Measurements with Heated Films—Analysis and Experiment	Univ. of Nebraska, Lincoln	ASME	20,000
RI-A-88-9	1988-	J. G. Georgiadis	Fundamental Studies of Transport in Two-Phase Media of Disordered Structure	Duke Univ.	ASME	20,000
RI-A-88-10	1988-	T. M. Egan	Scanning Tunneling Spectroscopy of Si-Si02 Interfaces	Carnegie Mellon Univ.	IEEE	20,000
RI-A-88-11	1988-	J.-P. Tsai	A Knowledge-Based Approach for Debugging Real-Time Software Systems	U. of Illinois	IEEE	20,000
RI-A-88-12	1988-	J. Dunn	A Novel Antenna and Resonator for Mic's Using Guided Waves	U. of Colorado	IEEE	20,000
RI-A-88-13	1988-	K. Y. S. Ng	In-Situ Raman Characterization of Superconducting Y1 Ba Cu3 FxOy Material as a Function of Process Parameters	Wayne State U.	AIChE	20,000
RI-A-88-14	1988-	S. T. Oyama	Synthesis and Reactivity of High Surface Area Epitaxial Structures	Clarkson U.	AIChE	20,000

TABLE 5 Committees of the Engineering Foundation

#	*Committee Name*	*Chair's Term*	*& Name*	*Committee Members' Names & Period(s) of Service†*
1	on Applications	1915	A. R. Ledoux	M. I. Pupin, J. W. Smith, A. C. Humphries
		1916	M. I. Pupin	
2	Executive	1918–19	M. I. Pupin	E. D. Adams 18–25; W. F. M. Goss 18, 24; S. H. Woodard 18–19;
		1919–20	W. F. M. Goss	C. W. Hunt 18; D. S. Jacobus 19; H. H. Porter 19–23;
		1920–25	C. F. Rand	G. H. Pegram 20–23; F. B. Jewett 20–24; E. A. Sperry 24–26;
		1925–29	L. B. Stillwell	G. A. Orrok 25–29; A. L. Walker 25; J. V. Davies 26–27 A. M. Greene, Jr. 26–27; A. D. Little 27–29; R. Ridgway 28–29; J. V. N. Dorr 28–29, 33–34;
		1929–33	H. H. Porter	O. E. Hovey 31, 34–37; G. H. Clevenger 31; G. L. Knight 31; H. C. Bellinger 31, 33; G. H. Fuller 32; C. E. Skinner 32;
		1933–34	G. W. Fuller	G. D. Barron 32; D. R. Yarnall 32–37; H. P. Charlesworth 33;
		1934–36	H. P. Charlesworth	E. R. Fish 34–35; A. L. J. Queneau 35–40, 42;
		1936–39	F. M. Farmer	H. I. Schlichter 36–40, 42–43; G. E. Beggs 37–39; F. F. Colcord 38, 40–41; J. H. R. Arms (1) 37–61;
		1939	G. E. Beggs	K. H. Condit 38–40, 42, 43; J. D. Justin 40–41;
		1939–42	O. E. Buckley	J. P. H. Perry 41–44; C. E. Stephens 41–44; J. S. Casey 44–47;
		1942–45	A. J. L. Queneau	R. H. Chambers 44–49; R. E. Dougherty 44–45;
		1945–48	A. B. Kinzel	A. B. Kinzel 44–45, 48–52; C. R. Jones 45–47;
		1948–49	J. D. Justin	O. E. Buckley 46–49; G. L. Knight 47–48;
		1949–51	B. A. Bakhmeteff	J. W. Barker 47–48, 52–53; H. Weisberg 48–51, 52–53;
		1951	C. G. Suits	D. A. Quarles 49–52; C. G. Suits 49–52; L. G. Holleran 51–58;
		1951–53	H. Weisberg	W. M. Peirce 52–53, 54–57; E. M. Barber 54–55, 57–61;
		1953–55	J. W. Barker	C. S. Purnell 54–55; J. F. Fairman 55–56, 58–59;
		1955–57	E. M. Barber	R. J. Wiseman 55–56; C. B. Molineaux 56–69, 61, 63–64;
		1957–59	W. M. Peirce	E. I. Green 56–58; G. O. Curme, Jr. 57–59; A. M. Gaudin 58–59;
		1959–61	A. M. Gaudin	J. D. Tebo 60–62, 65–68; C. W. Borgmann 60;
		1961–62	A. H. Kidder	W. E. Lobo 61–62, 74–76; T. J. Fratar 62; E. L. Robinson 62;
		1962–66	W. C. Schreiner	L. F. Hickernell 62; S. S. Cole 63–64; J. A. Hitchcock 63–64; A. V. Loughren 63–64, 69–75; E. Kirkendall (1) 63–64;

TABLE 5 Committees of the Engineering Foundation (continued)

#	*Committee Name*	*Chair's Term*	*& Name*	*Committee Members' Names & Period(s) of Service†*
		1966–67	W. D. Alexander, III	S. W. Marras (1) 60–62; M. A. Mason 65; J. S. Smart, Jr. 65–68; C. L. Tutt, Jr. 63–67, 74–75; J. A. Zecca (1) 65–75;
		1967–73	C. L. Tutt, Jr.	W. C. Schreiner 66, 68–72; W. D. Alexander, III 68–73, 75; R. P. Genereaux 69; W. J. Harris, Jr. 70; W. K. MacAdam 70;
		1973–75	P. E. Queneau	W. E. Glenn 71; P. E. Queneau 72–73, 76; H. D. Guthrie 73, 77;
		1975–77	R. W. Sears	T. W. Mermel 74; A. B. Giordano 74, 76, 78, 82–84;
		1977–79	M. A. Cordovi	R. H. Tatlow III 76; O. S. Bray 77–78; G. F. Habach 77;
		1979–80	A. B. Giordano	R. M. McGeorge 77–80; E. W. Herold 78; C. E. Ford 78–80;
		1980–82	R. P. Genereaux	J. A. Haddad 79, 85–86; M. A. Cordovi 80; F. F. Aplan 81–85;
		1982–83	J. A. Haddad	M. C. Shaw 81; R. K. Hellmann 81–84; W. H. Manogue 81–83;
		1983–85	I. M. Viest	H. Bieber 84–86; A. E. Brandt, Jr. 85; D. Swan 86; H. Shaw 85–86;
		1985–87	F. F. Applan	I. M. Viest 86; R. S. Fein 86; P. Somasundaran 86;
		1987–	H. Shaw	A. S. West 86; A. D. Korwek (1) 82–86; H. A. Comerer (2) 86;
3	Policy as to Patents & Standard Form for a Grant	1920–21	H. H. Porter	E. W. Rice, Jr., C. Herschel, W. F. M. Goss
4	Mental Hygiene of Industry	1920	W. F. M. Goss	E. W. Rice, Jr., J. P. Channing, T. T. Read
5	Hydraulic Research	1920–23		J. W. Smith, S. H. Woodard
6	Industrial Education& Training	1920	W. F. M. Goss	G. H. Pegram, I. N. Hollis, J. W. Richards, A. W. Berresford G. B. Pegram
7	Declaration of Trust & Form of Deed of Gift	1920		E. D. Adams
8	Control of Publication of Reports	1920–23		E. D. Adams
9	Endowment Increase	1920–23		C. F. Rand 20–23;
		1924	C. F. Rand	E. D. Adams, W. L. Sanders & G. A. Orrok 24;
10	Steam Tables Research	1921		D. S. Jacobus
11	Paint-on-Wood Research	1921		A. H. Sabin, G. H. Pegram & A. D. Flinn 21–24;
	Wood Finishing Research	1922–24		

TABLE 5 Committees of the Engineering Foundation (continued)

#	*Committee Name*	*Chair's Term & Name*		*Committee Members' Names & Period(s) of Service†*
12	Relations with NRC Div. of Engineering**	1921–22		G. H. Pegram
13	Arch Dam Investigation	1923–24	C. Derleth, Jr.	F. A. Noetzli (1), D. C. Henny, M. M. O'Shaughnessy, H. H. Porter, F. E. Weymouth, S. H. Woodard, P. Bailey & W. Meredith (a) 23–33; W. F. McClure 23–25; R. P. McIntosh (a) 23–27; R. E. Davis (a) & J. Hinds (a) 24–33;
		1925–33	C. D. Marx	A. D. Flinn 25–28; H. W. Dennis & C. Derleth, Jr. 25–33; J. L. Savage & I. E. Houck (a) 26–33; G. S. Binckley 30–33;
14	Engineering Encyclopedia	1923		S. H. Woodard
15	Publication	1924		E. D. Adams, A. D. Flinn
16	Plan	1924	C. F. Rand	E. D. Adams, J. V. Davies, B. Gherardi; W. F. M. Goss, A. L. Walker
17	Relations with Founder Societies	1924		W. F. M. Goss
18	Iron Alloys	1929–44	G. B. Waterhouse	C. H. Herty, Jr. (a) & G. K. Burgess 29–31; S. Turner 29–33; L. Jordan (r) & H. W. Gillett 29–34; J. A. Mathews 29–35; R. E. Kennedy 29–36; J. Strauss, B. Stoughton, T. H. Wickenden & J. H. Critchett 29–46; J. T. MacKenzie 31–46;
		1945–47	H. W. Gillett	R. S. Dean (a) 32–37; J. W. Finch 34–39; J. G. Thompson 36–44; J. L. Briggs 34–46; J. Johnston 36–46; W. Sykes 37–46; R. R. Sayers 40–46; J. S. Marsh (ed.) 41–42;
		1948–58	J. L. Gregg	F. T. Sisco (ed.) 32–58;
19	Education Research	1931–34	H. N. Davis	H. B. Smith & W. B. Plank 31; W. F. Whittemore, R. I. Rees, A. H. White & A. R. Stevenson, Jr. 31–34; W. S. Rodman, T. T. Read & J. W. Baker 32–34; H. P. Hammond (a), A. D. Flinn (1) & D. H. Killeffer (ed.) 32;
20	Research Procedure	1931–37	D. R. Yarnall	T. Merriman 31–37; F. M. Becket 31–33, 37; W. H. Fulweiler 31–41; L. W. Chubb 31–43; E. DeGolyer 31–34; A. D. Flinn 32–37; J. A. Mathews 33–35; S. Tour 37–42;
		1938–39	F. F. Colcord	K. H. Condit 38–40; F. C. Scobey 38; H. E. Wessman 39–46;

TABLE 5 Committees of the Engineering Foundation (continued)

#	*Committee Name*	*Chair's Term & Name*		*Committee Members' Names & Period(s) of Service†*
		1940–42	O. E. Buckley	E. M. T. Ryder 40–44; W. Trinks 42–46;
		1942–44	K. H. Condit	W. M. Peirce 43–46; A. L. J. Queneau 43–45; W. A. Lewis 44–46;
		1944–48	L. W. Chubb	E. A. Prentis 45–46; A. B. Kinzel 46; B. A. Bakhmeteff 47–51; J. F. D. Smith 47–48; E. F. Colcord 47–48;
		1948–51	B. A. Bakhmeteff	E. R. Kaiser 50–51; W. A. Newman 49; C. G. Suits 49–51; H. Weisberg 50–51;
		1951–52	H.Weisberg	A. M. Gaudin 51–53, 56–57; T. Saville & C. S. Purnell 51–54;
		1952–53	T. Saville	H. Weisberg 52–53; E. M. Barber 52–53, 55–57;
		1953–56	A. M. Gaudin	J. W. Barker 53–55; E. L. Robinson 53–57; E. K. Timby & E. I. Green 54–56;
		1956–57	E. K. Timby	R. J. Wiseman 56–58;
		1957–59	A. M. Gaudin	M. A. Mason, R. C. Allen, W. C. Schreiner & W. M. Peirce 57–59; H. A. Kidder 58–59;
21	Study of Industrial System	1932	J. H. Fahey	H. Aaron, W. Green, M. E. Leeds, W. C. MacFarlane, R. V. Wright
22	Welding Research	1935–43	C. A. Adams	H. C. Boardman, E. Chapman, J. H. Critchett, J. C. Crowe, C. L. Eksergian, A. J. Ely, D. S. Jocobus, G. F. Jenks & C. L. Eksergian, 35–43; J. Jones, F. T. Lewellyn & A. Weymouth (a) 35–39; H. M. Hobart & P. G. Lang, Jr. 35–41; A. S. Douglass 35–42; F. H. Frankland & R. E. Zimmerman 39–43; A. E. Pew & A. R. Wilson 40–43; O. U. Cook, T. S. Fuller & A. C. Weigel 41–43; A. S. Albright, D. Arnott, I. Harter & C. W. Williams 42–43;
23	Research	1960–62	W. C. Schreiner	B. G. Ballard, A. Oblad & E. P. Partridge 60; E. M. Barber, A. M. Gaudin, R. E. Fadum & H. K. Work 60–61; S. W. Marras 60–62; A. B. Kinzel 60, 64; S. S. Cole 60–65; R. A. Kinckiner 61–62; S. R. Beitler 61, 65–70; E. A. Walker 62; N. M. Newmark 62–63; R. N. Faiman 62–63, 65; C. L. Tutt, Jr. 62, 68–71;
		1963–64	R. A. Kinckiner	H. L. Mason & E. Kirkendal 63–64; M. A. Mason 63–67, 71; E. H. Ohsol 63–71; J. A. Logan 64–65;

TABLE 5 Committees of the Engineering Foundation (continued)

#	*Committee Name*	*Chair's Term & Name*		*Committee Members' Names & Period(s) of Service†*
		1965–67	J. D. Tebo	W. D. Alexander, III 65; J. A. Zecca 65–71; W. J. Harris, Jr. 66–68; T. W. Mermel & C. Concordia 66–71;
		1968–70	M. A. Mason	W. K. MacAdam 68–69; E. F. von Wettberg, Jr. 68–70; D. A. Dahlstrom 69; F. W. Luerssen 70–71;
		1970–71	E. F. von Wettberg, Jr.	A. B. Giordano & S. Way 71;
24	Public Relations	1960–64	J. H. Hitchcock	M. O. Chenoweth & C. W. Swartout 60–62; C. Bayer, G. W. Bower & L. S. Dennegar 60–64; H. J. Kostkos 61; C. R. Landgren 63–64; H. R. Hands 64;
25	Financial Development	1960	G. O. Curme, Jr.	P. Rutledge 60–61; J. F. Thompson 60–61; H. E. Martin 60–61;
		1961	vacant	M. J. Kelly 60–61; R. Landau 60–61;
26	International Relations	1961	W. E. Lobo	B. F. Dodge, R. A. Morgen, W. V. Binger, W. G. Bowman, W. H. Wisely, G. Browne
27*	Coordinating on Air Pollution	1961–68	W. T. Ingram	H. E. Furgeson & L. B. Hitchcock 62; W. L. Winkelman & L. Dolkart 62–64; A. B. Mindler 63; C. K. Stoddard 63–65; H. I. Wolff 64–68; E. Wilson 65–68; A. Fletcher 66–68;
28*	Engineering Foundation Research Conferences	1962	W. C. Schreiner	H. K. Work 62; W. J. Harris, Jr. & T. O. Paine 62–64;
		1963–64	H. K. Work	R. A. Kinckiner & L. K. Wheelock 63–64; D. C. Taylor 64, 66–67;
		1965–67	W. J. Harris, Jr.	W. Hausz, T. P. Meloy & S. S. Cole (2) 66–67;
29*	Composite Materials Advisory	1963–68	J. H. Jackson	A. M. Blamphin, C. M. Flanagan, R. B. Mears, H. Perry, J. J. Soroka & H. S. Wolko 63–68;
30	Policy Review	1962	A. H. Kidder	H. K. Work, C. B. Molineaux, J. S. Smart, H. E. Martin, L. F. Hickernall, W. E. Lobo
31	Finance	1962–64	W. E. Lobo	A. Fletcher, J. D. Tebo, H. K. Work
32	Rules and Policy	1963–64	vacant	M. A. Mason 63–64; J. S. Smart, Jr. 63–65; A. B. Kinzel 63–64;
		1965	W. D. Alexander, III	C. L. Tutt, Jr. 63–64; J. H. Hitchcock 63–64; A. V. Loughren 63–64; W. E. Lobo 63–64;
33	Conferences	1968–70	W. Hausz	D. C. Taylor 68; W. J. Harris, Jr. 68–69; T. P. Meloy 68–74; D. E. Jones, Jr. 68–72; S. S. Cole (2) 68–84; C. Harrison 69;

TABLE 5 Committees of the Engineering Foundation (continued)

#	*Committee Name*	*Chair's Term & Name*		*Committee Members' Names & Period(s) of Service†*
				H. Bieber 68–70, 73, 75–79, 83–86; C. O. Harris 69–73; G. B. Devey 69–72, 74;
		1970–72	H. Bieber	W. Hausz 71; E. F. von Wettberg, Jr. 71–74, 77–85; L. G. Mayfield 72–73; T. W. Mermel 72, 75–77;
		1972–74	T. W. Mermel	V. W. Uhl 73–77; L. H. Fink 73–81; P. E. Queneau 74–75; M. C. Shaw & J. A. Zecca (1) 74–81; G. P. Fisher 74–82, 86;
		1974–76	E. F. von Wettberg, Jr.	H. H. Kellogg, A. V. Loughren & W. E. Reaser (3) 75; R. P. Genereaux 75–76, 80–82; R. W. Sears 76–77; R. W. Mann 76–81; M. A. Cordovi 76, 78–79, 81; F. F. Applan 76–77, 80–83, 85–86; H. A. Comerer (3)(2) 76–86;
		1976–79	R. P. Genereaux	J. R. Brand 78–79; A. B. Giordano 78–80; M. M. Denn 78–81; P. T. Luckie 78–82;
		1979–82	H. Bieber	R. W. Bryers 80–82; 84–85; J. A. Haddad 81–84; S. Kahne 82–83; D. E. Keairns, M. N. Salgo & A. D. Korwek (1) 82–86; D. Childress 82–85;
		1982–85	G. P. Fisher	G. F. Habach, R. K. Hellman & W. O. Winer 83; I. M. Viest & Somasundaran 83–85; H. Freitag & W. M. Sangster 84; F. Landis 84–86; E. W. Ernst & W. H. Taylor 85;
		1985–	P. Somasundaran	R. S. Fein, H. Shaw, C. B. Stott, A. I. Laskin, J. B. Reswick & A. S. West 86;
34	Projects	1972–73	E. F. von Wettberg, Jr.	E. O. Ohsohl & C. L. Tutt 72–73; M. A. Mason, F. W. Luerssen, S. Way & C. Concordia 72–74; A. B. Giordano 72–73, 77–80; J. A. Zecca (1) 72–81; T. W. Mermel 72–74, 76–78, 83;
		1973–77	A. B. Giordano	E. F. von Wettberg, Jr. 73–84; P. E. Queneau 74–75; M. A. Cordovi 74–79; A. V. Loughren 75; S. R. Beitler 75–76; W. H. Manogue 75–84, 86; J. Arthur 76; H. H. Kellogg 76–77; R. W. Sears 76–77, 79; R. K. Hellmann 76–80; J. Kestin 77;
		1977–78	R. W. Sears	G. F. Habach 77–78, 83–84; E. O. Pfrang 78; M. P. Freeman & S. Gratch 78–79; S. J. Angello 78–80; W. G. Prast 78–86;

TABLE 5 Committees of the Engineering Foundation (continued)

#	*Committee Name*	*Chair's Term & Name*		*Committee Members' Names & Period(s) of Service†*
		1978–80	C. B. Molineaux	C. G. Clyde & W. E. Lobo 79–82; J. H. Kelly 79–86; S. A. Newman 80–81; R. P. Genereaux 80–82; G. Reethof 80–83; I. M. Viest 80, 84–85;
		1980–83	I. M. Viest	J. A. Haddad 81–84; H. Shaw & A. D. Korwek (1) 82–86; F. H. Blecher 83–86;
		1983–	G. Reethof	A. E. Brandt 84; D. H. Hall & H. A. Comerer (2) 84–86; F. F. Aplan 85–86; R. S. Fein, G. F. Habach, R. B. Finch, R. L. Mattingly, D. Swan & A. S. West 86;
35	Development	1983–85	H. Bieber	W. H. Manogue 83–84; I. M. Viest 83–85; A. E. Brandt, Jr., H. A. Comerer (2) & A. D. Korwek (1) 82–86;
		1985–	F. F. Aplan	A. S. West 86;

* Subcommittee of Research Committee

** Committee on Relations with the NRC was appointed in 1919 with W. F. M. Goss as chairman and C. W. Hunt, S. M. Woodard and F. B. Jewett as members. The annual reports list this committee only for 1921–22.

†When omitted, period of service identical with chair's term.

(1) secretary; (a) alternate; (r) representative; (2) director; (3) assistant director

TABLE 6 Contributions to Engineering Foundation Endowment

Date of Gift	*Donor*	*Amount of Gift, $ †*	*Source of Information*
2/24/1915	Ambrose Swasey	200,000	Reference 1.5, p. 12
9/28/1918	Ambrose Swasey	100,000	Reference 1.5, p. 13
10/25/1920	Ambrose Swasey	200,000	Reference 1.10, p. 13
6/30/1925	Henry R. Towne estate	50,000	Reference 1.17, p. 5; Reference 1.10, p. 5; Board Minutes Dec. 16, 1924
1/20/1927	Edward Dean Adams	50,000	Reference 1.19, pp. 6 & 24–26; Board Minutes Febr. 17, 1927
1928	Seeley W. Mudd estate	9,500	Reference 1.22, p. 5; Board Minutes Febr. 24, 1928
1928	Central Hudson Gas and Electric Corporation	10,000	Reference 1.22, p. 5; Board Minutes Febr. 24, 1928
1931	Ambrose Swasey	250,000	Reference 1.34, p. 10
1936	Harry de Berkeley Parsons	4,500	Reference 1.50, p. 10; Board Minutes Mar. 12, 1936
1937	Ambrose Swasey estate	89,000	Reference 1.37, p. 13; Board Minutes Oct. 14, 1937
10/19/39	Sophie M. Gondron estate	25,000	Reference 1.39, p. 7
1939	Karl Emil Hilgard estate	1,800	Reference 1.39, p. 7; Board Minutes June 15, 1939
1939	Richard Khuem, Jr. estate	2,000	Reference 1.39, p. 7; Board Minutes Oct. 19, 1939
1949*	George D. Barron	122,600	Reference 1.50, p. 10
1949*	W. S. Barstow	12,500	Reference 1.50, p. 10
1949*	G. L. Knight	1,000	Reference 1.50, p. 10
1949*	Alfred D. Flinn	1,000	Reference 1.50, p. 10
1949*	miscellaneous gifts	600	Reference 1.50, p. 10
1953	John C. Parker	1,000	Reference 1.54, p. 7; Board Minutes June 18, 1953
1953	James F. Fairman	700	Reference 1.54, p. 7; Board Minutes June 18, 1953
1956	Blanche H. McHenry Memorial Fund	408,400	Reference 1.34, p. 10; Reference 1.57, pp. 10–11; Board Minutes May 16, 1957
1956	Orlan W. Boston	4,400	Reference 2.118, p. 1; Board Minutes May 3, 1956
1963	Ernest L. Robinson	1,000	Board Minutes May 1, 1963
1975	Jules Breuchaud	60,000	Board Minutes May 8, 1975
	Total	1,605,000	

* Gift received in the year shown or earlier.
† Rounded to the nearest $100.

TABLE 7 Financial Contributors to Individual Projects

*Arch Dam Investigation (19)**

Allis-Chalmers Company
Aluminum Company of America
American Society of Civil Engineers
Bankers' Group:
  Blyth, Witter & Company
  Bond & Goodwin & Tucker
  Coffin & Burr, Inc.
  First Securities Company
  Harris, Forbes & Company, Inc., Boston
  Harris, Forbes & Company, Inc., New York
  Harris Trust & Savings Bank
  Mercantile Trust Company
  National City Company
  E. H. Rollins & Sons
  Security Trust & Savings Bank
  Wm. R. Staats & Company
Bent Brothers
Blaw-Knox Company
Bureau of Standards
Burns & McDonnell Engineering Company
Byllesby Engineering & Management Corporation
William Cain
State of California
California Institute of Technology
California-Oregon Power Company
California Portland Cement Company
Coast and Geodetic Survey
Hugh L. Cooper & Company
County of Los Angeles
A. S. Crane
W. P. Creager
Arthur P. Davis
Electric Bond and Share Company
  Carolina Power & Light Company
  Idaho Power Company
  Minnesota Power & Light Company
  Pacific Power & Light Company
  Pennsylvania Power & Light Company
  Utah Power & Light Company
Electro-Metallurgical Company (Union Carbide Company)
Engineering Foundation
Alfred D. Flinn
Foundation Company
Gannett, Seelye & Fleming
General Electric Company
Giant Powder Company
Great Western Power Company
Halsey, Stuart & Company, Inc.
Ingersol-Rand Company
Insley Manufacturing Company
International Paper Company
M. W. Kellogg Company
The A. Lietz Company
Massachusetts Institute of Technology
Middle West Utilities Company
Monolith Portland Cement Company
Fred A. Noetzli
State of Oregon

TABLE 7 Financial Contributors to Individual Projects (continued)

Arch Dam Investigation (19) (continued)*	
Pacific Coast Steel Company	Sugar Pine Lumber Company
Pacific Gas & Electric Company	George F. Swain
H. deB Parsons	Arthur N. Talbot
Pelton Water Wheel Company	U. G. I. Contracting Company
Portland Cement Association	University of California
Portland Electrical Power Company (Portland Railway, Light and Power Company)	University of Colorado
	U.S. Bureau of Reclamation
Princeton University	U.S. Department of Agriculture
Rensselaer Valve Company	U.S. Indian Service
Riverside Portland Cement Company	Utica Gas & Electric Company
Dwight P. Robinson & Company, Inc.	H. D. Walbridge & Company
John A. Roebling's Sons Company	Washington Water Power Company
Sanderson & Porter	Water Works Supply Company
San Joaquin Light & Power Company	Western Pipe & Steel Company
Southern California Edison Company	Westinghouse Electric & Manufacturing Company
Southern Pacific Railway	William Wheeler
Stone & Webster, Inc.	J. G. White Engineering Company
Blast Furnace Slags Investigation (26)	
Colorado Fuel and Iron Company	International Harvester Company
Engineering Foundation	Jones & Laughlin Steel Company
Illinois Steel Company	Tennessee Coal, Iron & Railroad Company
Inland Steel Company	The Youngstown Sheet and Tube Company
Dielectric Absorption: Electric Insulation (27)	
The Acme Wire Company	American Steel and Wire Company
American Brass Company	Bakelite Corporation

TABLE 7 Financial Contributors to Individual Projects (continued)

Dielectric Absorption: Electric Insulation (27) (continued)	
Bell Telephone Laboratories, Inc.	Simplex Wire & Cable Company
Boston Insulated Wire & Cable Company	Standard Oil Development Company
Engineering Foundation	Standard Underground Cable Company
General Electric Company	Wagner Electric Company
The Okonite Company	Westinghouse Electric & Manufacturing Company
John A. Roebling's Sons Company	
Engineering Education Investigation (28)	
Adams, Edward Dean	Samuel Insull
American Brown Boveri Electric Corporation	International Nickel Company
American Institute of Electrical Engineers	Samuel Lewisohn
American Institute of Mining and Metallurgical Engineers	Jas. H. McGraw
American Society of Civil Engineers	Sangamo Electric Company
American Society of Mechanical Engineers	Henry D. Sharpe
American Telephone and Telegraph Company	Sullivan Machinery Company
American Water Works and Electric Company	Union Carbide Company
Babcock & Wilcox Company	United Shoe Machinery Corporation
Brown & Sharpe Manufacturing Company	Western Union Telegraph Company
Detroit Edison Company	Westinghouse Electric & Manufacturing Company
General Electric Company	Stone & Webster
Guggenheim Fund for Promotion of Aeronautics	
Alloys of Iron Research (32)	
Acme Steel Company	American Foundrymen's Association
Alan Wood Steel Company	American Locomotive Company
Algoma Steel Corporation, Ltd.	American Radiator Company
American Chain & Cable Company, Inc.	American Rolling Mill Company

TABLE 7 Financial Contributors to Individual Projects (continued)

Alloys of Iron Research (32) (continued)	
American Society for Metals	Jones & Laughlin Steel Corporation
Anaconda Copper Mining Company	Kennecott Copper Corporation
Atlantic Steel Company	Latrobe Steel Company
Babcock & Wilcox Company	Lukens Steel Company
Baldwin Locomotive Works	Mesta Machine Company
Battelle Memorial Institute	The Midvale Company
Bethlehem Steel Company	Morgan Construction Company
Buckeye Steel Castings Company	National Bureau of Standards
Bucyrus Erie Company	National Cash Register Company
Canadian Atlas Steels, Ltd.	National Malleable & Steel Castings Company
Carpenter Steel Company, The	Pacific Coast Steel Corporation
Climax Molybdenum Company	Phelps Dodge Corporation
Colonial Steel Company	Pittsburgh Steel Company
Colorado Fuel & Iron Company	Pullman Car & Manufacturing Corporation
Columbia Steel Company	Republic Steel Corporation
Columbia Tool Steel Company	John A. Roebling's Sons Company
Crane Company	Sharon Steel Corporation
Crucible Steel Corporation of America	Simonds Saw and Steel Company
Henry Disston & Sons	Singer Manufacturing Company
Dominion Steel and Coal Corporation, Ltd.	Souther California Iron & Steel Company
Electro Metallurgical Company	Spang, Chalfant & Company, Inc.
Firth Sterling Steel Company	Taylor-Wharton Iron & Steel Company
Follansbee Brothers, Inc.	Timken Steel & Tube Company
General Electric Company	U.S. Steel Corporation
General Motors Corporation	Vanadium-Alloys Steel Company
Great Lakes Steel Corporation	Vanadium Corporation of America
Harrisburg Steel Corporation	Vulcan Crucible Steel Company
Heppenstall Company	Warman Steel Castings Company
Inland Steel Company	Weirton Steel Company
The International Nickel Company, Inc.	Western Electric Company

TABLE 7 Financial Contributors to Individual Projects (continued)

Alloys of Iron Research (32) (continued)	
Westinghouse Electric & Manufacturing Company	Wisconsin Steel Company
Wheeling Steeling Corporation	
Welding Research Council (62)	
Air Reduction Company	Beech Aircraft Corporation
Air Reduction Sales Company	Bethlehem Shipbuilding Corporation, Ltd.
Alabama Drydock and Shipbuilding Company	Bethlehem Steel Corporation
Allis-Chalmers Manufacturing Company	Biggs Boiler Works Company
Alloy Fabricators, Inc.	Black, Sivalls & Bryson, Inc.
Aluminum Company of America	Blaw-Knox Company
The American Brass Company	S. Blickman, Inc.
American Bureau of Shipping	Boston Edison Company
American Gas and Electric Service Corporation	E. G. Budd Manufacturing Company
American Institute of Electrical Engineers	Bureau of Public Roads
American Institute of Steel Construction, Inc.	J. I. Case Company
American Iron and Steel Institute	Caterpillar Tractor Company
American Locomotive Company	Champion Rivet Company
American Petroleum Institute	Chase Brass & Copper Company
American Society for Metals	Chattanooga Boiler & Tank Company
American Society for Testing Materials	Chicago Bridge & Iron Company
American Society of Civil Engineers	Clark Controller Company
American Transit Association	Cleveland Electric Illuminating Company
American Welding Society	Climax Molybdenum Company
Arcos Corporation	Combustion Engineering Company, Inc.
Association of American Railroads	Consolidated Aircraft Corporation
Austin Company	Consolidated Gas Electric & Power Company of Baltimore
Babcock and Wilcox Company	Consolidated Edison Company of New York, Inc.
Baldwin Locomotive Works	Consolidated Steel Corporation, Ltd.
Bastian-Blessing Company	Consolidated Vultee Aircraft Corporation

TABLE 7 Financial Contributors to Individual Projects (continued)

Welding Research Council (62) (continued)

Copper and Brass Research Association
Crane Company
Curtiss-Wright Corporation—Airplane Division
Cutler-Hammer, Inc.
Detroit Edison Company
Downingtown Iron Works Company
Dravo Corporation
Du Pont de Nemours and Company
Duquesne Light Company
Edge Moore Iron Works, Inc.
Electric Arc Cutting & Welding Company
Engineering Foundation
Fairbanks, Morse & Company
Federal Shipbuilding and Dry Dock Company
Foote Mineral Company
Foster Wheeler Company
General American Transportation Corporations
General Electric Company
General Motors Corporation—Fisher Body Division
Girdler Corporation
Graver Tank & Manufacturing Company, Inc.
Griscom-Russell Company
Grumman Aircraft Engineering Corporation
Handy & Harman Company
Harnischfeger Corporation
Ingalls Shipbuilding Corporation
International Business Machines Company
International Harvester Company
International Nickel Company, Inc.
Kaiser Company, Inc.
Kelite Products, Inc.
The M. W. Kellogg Company
Leeds and Northrup Company
Lenape Hydraulic Pressing and Forging Company
R. G. Le Tourneau, Inc.
Lima Locomotive Works, Inc.
Lione Corporation
Lukens Steel Company
Metal and Thermit Corporation
Midwest Piping and Supply Company, Inc.
National Cash Register Company
National Cylinder Gas Company
National Supply Company
New York Shipbuilding Corporation
John Nooter Boiler Works Company
Northwest Engineering Company
Oakite Products, Inc.
Ohio Public Service Corporation
Oregon Shipbuilding Corporation
Pennsylvania Crusher Company
Permanente Metals Corporation
Philadelphia Electric Company
Phillips Petroleum Company
Pittsburgh-Des Moines Company
Henry Pratt Company
Public Roads Administration, Federal Works Agency
Public Service Electric and Gas Company
Pullman-Standard Car Manufacturing
Resistance Welder Manufacturers Association
Revere Copper & Brass, Inc.

TABLE 7 Financial Contributors to Individual Projects (continued)

Welding Research Council (62) (continued)	
Ryan Aeronautical Company	Union Carbide and Carbon Company
Scaife Company	United Engineering and Foundry Company
Shawinigan Products Corporation	United States Gauge Company
Singer Manugacturing Company	U.S. Maritime Commission
S. Morgan Smith Company	U.S. Navy—Bureau of Construction & Repairs
A. O. Smith Corporation	U.S. Navy—Bureau of Ships
Solar Aircraft Company	U.S. Navy—Bureau of Yards and Docks
Southern California Gas Company	U.S. Steel Corporation
Steel Cooperage Company	Utilities Research Commission
Steel Founders' Society of America	Henry Vogt Machine Company
Steel Products Company	Wagner Electric Corporation
Struthers-Wells	Warner & Swasey Company
Sun Shipbuilding & Dry Dock Company	Wellman Engineering Company
Tennessee Eastman Corporation	Western Pipe & Steel Company of California
Timken Roller Bearing Company	Westinghouse Electric & Manufacturing Company
Titanium Alloy Manufacturing Company	Worthington Pump and Machinery Corporation
Todd Shipyards Corporation	Wyatt Metal & Boiler Works
Treadwell Construction Company	York Corporation
Tube Turns, Inc.	York Safe and Lock Company
United Aircraft Corporation	
Impregnated Paper Insulation (66)	
American Gas & Electric Service Corporation	Hartford Electric Light Company
American Institute of Electrical Engineers	Insulated Power Cable Engineers Association
Boston Edison Company	Nebraska Power Company
Cincinnati Gas & Electric Company	Northern States Power Company
Consolidated Edison Company of New York, Inc.	Pennsylvania Water & Power Company
Detroit Edison Company	Philadelphia Electric Company
Duquesne Light Company	Potomac Electric Power Company

TABLE 7 Financial Contributors to Individual Projects (continued)

Impregnated Paper Insulation (66) (continued)	
Riegel Paper Company	Utilities Research Commission
Rochester Gas & Electric Corporation	Utilities Coordinating Research, Inc.
Union Mills Paper Manufacturing Company	
Insulating Oils and Cable Saturants (74)	
American Gas & Electric Service Corporation	Nebraska Power Company
American Institute of Electrical Engineers	Northern States Power Company
American Steel & Wire Company	Ohio Brass
Anaconda Wire & Cable Company	Okonite Company
Boston Edison Company	Pennsylvania Water & Power Company
Buffalo, Niagara & Eastern Power Corporation	Philadelphia Electric Company
Cincinnati Gas & Electric Company	Potomac Electric Power Company
Consolidated Edison Company of New York, Inc.	Rochester Gas & Electric Company
Consolidated Gas Electric Light & Power Company	John A. Roebling's Sons Company
Dayton Power & Light Company	Shell Development Company
Detroit Edison Company	Shell Oil Company
Duquesne Light Company	Sinclair Refining Company
Ebasco Services, Inc.	Socony-Vacuum Oil Company
General Cable Corporation	Standard Oil Company of California
General Electric Company	Standard Oil Company of Indiana
Gulf Research and Development Company	Standard Oil Development Company
Hartford Electric Light Company	Sun Oil Company
Kansas City Power & Light Company	The Texas Company
Kennecott Wire & Cable Company	Utilities Research Commission
National Electrical Manufacturers' Association	

* The listing includes contributors and co-operators.

TABLE 8 Financial Sponsors of Engineering Foundation Conferences

Year	*Name of Sponsor*	*Name of Sponsor*
1970	Bristol-Meyers International	Searl-Medidata, Inc.
	F. Hoffmann-La Roche & Company, Ltd.	Swiss Research Institute
	National Science Foundation	
1971	National Science Foundation	
1973	National Science Foundation	U.S. Department of Transportation
	U.S. Department of Health, Education and Welfare	
1974	National Institute of Cancer	U.S. Department of Health, Education and Welfare
	National Science Foundation	
1975	National Science Foundation	U.S. Department of Interior
	U.S. Department of Health, Education and Welfare	U.S. Department of Transportation
1976	Environmental Resources Development Administration	U.S. Department of Health, Education and Welfare
	National Science Foundation	U.S. Department of Transportation
	Contributions were made also by several industrial firms	
1977	Electric Power Research Institute	National Science Foundation
	Environmental Resources Development Administration	U.S. Department of Health, Education and Welfare
	Government of Federal Republic of Germany	U.S. Department of Transportation
	Contributions were made also by several industrial firms	
1978	Deutsche Forschungsgemeinschaft	National Science Foundation
	Deutsche Krebforschungszentrum	New York Academy of Sciences
	Electric Power Research Institute	U.S. Department of Energy
	Environmental Protection Agency	U.S. Department of Health, Education and Welfare
	Environmental Resources Development Administration	U.S. Department of Transportation
	Ministry of Health	
	Contributions were made also by several industrial firms	
1979	Abbott Laboratories, Inc.	Marquette Electronics Company
	Air Products and Chemical Company	Miles Laboratories, Inc.

TABLE 8 Financial Sponsors of Engineering Foundation Conferences (continued)

Year	*Name of Sponsor*	*Name of Sponsor*
	Alfa Laval AB	National Science Foundation
	Allied Chemical Foundation	Novo Laboratories, Inc.
	American Can Company	Ortho-Diagnostics
	Amersham Corporation	Owens-Illinois Company
	Anheuser-Busch Brewing Company	Pennsylvania Power and Light
	Astra Lakimedelk	Pfizer, Inc.
	BASF Wyandotte	Pharmacia Diagnostics
	Bayer AG	PPG Industries
	Bio-Dynamics Corporation	Snamprogetti SPA
	Boehringer Mannheim GmbH	Technicon Instruments Corporation
	Ciba-Geigy AG	Telemed Company
	Corning Glass Works	Union Carbide Corporation
	Dow Chemical Company	U.S. Department of the Army
	Electric Power Research Institute	U.S. Department of Energy
	Environmental Protection Agency	U.S. Department of Interior
	Fisher Scientific Company	U.S. Water Resources Council
	Hewlett-Packard Company	United States Steel Corporation
	International Business Machines Corporation	Upjohn Diagnostics
	Japanese Industry	Water and Wastewater Equipment Manufacturers Association
	Kaiser Aluminum Company	
1980	Abbot Laboratories, Inc.	National Science Foundation
	American Iron and Steel Institute	New York Academy of Sciences
	Beekman Instruments, Inc.	Pharmacia Diagnostics
	Carter-Wallace, Inc.	U.S. Department of Energy
	Hewlett-Packard Company	U.S. Department of Interior
	International Diagnostic Technology	Wampole Laboratories Division
1981	Abbott Laboratories, Inc.	Kraft, Inc.
	Ajinomoto Company, Ltd.	Kyowa Hakko Kogyo Company, Ltd.
	Assoreni	McNeil Pharmaceutical

TABLE 8 Financial Sponsors of Engineering Foundation Conferences (continued)

Year	*Name of Sponsor*	*Name of Sponsor*
	Bayer, A. G.	Marquette Electronics
	Beecham Group, Ltd.	Meiji Seika Kaisha, Ltd.
	Behringwerke, A. G.	Merk & Company Inc.
	Boehringer Mannheim, GmbH	Micromedic Systems, Inc.
	British Columbia Hydro and Power Authority	Miles Kali-Chemie, GmbH
	British Petroleum Company	Miles Laboratories
	Burdick Corporation	Miles Laboratories/Bayer/Cutter/Miles
	Cambridge Instruments, Inc.	Mixing Equipment Company, Inc.
	Celanese Research Company	Monsanto Company
	Cetus Corporation	Nihon Kohden Kogyo Company
	Chevron Research Company	National Science Foundation
	Ciba-Geigy, Ltd.	Novo Industri
	Corning Glass Works	Ortho Diagnostics
	CTIP International, Inc.	Pfizer, Inc.
	Dorr-Oliver, Inc.	Pharmacia, Inc.
	Dow Chemical U.S.A.	Pharmacia Diagnostics
	DSM, N.V.	Pharmacia Japan Kabushikikaisha
	E.I. du Pont de Nemours Company	P.P.G. Industries, Inc.
	Eastman Kodak Company	The Proctor & Gamble Company
	Exxon Research and Engineering Company	Scientific Systems, Inc.
	F.M.C. Corporation	G. D. Searle & Company
	Gist-Brocades	Shell Development Company
	Genex Corporation	Smith-Kline Instruments
	Glaxo Operations	A.E. Staley Manufacturing Company
	W. R. Grace & Company	Stauffer Chemical Company
	CHR Hansen's Laboratorium	Sun Company, Inc.
	Hewlett-Packard Company	Takeda Chemical Industries, Ltd.
	Hoechst, A. G.	Tanabe Seiyaku Company, Ltd.
	Hoffmann-La Roche, Inc.	Telemed Cardio Pulmonary Systems

TABLE 8 Financial Sponsors of Engineering Foundation Conferences (continued)

Year	*Name of Sponsor*	*Name of Sponsor*
	Imperial Chemical Industries	Toray Inc., Company
	International Business Machines	Toyo Jyozo Company, Ltd.
	International Minerals & Chemical Corporation	U.S. Department of Transportation
	Japanese Society of Enzyme Engineering	U.S. Department of Energy
	Kanegafuchi Chemical Industry Co., Ltd.	Warner Lambert Company
1982	American Fly Ash Company	Hewlett Packard
	American Society of Appraisers	Homer Hoyt Institute
	Becton Dickinson FACS Systems	Instruments for Cardiac Research, Inc.
	Biocon Limited	International Business Machines Corp.
	BioTechnica International	Ionics, Inc.
	Bio-Technical Resources, Inc.	Kellogg Company
	Boehringer Mannheim, GmbH	Lone Star Industries, Inc.
	Anna & Harry Borun Foundation	LSL Biolafitte, Inc.
	The Burdick Corporation	Marquette Electronics, Inc.
	Cambridge Medical Instruments-Picker International	Merck & Company, Inc.
	Cetus Corporation	Miles Laboratories/Biotechnology Group
	Chemapec, Inc.	Miller Brewing Company
	Comp-U-Med, Inc.	National Science Foundation
	Corning Glass Works	New York Academy of Sciences
	Coulter Electronics, Inc.	Novo Industri
	Environmental Protection Agency	Pharmacia, Inc.
	Exxon Minerals Company	Phone-A-Gram Systems, Inc.
	Fermco Biochemics, Inc.	Procter & Gamble Company
	Frito-Lay, Inc.	Stanley Sapiro
	Genentech, Inc.	Scientific Systems, Inc.
	Geometric Data	Technicon Instruments Corp.
	Gist Brocades, N.V.	U.S. Department of Energy
	W. R. Grace & Company	University of Oklahoma Foundation
	H. J. Heinz Company	Worthington Diagnostic Systems, Inc.

TABLE 8 Financial Sponsors of Engineering Foundation Conferences (continued)

Year	*Name of Sponsor*	*Name of Sponsor*
1983	AC Biotechnics	International Business Machines
	Air Products & Chemicals, Inc.	Iwatami Foundation
	Ajinomoto Company, Inc.	Japanese Society of Enzyme Engineering
	Alfa Laval Separation AB	Kashima Foundation
	American Society of Mechanical Engineers	Edwards Kniese & Company
	Asahi Glas Foundation	KVB
	Association of Microbial Food Enzyme Producers	Kyowa Hakko Kogyo
	B. Braun Melsungern, A. G.	Marquette Electronics, Inc.
	BASF Wyandotte Corporation	E. Merck
	Bayer, A. G.	Merck & Company, Inc.
	Beecham Pharmaceuticals	Miles Kali-Chemie GmbH & Company
	Biochemie GmbH	Miles Laboratories
	Goehringer Mannheim GmbH	Monsanto Company
	Calgon Corporation	Naamloze Vennootschap D.S.M.
	Carnation Company	National Bureau of Standards
	Cetus Corporation	National Institute of Health
	Chemap AG	National Science Foundation
	Ciba-Geigy	New York Academy of Sciences
	Conoco, Inc.	Novo Laboratories, Inc.
	Corning Glass Works	Occidental Research Corporation
	E. I. du Pont de Nemours & Company	Otto Rohm, GmbH
	Electric Power Research Institute	Phone-A-Gram Systems, Inc.
	Enka AG	Picker International
	Environmental Protection Agency	Schering AG
	Exxon Chemical Company	Shell Development Company
	Exxon Corporation	Siemens-Elema AB
	Exxon Research & Engineering Company	Sudzucker
	Fukuda Denshi Company	TRW
	Giovanola Freres SA	Union Carbide Corporation

TABLE 8 Financial Sponsors of Engineering Foundation Conferences (continued)

Year	*Name of Sponsor*	*Name of Sponsor*
	Gist Brocades NV	U.S. Department of Energy
	Haarman & Reiner GmbH	U.S. Department of the Army
	Heinz, U.S.A.	UOP Foundation
	Henkel KG AA Chemische	Westvaco
	Hewlett-Packard	Westfalia Separator
	Hoechst-Roussel Pharmaceuticals, Inc.	Worthington Diagnostic Systems, Inc.
	Imperial Chemical Industries	
1984	Abbott Laboratories	Hoffman-LaRoche, Inc.
	Ajinomoto Company	Imperial Chemical Industries
	Alfa-Laval	Japanese Society of Enzyme Engineering
	American Society of Heating, Refrigeration &	Kraft, Inc.
	Air Conditioning Engineers	Kyowa Hakko Kogya Company
	Amicon	Marquette Electronics, Inc.
	Amoco Oil Company	Merck, Sharp & Dohme
	Amoco Research-Standard Oil Company	Miles Laboratories
	Arthur D. Little Company	Millipore
	Battelle Memorial Institute	Monsanto Company
	Beecham Pharmaceuticals	National Aeronautics and Space Administration
	Biogen Corporation	National Science Foundation
	Bio Information Associates, Inc.	New York Academy of Sciences
	Bio Technical Resources	NortonCompany
	Cardio Data	Novo Industri
	Celanese Corporation	Pall Ultrafine Filtration Corporation
	Cetus Corporation	Phone-A-Gram System
	Chemapec, Inc.	Pfizer, Inc.
	Ciba-Geigy, Ltd.	Picker International
	Conoco, Inc.	PPG Industries
	Dorr-Oliver, Inc.	Quinton Instrument Company
	Dow Chemical Company	Rhone-Poulenc, Inc.

TABLE 8 Financial Sponsors of Engineering Foundation Conferences (continued)

Year	*Name of Sponsor*	*Name of Sponsor*
	E.I. duPont de Nemours & Company	Rohm & Haas
	Electric Power Research Institute	Schering-Plough
	Elf Acquitaine	Scole Engineering Company
	Engenics	Siemens-Elema AB
	Environmental Protection Agency	A. E. Staley Company
	Fujisawa Pharmaceutical	Standand Oil Company
	W. R. Grace & Company	3M
	Genentech	Texaco, Inc.
	Genetics Institute	Union Carbide Corporation
	Gist Brocades NV	Upjohn Company
	Heinz, U.S.A.	U.S. Department of Energy
	Hewlett-Packard	
1985	Agway, Inc.	Hoffman-La Roche, Inc.
	Air Products & Chemicals, Inc.	Imperial Chemical Industries, PLC
	AM Gen	Jacobs Engineering Group, Inc.
	Bio-Technical Resources	Kabigne AB
	Anna Borun & Harry Borun Foundation	Kraft, Inc.
	Brown-Boveri	Lord Corporation
	Cetus Corporation	Merck, Sharp & Dohme Research Laboratories
	Chemineer, Inc.	New York Academy of Sciences
	Colgate-Palmolive Company	Novo Industri
	International Society for Computerized ECG	Occidental Chemical Company
	Data General Corporation	Perstrop Biotech AB
	Dorr-Oliver, Inc.	Pharmacia AB
	Dow Chemical Company	Phone-A-Gram System
	DSM, N. V.	Quinton Instruments Company
	E. I. duPont de Nemours & Company	Redlake Corporation
	Electric Power Research Institute	Rohm & Haas Company
	Japanese Society for Enzyme Engineering	G. D. Searle & Company

TABLE 8 Financial Sponsors of Engineering Foundation Conferences (continued)

Year	Name of Sponsor	Name of Sponsor
	Exxon Corporation	Shell Development Company
	Exxon Research & Engineering Company	Signal UOP Research Center
	Fermenta	Sockerbolaget
	Genencor, Inc.	The Standard Oil Company of Indiana
	General Mills, Inc.	The Standard Oil Company of Ohio
	Genex Corporation	Svenson Utuecklings Bolaget
	Gist-Brocades	Technicon Instruments
	W. R. Grace & Company	Townley Manufacturing Company
	Henkel Corporation	Union Carbide Corporation
	Hewlett-Packard	Westvaco
1986	Aalborg Vaerft A/S	Geschaftfuhrer
	A. C. Biotechnics AB	Gist Brocades BV
	Air Force Office of Scientific Research	W. R. Grace & Company
	Air Products & Chemicals	Grunenthal GmbH
	Ajinimoto	Haarman & Reimer GmbH
	Alfa-Laval	Henkel KGaA
	Alko Ltd.	H. J. Heinz Company
	AMF Cuno Division Europe	Hoffman La Roche & Company
	AMF Deutschland	U.S. Department of Interior, Bureau of Reclamation
	Amicon GmbH	Kabi Vitrum AB
	Amoco Corporation	Kraft, Inc.
	U.S. Army Research Office	Lauritz Andersens Foods
	Artisan Industries, Inc.	Mees & Hope NV
	A/S Cheminova	E. Merck
	Asahi Glass Company, Ltd.	Miles Laboratories
	Bayer (Swerige) AG	Millipore Foundation
	Beecham Pharmaceuticals	Millipore GmbH
	Bio-Information Associates	Monsanto
	BioTechnica International	National Science Foundation

TABLE 8 Financial Sponsors of Engineering Foundation Conferences (continued)

Year	*Name of Sponsor*	*Name of Sponsor*
	Bio-Technical Resources, Inc.	Office of Naval Research
	Boehringer-Mannheim GmbH	New York Academy of Sciences
	Bran & Lobbe GmbH	Nordisk Insulinlaboratorium
	B. Braun Melsungen	Norsk Biotech A/S
	Cetus Corporation	Norsk Hydro
	Chemap AG	Novo Industries A/S
	Chevron Oil Field Research	Pall Corporation
	Ciba-Geigy	Pall Filtrationstechnik GmbH
	U.S. Army Corps of Engineers	Pfeifer & Langen
	Dansk Sukker Fabrikker	Pfizer, Inc.
	DDS-RO Division	Pharmacia
	Diessel GmbH & Company	The PQ Corporation
	Diosynth BV	Rhone-Poulenc, Inc.
	Dorr-Oliver, Inc.	Rohm GmbH
	Eastman Kodak Company	Rohm & Haas Company
	Edwards, Kniese & Company	Sandos Forschungsinstitut GmbH
	Enka, AG	Schering Corporation
	U.S. Environmental Protection Agency	Suddeutsche Zucker AG
	Exxon R & E Company	Sumitomo Chemical Company, Ltd.
	Faxe Kalk	Toray Industries
	Fermenta Products AB	Unilever Research Laboratorium
	Forsk. Sekretariat Kobenhaven	Union Carbide Corporation
	Gebruder Sulzer AG	UOP Foundation
	Genencor, Inc.	Vogelbusch
	Genentech, Inc.	Westfalia Separation AG
	Genzyme	Westvaco

TABLE 9 Engineering Foundation Representatives

Organization	Representative	Years of Service
American Bureau of Welding	D. S. Jacobus	1921–23
Advisory Board on Rock Drill Steel	Herbert M. Boylston	1921–23
Personnel Research Foundation	Alfred D. Flinn	1921–23
Advisory Board on Highway Research	Alfred D. Flinn	1921–23
Highway Research Board	O. E. Hovey	1940–41
Steam Table Research	D. S. Jacobus	1922–23
National Research Council		
Executive Board	F. M. Farmer	1936–39
	George E. Beggs	1939
	O. E. Buckley	1939–42
	A. L. J. Queneau	1942–45
	A. B. Kinzel	1945–46
	F. T. Sisco	1948–52
	J. W. Barker	1952–55
	E. M. Barber	1955–57
	W. M. Peirce	1957–59
	G. O. Curme, Jr.	1960
	W. Schreiner	1961–64
	A. Loughren	1965–70
	E. F. von Wettberg, Jr.	1970–73
Metal Properties Council	Charles L. Tutt, Jr.	1967–70

APPENDIX B

Selected Documents

Contents:

INFORMAL LETTER SUGGESTING FORMATION OF AN ENGINEERING FOUNDATION

The American Society of Mechanical Engineers
29 West 39th Street, New York

April 21, 1914

Personal and Informal

Mr. Gano Dunn, President,
Board of Trustees, U.E.S.,
c/o J. G. White & Co.,
43 Exchange Place, New York.

My Dear Mr. Dunn,

One of the members of the American Society of Mechanical Engineers has long cherished the desire to benefit mankind through the instrumentality of the engineer. His ideal has been taking shape and I am now authorized to state confidentially that our friend is prepared to make a donation of possibly $250,000 to an Engineering Research Foundation.

Our friend intended to make the American Society of Mechanical Engineers custodian of this Foundation, but in consultation with a number of our members, all agreed that we have in our community of interests through the agency of the United Engineering Society, the machinery for even larger and broader vision and that it might be more conducive to what is best if the trust were undertaken by the United Engineering Society, as representing the three Founder Societies. The object of this note is to inquire informally if the United Engineering Society would undertake such a trust were it offered to them.

The Charter appears to be adequate and the organization of the Library Board indicates the manner in which the Research Board might be formed and made the Executive of the Trustees in the matter of research.

I will bring up the matter either before the meeting of the Trustees or after we adjourn and learn the attitude of the gentlemen present. Copy of this letter is sent to each member of the Board for their information and thought. All are requested to treat the matter confidentially for the present.

Yours truly,

(signed) CALWIN W. RICE
Secretary.

FIRST FORMAL MENTION OF AN ENGINEERING FOUNDATION

MINUTES, UNITED ENGINEERING SOCIETY, MAY 28, 1914: The President reported a conference with an eminent engineer, who had expressed his desire to present to United Engineering Society a considerable fund for the advancement of the profession of engineering. It was the recommendation of the President that a special committee be appointed to consider and report upon the best means to accept the generous gift and administer and establish an engineering research foundation under the broad terms of the donor's expressed wishes. Creation of such a Special Committee was approved, two from each of the Founder Societies.

LETTER SUBMITTING PLAN OF ORGANIZATION TO MR. SWASEY

UNITED ENGINEERING SOCIETY

November 20, 1914

Mr. Ambrose Swasey,
7808 Euclid Avenue,
Cleveland, Ohio.

Dear Mr. Swasey:

At the meeting of the Board of Trustees on November nineteenth, a specially qualified committee, of which Dr. Alex. C. Humphreys is Chairman, reported its recommendations or by-laws, which should administer under the Board of Trustees the Engineering Foundation, for which you have so generously provided.

It was suggested that the proposed organization of the Engineering Foundation Board should be submitted to you for your approval before it became the law of the Trustees of United Engineering Society. We would be glad also to have any further suggestions as to the proposed by-laws of which I enclose you a copy. I need not add that any comment will be gladly welcomed, if you think they can be improved for the purpose in hand. . . .

Very truly yours,

F. R. HUTTON,
Secretary.

FIRST BYLAWS OF THE ENGINEERING FOUNDATION
Adopted December 17, 1914

101. The Board of Trustees shall establish a fund to be known as The Engineering Foundation.

101a. The Engineering Foundation shall be controlled and administered by a board to be known as The Engineering Foundation Board.

101b. The Engineering Foundation Board shall consist of eleven members as follows:

The President of the United Engineering Society, ex officio.

One member of the Board of Trustees of United Engineering Society from each Founder Society; elected by the Board at the annual meeting; serving until the next annual meeting or until his successor is elected and eligible for re-election while a member of the Board of Trustees of the United Engineering Society. Should the Board for any reason fail to elect such members or any of them at the annual meeting, then the election shall be made as soon after such annual meeting as may be practicable.

One member from each Founder Society to be elected by its governing body prior to each annual meeting of the United Engineering Society and to serve until the next annual meeting. Should the election of any such member not be made prior to the annual meeting of the United Engineering Society the election on the approval of the date therefor by the Board of Trustees of the United Engineering Society may be made at a subsequent date.

Two members to be elected by the Board of Trustees of the United Engineering Society upon the nomination of the American Society of Civil Engineers prior to the annual meeting of United Engineering Society, and to serve until its next annual meeting. Should the Election of such members not be made prior to the annual meeting of the United Engineering Society, the election on the approval of the date therefor by the Board of Trustees of the United Engineering Society may be made at a subsequent date.

Two members at large selected by the Board of Trustees of the United

Engineering Society at its annual meeting and to serve until its next annual meeting, or until their successors are elected. Should the Board for any reason fail to elect such members at the annual meeting, then the election shall be made as soon after such annual meeting as may be practicable.

Any vacancy occurring in The Engineering Foundation Board may be filled by the Board of Trustees of the United Engineering Society.

101c. Regular meetings of the Engineering Foundation Board shall be held in the months of February, May, September and December of each year. Special meetings may be called at the option of the Chairman, on not less than seven days' notice, and must be called by the Chairman or Secretary on the written request of three or more members. No business shall be transacted at a special meeting beyond that stated in the call. A quorum shall consist of three or more members.

101d. At the regular meeting in February of each year, The Engineering Foundation Board shall elect one of its members Chairman to serve for one year or until his successor is elected. At the same meeting the Board shall select a Secretary to serve for one year or until his successor is selected. The Secretary unless he be a member of the Board shall have no vote.

101e. The Engineering Foundation shall have authority to receive, conserve, invest, hold in trust or administer such monies, or other property, as it may elect to receive from the United Engineering Society or any other society or organization, or from any person, or the estate of any deceased person, or from any other source.

101f. The Engineering Foundation Board shall have discretionary power under the by-laws in the disposition of the funds received by it.

The Board may use any part of its funds, and in any manner, which it deems proper for the furtherance of research in Science and Engineering, or for the advancement in any other manner of the Profession of Engineering and the good of Mankind.

The Board may, by publications or public lectures or by other means, in its discretion, make known to the world the results of its undertakings.

101g. The Engineering Foundation Board shall present to each annual meeting of the Trustees of the United Engineering Society a complete report for the year of its work and finances. The Treasurer of the United Engineering Society shall act as Treasurer of The Engineering Foundation.

101h. That the fund known as "The Engineering Foundation" as now or hereafter constituted shall at all times be subject to the by-laws of the United Engineering Society or any amendments which may be made thereto, and any of the provisions of the by-laws respecting said Fund of The Engineering Board or any matters respecting the Administration of said fund may be altered from time to time by the by-laws of the United Engineering Society.

LETTER OF GIFT

November 30, 1914.

PROFESSOR F. R. HUTTON, *Secretary*,
United Engineering Society,
New York City.

DEAR PROFESSOR HUTTON:

I am pleased to acknowledge receipt of yours of the 20th instant, also copy of the minutes of the Board of Trustees of November 19th, and I appreciate the courtesy of the Board in submitting them to me before final approval.

The name adopted, "The Engineering Foundation," is ideal, and the plan of organization and administration, as given in the minutes, is along the broadest lines and most admirable in every respect. I have no suggestions or recommendations to offer.

As soon as I am advised that the plan of organization of the Engineering Foundation, submitted, has become the law of the United Engineering Society, I will be pleased to transmit to the officer designated by the Society, the two hundred thousand dollars ($200,000) which constitutes my gift to the Society for the Engineering Foundation; the income only of which is to be used for the purposes of the Foundation.

As to the date of the general meeting when the plan of the Foundation is to be made public; if agreeable to the Board, it seems to me it would be well to have it some time during the last week in January.

I want the members of the Board to know how much I appreciate the interest they have manifested in working out the problems relative to the Foundation, and the pleasure it has given me to be associated with them in their splendid undertaking.

With all best wishes,

Very truly yours,

(Signed) AMBROSE SWASEY.

RECORD OF RECEIPT OF FIRST GIFT

MINUTES, UNITED ENGINEERING SOCIETY, FEBRUARY 25, 1915: The Secretary stated that the President, the Treasurer, the Secretary, and Messrs. H. H. Barnes, Jr., and Jesse M. Smith of the Finance Committee, had met Mr. Swasey February 24th, in the private room of the Astor Trust Company; that in their presence representatives of the Bankers' Trust Company, as Mr. Swasey's fiscal agents, had turned over to him two packages of bonds of the City of New York, aggregating $200,000; that Mr. Swasey had then handed to the President of the United Engineering Society an autograph letter, listing these bonds.

Upon transfer of the securities to United Engineering Society the President and Treasurer executed a receipt. These securities were then intrusted to the care of the Astor Trust Company, and a receipt for them was handed to the Treasurer.

The Secretary presented to the Board the formal announcement from the Founder Societies of their action in nominating under the by-laws of the United Engineering Society, the three members of The Engineering Foundation Board, to be nominated by this method.

LETTER ACCOMPANYING SECOND GIFT

September 28, 1918.

MR. CHARLES F. RAND, *President,*
United Engineering Society,
New York.

DEAR SIR:

The admirable manner in which the Engineering Foundation has been conducted, and the large measure of helpfulness it has extended to the engineering world, have been most gratifying to me and thoroughly justify its establishment.

In recognition of what had then been accomplished, it seemed to me in 1916 that it would be well to plan for its further extension, and I then made the United Engineering Society the beneficiary of a trust fund of $100,000, established with the Cleveland Trust Company. In 1917, from the income of this trust, $5,000 was contributed to the Foundation for the purpose of carrying forward its work.

The many vital problems created by war conditions give new opportunities to the Foundation, and impose added responsibilities upon it. In recognition of this I feel that the Foundation should now have the full benefit of the income from the fund established in 1916, and I am pleased to add to the previous gift of $200,000, the fund of $100,000, held in trust as above, to be on the same terms and conditions as my original gift.

The Board is at liberty to use its discretion as to whether the funds shall remain with the Cleveland Trust Company for a time, or be transferred to the custody of the United Engineering Society. In either case, I desire that the Foundation shall have the benefit of the income of this additional fund after September 1, 1918.

Very truly yours,

(Signed) AMBROSE SWASEY.

LETTER ACCOMPANYING MR. SWASEY'S THIRD GIFT

AMBROSE SWASEY,
Cleveland, Ohio.

October 25th, 1920.

MR. CHARLES F. RAND, *Chairman*,
The Engineering Foundation,
New York City.

DEAR MR. RAND:

The service to humanity which has accrued from successful research and development work in the several fields of engineering during the past decade would have been impressive even without the notable results achieved by engineers and scientists during the great war.

Thinking men throughout the world are giving constant recognition to this general fact, and emphasizing the opportunity which is presented by the stimulation thus given to technical men and technical organizations to still further advance knowledge and practical attainments in these fields. Therefore, the time seems to me to be peculiarly opportune to respond to the appeal which you have recently made for increased funds for the Engineering Foundation.

When I made my first gift of $200,000 in 1914, I had strong hopes that the establishment of an Engineering Foundation would be productive of helpful results. In 1918, the work already undertaken and the broad and practical spirit in which the Board of the Foundation had laid its plans made me feel that a further gift of $100,000 was fully justified. Now, I feel even more strongly that the Engineering Foundation has undertaken its mission at a time which gives it most unusual opportunities for helpfulness to the technical profession and consequently to mankind.

Therefore, I have sincere pleasure in informing you that I have to-day directed The Cleveland Trust Company to transfer to the United Engineering Society securities having a par value of $200,000, to be held by the United Engineering Society as a part of its permanent fund for the benefit of the Engineering Foundation, upon the same terms and conditions as my two previous gifts covered by my letters of November 30, 1914, and September 28, 1918. The income on these securities which has accrued since Jan. 1st, 1920, and which I have ordered to be transferred to the United Engineering Society, is available for use as income for the purposes of the Engineering Foundation, and is not denominated by me a part of the permanent fund.

Very truly yours,

AMBROSE SWASEY.

APPENDIX C

Biographies

Contents:

1. GANO DUNN

Gano Dunn was one of the most respected of America's engineers during the first half of the twentieth century. He served as the first chairman of the Foundation. Through his prestige and leadership ability he did much to bring founder Ambrose Swasey's plans for the Engineering Foundation into reality.

Dunn was born on October 18, 1870 at New York City. He received his early engineering education at the City College of New York from which he graduated with a B.S. degree in 1889. He then entered Columbia University where he studied under Michael Idvorsky Pupin, who would later remark that Dunn was the most gifted of all of his students. In 1891 Dunn graduated from Columbia University with a degree in electrical engineering. He then returned to the City College of New York where he completed his formal education by earning an M.S. degree in 1897.

Dunn began his working carrier in a humble manner when he secured a position as a night shift telegraph operator at Western Union's New York City office. After graduation from Columbia, he entered employment with the Crocker Wheeler Electric Manufacturing Company at Ampere, New Jersey. At Crocker Wheeler his talents as an electrical engineer began to flourish. During the next decade his research would result in over thirty patents including a revolutionary process which made it possible to wind copper strips edge-wise in a simple and economical manner. This process greatly reduced the costs of producing electrical generators and motors. He also played a prominent role in the development of the long distance telephone system. By 1898 Dunn had become the vice president and chief engineer of Crocker Wheeler, a position which he held for more than a decade.

In 1911 Dunn was elected vice president in charge of engineering and construction of the J. G. White & Company, and in 1913 he became president of the newly organized J. G. White Engineering Company. Dunn enjoyed a brilliant career at J. G. White. During World War I he planned and directed such notable projects as the construction of the original government research facilities at Langley Field in Virginia and the great electrical generating plant at Muscle Shoals, Alabama. His firm also constructed the first long distance natural gas transmission pipeline, a large number of transoceanic radio relay stations and several oil refineries. Its success continued during the 1920s and 1930s when it engineered railroads in Persia and dams in Ethiopia.

Dunn also played a prominent role in the development of professional engineering organizations and government advisory committees. During 1900–02 he served as the president of the New York Electrical Society, in 1911–12 as the president of AIEE and between 1913 and 1916 as president of the United Engineering Society. In the last capacity he played major part in the establishment of the Engineering Foundation which he served as the first chairman in 1915–17. He did much to establish the Foundation's strong emphasis on funding of research projects.

During World War I Dunn offered his services to the U.S. Government. He became a member of the War Department's Nitrite Commission which had the critical task of insuring an adequate supply of ammunition. He also played an important role in the activities of the Engineering Committee on National Defense. Perhaps his most important service to America was the leadership role that he played as chairman of the Joint Committee on Submarine Cables. This body insured the security of the primary means of transatlantic communications during the war years.

After World War I Dunn continued his volunteer service. He became the chairman of the visiting committee of the National Bureau of Standards. He also served as chairman of the National Research Council between 1923 and 1928. In 1936 he became a member of the executive committee of the World Power Conference. Throughout the 1930s he served as a member of the Patent Office advisory committee. Due to the many innovations in steel plant design that his firm pioneered, Dunn became a special consultant on the steel industry to the Office of Production Management. For this World War II federal agency he produced a special report on the productive capacity of the steel industry in the United States. Finally, it should be noted that Dunn served in 1944 as a member of New York Governor Thomas Dewey's Committee on Technical Industry Development.

Dunn also devoted much of his time to the betterment of educational institutions. For many years he served as a trustee of both Columbia University and Barnard College. During the 1930s he successfully served as the president and chairman of the Copper Union in New York City and did much to insure that this institution continued to prosper during the harsh economic climate of the Great Depression.

Dunn's varied achievements were recognized by many organizations during the course of his lifetime. He received the first Egleston Medal of the Columbia University engineering school's alumni association and was awarded the Townsend Harris Medal by the City College of New York. He was presented with the Edison Medal by AIEE in 1937, while he won the prestigious Hoover Medal of the Founder Societies in 1939. In 1940 the Republic of Haiti presented a special Medal of Merit to Dunn, while in 1945 he was awarded the Gold Insignia of the Pan American Society.

Dunn enjoyed a happy marriage to the former Julia Gardiner Gayley who died in 1937. Dunn's many labors ceased on April 10, 1953 in New York City.

2. ENGINEER NO. 1

"Would you like," asked the fat man as he tossed a piece of soap into the air and caught it—"would you like to have me explain the Einstein theory?" The group in the washroom of the Pullman car eyed him sleepily.

"That soap," continued the washroom scientist, "fell into my hand exactly underneath the highest point that it reached in its trip into the air. But the train in which we are riding is going sixty miles an hour. Why didn't the train, this room, my hand, move away from the downward path of the soap?"

Nobody knew. "Neither do I," said the fat man. "Neither does Einstein. But," he mused, "Einstein pointed out the fact that the train, my hand, and the soap don't change positions relatively. And that, my friends, is all I know about relativity."

The audience looked dashed. The demonstration seemed completely pointless until a little man in the corner muttered, "You've got something there. I've just spent a week in Washington trying to find out from the OPM and the SPAB how I'm supposed to go on employing people and turning out goods when I can't get raw materials. And I tell you that moving our production machine at a faster and faster pace for defense while we try to get the soap to fall down in the same place—wait a minute, I mean while we try to keep the same relative economic situations we've always had in this country—well, that's something for Einstein to dope out. They ought to get that guy down in Washington."

A Washington SOS

Washington seems to have had some such idea months back. The great scientist himself was not summoned, of course. But President Roosevelt did call in Gano Dunn, an old friend of Einstein's, one of the eleven men once reputed to understand the Einstein theory. And to Dunn, early this year, the President put the most crucial of all how-to-get-the-economic-soap-to-fall-down-in-the-same-place questions. He asked Dunn to advise him on the capacity of the steel industry.

Everyone who has even the dimmest notion of the mechanics of defense production knows that steel capacity, intelligently used, is the cornerstone of the defense effort. What is not generally known is that for many months now there has been a raging controversy about whether there should or should not be a vast expansion of the steel industry. Needless to say, the arguments on each side involve considerations of defense. But they also involve calculations of what will happen to this basic American industry and the industries that depend on it when the defense effort is over. Which means that they involve critical questions about the political and economic future of us all.

Gano Dunn, to whom the President turned for advice, is the head of the J. G. White Engineering Corporation. As such, he has directed great construction projects all over the world—among them the steam plant at Muscle Shoals now operated by the TVA; the first long-distance natural-gas pipe line in California; enormous sulphur-extraction plants and oil refineries; thirteen transoceanic radio stations; the original Government aviation station at Langley Field; the naval oil base at Pearl Harbor; the 20,000-kilowatt power plant that pumps oil taken from under water off the shores of Lake Maracaibo in Venezuela. For the republic of Chile he has built three large irrigation and power dams; for the republic of Mexico, five; for the government of Colombia he has built power plants; for the republic of Panama, the Chiriqui railroad; for the Haitian government he is now carrying out an immense bridge-and-road-building project.

But it is as an engineer's engineer that Gano Dunn is best known. He and his firm are called in on difficult problems by the engineers of railroads, utilities, radio and telegraph companies, oil producers and refiners, in much the same way that a big law firm is called in by the legal divisions of its corporate clients. In accepting the President's assignment, Dunn might well have recalled Oscar Wilde's bitter half-truth: "It is always a silly thing to give advice, but to give good advice is absolutely fatal." On the other hand, as an engineer's engineer he has successfully survived a lifetime of giving good advice. So sound has his advice proved, in fact, that his signature on an engineering report is accepted the country over as a symbol of authority and integrity. To him have come the highest honors that his profession can give—among them the Hoover Medal and the Edison Medal.

Dunn's answer to the President's question on steel took the form of two reports, the first of which appeared late in February and the second of which appeared late in May. No one has ever made a survey that compares even remotely with the scope of his. Dunn began by appraising the physical capacity of the nation's steel plants—a job that involved the use of production figures always hitherto kept the closest of secrets by the steel people. These calculations he then extended, Einsteinwise, beyond the question of the capacity of the steel mills to produce. He studied the sources of the materials that go into steel—coke, ore, manganese, and the rest. He took into account the ability of the Great Lakes and other transportation systems to convey such materials to the mills and the amount of available labor in the face of the emergency. And, above all, he grappled with the theory that there ought to be a precipitate expansion of the steel industry now, in the midst of the armament program.

Struggle in Steel

For years it has been one of the basic tenets of the New Deal economists that steel could be much cheaper if steelmakers would produce more steel. The steel people, on the other hand, have insisted that producing steel is not like making and selling electric current or straw hats. Increased volume in producing such things results in lower costs of production. Increased volume in steel production results in lower costs to only a limited extent. This the steel people demonstrated with facts and figures in the hearings before the late Temporary National Economic Committee a year and a half ago. They convinced everyone but the New Deal economists.

Among those economists is a Mr. Melvin G. de Chazeau, who is now associated with the OPA. Before Dunn's first survey, Mr. de Chazeau was asked to make a report on steel capacity. Not surprisingly, it twanged the dear old New Deal refrain—expanded steel production. After estimating existing steel capacity, defense needs and probable civilian needs in terms of possible future national income, De Chazeau recommended an increase of 8,000,000 tons in steel capacity.

Dunn's survey proceeded to deal with the De Chauzeau findings dispassionately. But when he had finished with them, with words as gentle as eiderdown he had spread a completely sopping blanket over the contentions of the expansionists. The radical journals screamed with rage. The New Dealers themselves were temporarily silent. Dunn's findings permitted no further argument. Worse still, from the New Dealers' point of view, the President, in announcing that Dunn had completed the report, stated that it was the best thing of its kind done to date—a standard on which the Government could rely. So enthusiastic about it was he that he asked Dunn to prepare a second report at the end of three months.

If anyone who read Dunn's first report missed the fact that he was dealing with a controversy whose outcome might change the whole American scheme of things, Dunn's second report left no possibility of misunderstanding. Shortly before it was made, Mr. Stacy May's Bureau of Research and Statistics in the OPM produced an estimate of 1942 demands for steel which literally dwarfed the estimates of the De Chazeau report. The steel industry would have to expand its capacity from 91,300,000 tons to 120,000,000 tons, or nearly 30,000,000 tons, it said. Dunn turned to a consideration of this new estimate and plucked it to pieces with the same benign detachment that marked his first report.

The increase for which the Stacy May estimate asked was not to meet the demands of the armament program, Dunn showed. For armament purposes not more than one quarter of the steel industry's existing capacity was needed. It was for civilian uses that the expansion was demanded. Moreover, to produce the amount of steel called for by the Stacy May estimate, Dunn showed, would require that 6,047,200 employees be added to the 7,591,500 employees of the steel and steel-consuming industries in 1940—an increase of 77 per cent in the labor required. Such an increase would be impossible. There was not only not enough available labor to produce the 30,000,000-ton expansion for which May's estimate called: there was not enough to consume it if it could conceivably be produced.

Pricking a Bubble

Even to expand the nation's steel capacity by, say, a comparatively modest 10,000,000 tons, Dunn observed, would be a grave step. The cost would be staggering—$1,250,000,000 for increased transportation facilities, ships and mining equipment, for mine development and coke-oven construction, for added furnaces and finishing mills. The time required would be two years.

More ominous still, the consequence would be to divert 4,160,000 tons of finished steel urgently needed for immediate use to the building of facilities to increase steel production in the future.

This final consideration is what momentarily stumped the New Dealers. For their theories of expansion simply had not taken it into account. By way of emphasis, when he presented his report to the President, Dunn told him the story, variously credited to Mark Twain and Lincoln, of the little Mississippi steamboat whose whistle was so big that it had to stop its engine each time it tooted. The President was delighted. When he went into his press conference he repeated the story with embellishments. Again he hailed the monumental character of Dunn's findings. Again he set the figurative seal of his approval upon them. The New Dealers' stubbornly fought expansion cause seemed, for the moment, to be lost.

But zeal such as theirs can survive many a blow. And so, ironically, within a few days after Dunn's second report, the New Dealers seem to have got authorization—how, why and with what powers of persuasion remain a mystery—to plan for a 10,000,000-ton steel expansion. Jesse Jones called in the amazed steel people, told them to expand, and announced that the Government would put up 90 per cent of the cost. Before long there was talk of a 15,000,000-ton expansion.

A Ninety-Seven-Cent Year

Then, suddenly, around mid-July, fragments of news began to suggest that, New Dealers or no New Dealers, the ambitious program of expansion was running into trouble at every turn. Expansion on the Pacific Coast was delayed because of fears that the Coast might be vulnerable to attack. Expansion elsewhere was slowed down by difficulties in getting skilled labor, by difficulties in providing for adequate railroad and ship facilities, and by each of the dozen other intricate factors that Gano Dunn had calmly warned about.

This is not to say that Dunn's advice will finally prevail. It is true that there are inexorable processes in motion that may bring themselves to bear conclusively before it is too late. But Russia's entry into the war may fundamentally increase the demands on the steel industry and alter the timetable of those demands. More, it would be a mistake to underestimate the determination of the New Dealers.

In any event, the decisions forced by the determination of the New Dealers will never be attributed to Dunn. For he assumed no further responsibility in the matter. Shortly after the steel expansion was announced, his resignation as Senior Consultant to the OPM, the official title he had held for nearly a year, was submitted to the President. This was done, he said, in line with a decision he had reached weeks before. His reasons for it had nothing to do with the fate of his reports. His business and many other personal responsibilities made his departure from Washington necessary, and his strength, at seventy, had been heavily taxed by the year's grueling work. Cordial and affectionate letters were exchanged with the President. Dunn received a ninety-seven-cent check from the Government for his services and went off for a vacation on his sailing boat. He will make no more steel reports for the Government.

These are the facts about Dunn's resignation. But what the New Deal expansionists have had to say about it is another story. In a manner now well established in contentious Washington, they have fired parting volleys at him. There has been bitter unofficial talk by Government officials. The writers for the left-wing daily and weekly press have echoed this talk. The Dunn reports, it is said, were "full of misconceptions." They used "misleading arguments." They have been "discredited" by the course of events, for is the expansion policy not proof that he "minimized" an impending steel shortage? Dunn's resignation is more "reassuring" than his reports. And, for good measure, one commentator has expressed the hope that Dunn's "spiritual brothers on the OPM" will follow him into retirement.

Dunn is not worried about such attacks. But he is concerned about the Governmental policy that disregards the economic consequences of a big steel expansion for civilian needs. For Dunn realizes what may happen when, after the precipitate expansion of steel capacity takes place, the abnormal demand for steel falls off. He knows, and has demonstrated in his reports, a transcendently important fact about the steel industry. The industry's expansions have largely taken place in periods of depression.

Consider what that means. It means that when the demand for steel on the part of big steel users has been slack, the steel industry itself has taken up some of that slack by producing the steel to expand itself. The effect of that policy has been to even out the production curve, to a degree, in this feast-and-famine industry, and to provide employment in slack times.

The New Dealers, on the other hand, want the steel industry to expand not in slack times but in piping times. And do they now ask it to do this in order to meet the requirements of the armament program? Certainly not. It is to meet a theoretical civilian demand when a theoretical economic concept called "national income" is rising, and to give their greater-volume-lower-cost theory a tryout later when the emergency is over.

In short, they propose a stupendous gamble—a gamble which, if unsuccessful, must be paid for by everyone—by the industry, by labor, by taxpayers, by everyone except those safely ensconced in Government jobs. And when the steel industry falters in the inevitable deflation to come, it may find itself, at long last, in the hands of the Government.

An American Lord Bacon

Time will tell whether the expansionists were justified in the gamble. But, on the record of Dunn's career, laying down the chips against his advice is scarcely the safest bet in the world. For what he is, what he knows and what he has achieved carry formidable weight.

Dunn would probably be shocked by a comparison of his peculiar talents with those of Lord Bacon. But there is something about his influence that suggests the role of that man of universal learning in the court of Queen Elizabeth. There is the hint of the sorcerer about Dunn—not the sorcerer of Disney's Fantasia, but the Elizabethan master of subtle mysteries. In twentieth-century terms, Dunn is a man who might push back impeccable cuffs and, to amuse his dinner guests between courses, repair a lagging chronometer or whip together a complete radio set. More astonishing than even such bits of legerdemain are the color and flashing opulence of his knowledge. Explanations of atomic power, coaxial cables and the uses of manganese in steelmaking are the kind of intellectual fare he sets forth for his friends. His mind moves joyously among the miracles of an age that has never grown tired of producing miracles and exploring the miraculous.

One of Dunn's first devotees was the late Frank Munsey. Early in the 1900's Dunn met Munsey. Dunn was then chief engineer of an electrical-manufacturing concern. Munsey had some magazines whose formula was based on his interest in science and adventure. After three or four evenings of exciting talk, the great man summoned Dunn to his office.

A Munsey Tribute

"What does your present job pay you?" Munsey demanded.

"Six thousand dollars," Dunn answered.

"Good. I am offering you ten."

Startled, Dunn asked, "What for?"

"Last night you explained to me how a turbine works," Munsey said. "You made me understand it as I'd never understood it before, by comparing a turbine to a child's pin wheel. I know of no other scientist engineer with your gift of simplification for the layman. I am offering you the job of editor in chief of all my magazines."

Dunn was flattered, but he preferred to confine his talents for popularization to a more limited audience. The audience remains small but distinguished. Such men as Edward Stettinius, Jr., W. A. Harriman, Harold Stanley, Joseph Ripley, Robert Millikan and David Sarnoff are among those of a whole generation in the business and scientific worlds upon whom his influence has been profound. Now filling key posts in the OPM and elsewhere in the Administration, some of these men and scores of others less well known have an almost worshipful respect for Dunn's judgments. They come to him not only for professional guidance but for personal advice.

Dunn's relationship with Sarnoff is more or less characteristic. Sarnoff, a young radio operator, had, by 1913, become chief inspector of the Marconi Wireless Telegraph Company of America, or, as it came to be known, the American Marconi Company. One of his duties was to serve as a contact man for his company in connection with the big job of erecting the first transatlantic radio towers, which his company had given to Dunn's engineering firm. Sarnoff, at twenty-two, thus found himself dealing with Dunn, a man already at the top of his profession. "He was the Pope of my realm," says Sarnoff. "He first taught me to see the engineer as an organizer. He showed me how important it was for the technician to understand the applications and implications of what he knew. And, without patronizing me, he gave me advice about my career. It's only fair to say that he was the first person of importance who took any interest at all in what was to become of me, personally."

It was a proud moment for Sarnoff years later when, as president of the Radio Corporation of America, he could ask Dunn to serve as a member of his board of directors. "But," Sarnoff says, "I wanted him as a director not because he was kind to me and not because to me he symbolized what is greatest in engineering. I have profound respect for his radio knowledge. And even more, I wanted him because, more than any other man living, Dunn brings poetry into science and engineering. I can explain that best by saying that if I ask a good engineer to explain a new circuit of some sort, I get an accurate description in suitable technical terms. But if I ask Dunn to describe it, I get an explanation of the relationship of that circuit to other devices, an estimate of its possible uses and a prophetic glimpse of the developments to which it's apt to lead."

Casually, after a board meeting, Dunn looks down shimmering vistas of science with such men as Sarnoff and describes marvels still to come. It was the result of such a quiet moment as this, for instance, that Dunn, in 1913, pulled a young unknown named Edwin H. Armstrong out of the figurative hat for Frederick M. Sammis, the chief engineer of the American Marconi Company. Dunn told Sammis that he'd do well to look into the work Armstrong was doing in one of the laboratories at Columbia University. Armstrong, later the inventor of the superheterodyne and frequency modulation, turned out to be working on the regenerative circuit, or "feed-back." His work was to revolutionize the radio industry.

Armstrong tells an illuminating story of the encounter that led to that momentous tip. He was sitting in his laboratory when Dunn, tall, handsome and clad in the frock coat and high hat of that period, asked permission to enter. He had heard, he said, that Armstrong had a device capable of amplifying wireless signals so greatly that he was receiving signals from Europe with it. Would Armstrong show him the device and explain how it worked?

Armstrong not only would and did but was so filled with enthusiasm by Dunn's quick and sympathetic questions that he offered to build him a feed-back.

"If you don't mind," Dunn said, "I'd prefer to build it at home myself. Then I can really fix the principle of the thing in my mind."

So, Armstrong recalls, he and Dunn set about collecting the materials Dunn would need to make a feed-back at home. When they came to the fiber tubing around which the coils were to be wound, they found none of the required length available. Dunn fell to and cut some. "And I will never forget the picture," Armstrong says. "Dunn, who was somebody in the engineering world, I can tell you, sawed away at that fiber tubing with his frock-coat tails flying, and asking more questions than one would think a president of the United Engineering Society would bother his head with." One out of a thousand engineers, Armstrong adds, would have had the patience to dig into the details of what he was talking about. "And of those who did, 99 per cent would have said I had a very nice toy. But Dunn saw the future of what I had in the feed-back instantly."

Down-to-Earth Science

There is a sense in which the word "popularize" means to jazz up the long words and abstruse ideas of others. But when Dunn quickly told the American Marconi people to look into Armstrong's work, he was popularizing science in a sense far beyond Munsey's comprehension of the term. He was describing the significance of what was known to three or four men in a way that would make it important to millions. He was bringing together the invention and the practical men who could employ it in terms that affected everyday living.

Only a man of vast insight can do this. A quack can be an excellent popularizer. The super-popularizer must keep the respect of the cognoscenti. That Dunn has, is attested by the fact that he has had literally dozens of offers from scientific and educational institutions. Herbert Hoover offered him the directorship of the Bureau of Standards. Once he was asked to consider the secretaryship of the Smithsonian Institution, and twice the presidency of the Massachusetts Institute of Technology.

Dunn has turned down all such invitations for as good a reason as any. His grandfather was a schoolteacher, and Gano was brought up on the dictum, "Now, don't be like your grandfather." The inference is that Grandpa Nathaniel Dunn, an "impractical visionary," was a kind of skeleton in the family closet. Actually, Gano is exceedingly proud of him.

Nathaniel Dunn was a member of the famous class of 1825 at Bowdoin which included Longfellow and Hawthorne. A successful schoolmaster from the first, he seems to have had only one vagary. Each time he saved enough money from his modest salary, he blew it in on an invention.

This career Grandpa Dunn summed up in a flash of rueful wit. "A rolling stone gathers no moss," said he. "That is not always true, however. It depends on how you roll. I rolled and gathered no moss. Peter Cooper rolled and gathered mountains of it. And the explanation of it in part is this: He was by nature a businessman, and I was not. Indeed, as respects business talents, I have been an unmitigated failure." Doubtless it would have comforted the "unmitigated failure" if he could have seen his grandson become president of the institution founded by Peter Cooper—The Cooper Union for the Advancement of Science and Art—for the education of poor boys and girls.

Grandpa Dunn ended his days lecturing on chemistry, philosophy, geology and mineralogy, with asides to his gaping classes on proper ventilation, animal heat and other "marvels of the animal, vegetable and mineral kingdoms." To young Gano, his stories of how chemicals behaved and how his inventions and "constructions" were made were a source of unfailing enchantment. Unquestionably it was he who made Gano decide to become an engineer. It was he who gave Dunn that consuming interest in what was new that saved him from the normal conservatism of engineers. And if Nathaniel Dunn left Gano permanently skeptical of the delights of a schoolmaster's life, he also passed on to him the rare gift of vivid exposition.

Autumn Romance

Yet time and chance had a part in Dunn's success too. Time and curious time relationships, as a matter of fact, have played a strange role in his life. A confirmed bachelor at forty, Dunn found himself dropping by with increasing frequency at the home of the lovely, spirited Mrs. Julia Gardiner Gayley. He and everyone else first thought that her three beautiful daughters were the reasons for his visits. A long time passed before it dawned on him that it was the mother with whom he was falling in love. The situation could hardly have been more difficult. Mrs. Gayley was older than he. Her sense of proprieties was a formidable obstacle. It took ten years of persistence to overcome it. Dunn himself was fifty when they married, and she lived only seventeen years longer. Still he believes himself blessed beyond all other men because he could share her last years.

Even now, he regards her daughters and granddaughters as his, portraits of her hang everywhere in his home, and her name recurs constantly in his conversation.

But if, when he thinks of her, like the hero of Berkeley Square, he grieves that he was born too late, the engineer in him must grant that he appeared on the scene at the most propitious of all possible moments. When Dunn was six years old the nation was celebrating its centennial at Philadelphia. There at the exposition, wholly overshadowed by the gigantic Corliss steam engine, were two little things which suggested that some tired workman had disposed of a few bits of wire by winding them on pieces of iron. They were dynamos. In that same year, in a barn in Wickliffe, near Cleveland, a young man named Charles F. Brush was working on a primitive electric dynamo that would create illumination through the agency of arc lamps. When Dunn was nine, Edison first produced light in a sealed glass bulb. By the time Dunn was entering New York's City College, William Stanley's historic local lighting system was operating in Great Barrington, Massachusetts, and queer horseless streetcars were soon to lurch uncertainly in city streets.

Those were the days when it was possible for a seventeen-year-old, working his way through college as a night telegrapher, to visit the shrine of Edison and talk with the wizard, man to man. Dunn told the story a half century later when he was awarded the Edison Medal. "Out of an almost infinite kindness for young men who were struggling, Mr. Edison received me in the midst of some laboratory work he was doing in the coating of laminated armature plates," he said. "Seeming to be interested in the questions I asked him, he drew me out in turn, and spent an hour personally showing me over his lamp works. . . . I still cherish two of the first one-candle-power lamps that were ever made, which Mr. Edison picked from a box and handed me as a present. . . . At the end of the visit, he offered me a job."

But Dunn decided to forgo the job and finish college. Edison afterward said that Dunn had decided well. Despite his early prejudice against book learning in students of electricity, he watched Dunn's career with the keenest interest. Twice the maestro visited the young engineer in his own laboratory to observe his experiments. In 1910 Edison himself took the lead in nominating Dunn for the presidency of the American Institute of Electrical Engineers.

A Columbia Pioneer

Dunn's studies at City College continued, sustained by his job with Western Union. Every night he sat behind a brass railing in the Murray Hill Hotel sending and receiving telegrams. And in those days that meant sitting close to the center of things. The famous and great of the Brownstone Decades passed through that lobby. Mark Twain, when he lived there, used to send and receive innumerable messages. One evening, in making change for him, Dunn inadvertently pushed a penny off the plate and nervously replaced it with an other from the till. The great humorist gravely pocketed the change and then, picking up the dropped penny from the floor, pocketed that too. "Thank you, sir, thank you," said he as he turned away, leaving behind him a boy baffled until the tomfoolery registered.

Dunn got his engineering degree from Columbia in 1891, after completing the first electrical-engineering course given by an American university. In fact, because his name was first in the alphabetical class roster, he happens to have been the first electrical engineer graduated by an American university. Two great pioneers, Francis Crocker and Michael Pupin, conceived Columbia's course in electrical engineering. Pupin was later to testify that Dunn was the most versatile and gifted of all his students. Crocker thought so well of him that he took him on as engineer in his Crocker-Wheeler Electrical Manufacturing Company.

When Dunn emerged from Columbia, the era of development and promotion had begun in the electrical industry. The job before him was to perfect the crude tools at hand. Almost without effort, it seemed, a stream of inventions flowed from his inquisitive brain. There was a process making it possible to wind strip copper edgewise simply and economically, which revolutionized the winding of generators and which is now used all over the world. There was the first noiseless machine for ringing telephones; the first long-distance telephone call from New York to Chicago was rung with one of Dunn's machines. And, finally, there were the designs for steelmaking machinery, some of which is still being used after more than forty years.

Stalking Capital

An electrical engineer was not simply an electrical engineer then; he had to be a raiser of capital too. In the new world that electricity was shaping, the part played by the marriage of capital to invention is easy to underestimate. The inventor sitting alone could not move mountains. He could only dream of moving mountains and draw blueprints for moving them. It took machines to move mountains, it took labor to make and operate the machines and it took management to direct labor. But money was needed to get these things—money in quantities beyond the dreams of all the ages, fast-moving, daring, reckless money. And money to develop the marvelous things born of electricity did not grow on street corners. It has to be got from people who had it or controlled it. Some inventors saw this. All engineers saw it. For engineers were the middle men between science and utility.

Dunn learned important things in the 90's. He learned, perhaps from his armory of inventive labor-saving ingenuity, how to conserve energy in the process of marrying invention and capital. Reduced to its simplest terms, the problem was how to walk streets and climb stairs with a minimum of effort. The solution was so obvious that it had escaped the notice of his colleagues. Clubs, societies, social functions saved shoe leather. As it happened, they also provided the kind of recreation that he really liked. But primarily they offered the opportunity to melt the stiffness of the banker's face with the magic flow of words; to talk, relaxed, with the usually crusty and suspicious inventor; to introduce, in an atmosphere of gracious urbanity, these reluctant parties to the promising union of science and venture capital.

Dunn's mastery of that job was almost immediately recognized. Any fair inventory of his gifts must include the fact that he is handsome, tall, with the controlled movements of a gentle man who is conscious of his powerful physique. Very early in the game he realized the importance of dressing well. The Lord Chesterfield in him first emerged in 1895, when he was sent to London, briefly, by his firm. Before that, he had had the indifference of most engineers to his own appearance, an attitude in which, sad to relate, he was encouraged by Edison himself. It seems that when he was visiting the Edison works in 1893, Edison, showing him a new construction shack, flung open the door upon a bathroom complete with tub, and whispered apologetically, "See that thing? My wife made me put it in."

London changed all that so far as Dunn was concerned. He began to see that while, in the realm of theory, the apparel did not proclaim the man, it certainly did get him into a lot of nice places. No more thereafter were black ties bought in dozen lots and suits acquired only when old suits wore out. Dunn became a model of fastidious elegance. And yet, characteristically, it was the engineer and not the Chesterfield in Dunn that chose the valet who has been serving him for thirty years. The valet, Mr. Hedquist to you, is an ex-armature winder. He spends much of his day in the tiny machine shop next to Dunn's bedroom, tinkering with motors, radios, clocks and Dunn's remarkable collection of microscopes.

Ultimately, long after social life had ceased to be even a partial means to an end, Dunn and his wife were to give intimate dinners that would be the talk of New York—dinners at which Albert Einstein played his violin, accompanied by virtuosi Harold Bauer and Ernest Schelling, and physicist Robert Millikan argued with astronomer George Ellery Hale over the merits of a play. But in the 90's and 1900's, Dunn's subtle social sense was put to more practical use. He had the capacity to pick from the conflict of personalities the most minute particles of agreement. Temper and pique were luxuries that the engineer, whose function was to articulate the elixir of capitalism with the genius of invention, could not afford.

Only once did he discard the velvet glove. Thirty years ago, when Dunn was president of the American Institute of Electrical Engineers, he invited Samuel Insull to read a paper before one of its meetings. Insull was at the height of his power, a power which has perhaps never been wielded so completely by one man in a great industry. Even Rockefeller had partners. Insull had only employees. Engineers, executives, bankers and public officials quailed before his arrogance. When Insull had finished reading his address to the institute, Dunn, in conformity with custom, asked for questions from the audience. Dr. Cary T. Hutchinson, a well-known engineer, rose and observed coolly that what Mr. Insull had said was remarkable if his figures on the cost of power production were correct. Would the speaker, Hutchinson continued, favor him with a description of the method by which those figures were arrived at?

Drama in One Act

Insull was not a man to be questioned. While Hutchinson was still speaking, he shouted irrelevantly, "That's not true! It's false!"

In the pandemonium that followed, Dunn, in an aside on the stage, asked Insull to withdraw the remark, since it was neither responsive nor parliamentary.

"I'll be damned if I will," Insull snapped.

For a fraction of a second Dunn considered the possible consequences of crossing Insull. And then, rising to his full height, he announced to the incredulous audience, "In the name of the American Institute of Electrical Engineers I apologize to Doctor Hutchinson for the rude conduct of Samuel Insull. The meeting is adjourned."

Adolph Lewisohn was one of the men who put capital into the Crocker-Wheeler Electrical Manufacturing Company in the early 1900's. Because of a sense of obligation to men like Lewisohn, who had invested their money with a confidence born of knowing him, Dunn turned down all outside offers for many years, including three from General Electric. His quiet refusals were so quixotic that Lewisohn was moved to reward them materially in 1911, when Dunn, after discharging his "obligation" many times over, eventually decided to resign. Lewisohn offered Dunn the presidency of the Crocker-Wheeler Company.

But Dunn had by this time determined on a quite different course. The problems of a manufacturing engineer no longer interested him. Above all, now that he felt free to choose what he would do, he wanted to keep his independence, to pick his own jobs, to work on projects of many kinds.

A Persian Comedy

In making that decision and staying by it, Dunn has become the member of a tribe now approaching extinction—the independent engineer. Over the years, as head of the J. G. White Engineering Corporation, he has seen scores of other big engineering firms disappear, their personnel swallowed up, more often than not, by the engineering departments of the utilities. Dunn's firm has kept its identity through wars, depressions, New Deals and utility-holding-company waxings and wanings. It has emerged, out of it all, as the best-known American Engineering firm carrying out construction in foreign countries. But of their numberless foreign jobs, Dunn and his associates like best to recall the two that gave them both the most trouble and the most fun—working on the railroad for the Shah of Persia and preparing to dam the Blue Nile at Lake Tana for Emperor Haile Selassie of Ethiopia.

In the late 20's, Dunn's firm was employed by the Shah of Persia, together with others of an engineering syndicate, to construct the southern section of the Persian railroad. Months before the line was to have been finished, the shah suddenly took it into his head to inaugurate it. To begin with, the rolling stock of the railroad had not yet arrived, and so the royal train consisted only of a few sorry flatcars and boxcars drawn by a small construction engine. The engine itself, being equipped with a Richardson pop valve, occasionally let out steam with an ear-splitting blast. The shah and his retinue appeared and prepared to cut the usual ribbons before boarding the train. Just as the shah poised his shears, the valve on the engine popped. The shah and his company were thrown into an uproar mistaking the terrifying pop of the valve for an assassin's gun.

From that moment, one disaster succeeded another. The valve persisted in popping until the shah, enraged, charged that the Americans were deliberately tormenting him. To stop it, the engineer drew the fire. Whereupon the engine refused to move the train. When a compromise had been worked out with the engine, the train moved on. But then a small triumphal arch was struck by a projection from the train and wobbled. One of the shah's dignitaries, the governor of Ahwaz, arrayed in a brilliant uniform, tripped as he was bowing himself backward on a dock and fell off into the mud. A crew of the shah's own men, who, at the shah's insistence, had been placed in charge of a launch used at the sea end of the railroad, forgot to cast off one of the lines that held the launch to the dock; the boat started and, while the engineers looked on in an agony of suppressed laughter, it stopped with a jerk that hurled the shah and his retinue flat on their faces. At the end of that mad journey, the American syndicate was told to leave the country.

The Blue Nile job was another cup of tea. For more than twenty years the British government had been trying to persuade the Ethiopian emperor to permit the damming of Lake Tana, whose waters they needed to develop some 3,000,000 acres of the Anglo-Egyptian Sudan. Haile Selassie just as persistently refused, apparently fearing that British engineers would be more tender of Anglo-Egyptian interests than they would of his. One day, wholly unexpectedly, his emissary appeared at Dunn's New York office. He wished, he said, to discuss the vast project. When he left the country he left behind him an appointment giving the job of surveying and other preliminary work to Dunn's firm.

At once the British Foreign Office and the British newspapers began to howl for Dunn's head. They all but implied that it was an affront to the British Empire for an American company to get the agreement that their twenty-odd years of statesmanship had failed to clinch.

Within a few days these complaints grew loud enough to be heard across the Atlantic. Suddenly it occurred to Dunn that the emperor's emissary had tossed in his lap the job of being a diplomat as well as an engineer, for without British co-operation it was hopeless to proceed.

A diplomat Dunn, therefore became. He went to the British ambassador and explained that he had never contemplated proceeding with the job unless the British approved. In fact, he had made a stipulation to that effect in the agreement with the emperor's representative. He visited London and spent weeks conferring with government officials. And by the time the British had decided that he was a gentleman and a scholar, preparations for the Ethiopian expedition had been completed. Two great expeditions went and came back from the wild Lake Tana region while Dunn kept officialdom purring. So far did Dunn's diplomacy succeed that only the outbreak of the Italo-Ethiopian War prevented actual construction at Lake Tana by Dunn's firm.

Stemming the New Dealers' expansionist ideas makes the problem of damming Lake Tana look like child's play. This, in 1941, brought to a critical test even Dunn's practiced hand as a diplomat. Dunn's initial Washington triumph was remarkable. It consisted in bringing out, with the enthusiastic acclaim of the President, a picture of the steel situation that was anathema to the New Deal economists.

That decisive victory, as we have seen, was followed by what looked like a defeat. But time works its own judgments, and time alone will write the end of the story. If, as Dunn fears, a program of steel expansion great enough for both "guns and butter" ultimately causes a profound change in our economic system, it will be well for all of us to remember that a great engineer raised a minatory finger when there was still time—an engineer whose life had been devoted to the task of bringing people the largest measure of the good things of life with the least effort and the smallest possible cost.

For Dunn's definition of an engineer is a product of his personal philosophy. Engineering he defined long ago as "the art of the economic application of science to the purposes of man." Stressing the importance of the word "economic" in this definition, he discarded soft speech on one occasion. "A real engineer," he said, "is a man who can do for one dollar what any fool can do for two." If any man ever lived his philosophy, that man is Dunn. It is not without significance that he was the inventor, forty years ago, of a noiseless charging machine for the battery of telephone transmitters. Think of Dunn when you make your next telephone call. He stands for less noise and more clarity.

3. MICHAEL IDVORSKY PUPIN

Michael Idvorsky Pupin was both a distinguished researcher and a gifted teacher. He played a prominent role in the development of radio, telephone and X-ray technologies. As the second chairman of the Engineering Foundation, he guided it through the pressure filled years of World War I.

Pupin was born on October 4, 1858 at Idvor, Banat in what is now Yugoslavia. Coming from a poor background, his scholarship and intelligence won for him a place at the Prague Technical Institute. He emigrated to the United States in 1874 where he initially worked as a laborer. He entered Columbia University in

1879 and graduated from that institution in 1883. Desiring to increase his scientific and mathematical knowledge, he returned to Europe for his graduate education. He studied advanced mathematics at England's Cambridge University and he held later the John Tyndall Fellowship of Columbia University while studying physics at the University of Berlin. In 1889 he received his Ph.D. from the University of Berlin and returned to the United States to become an instructor of mathematical physics at Columbia. In 1892 he became an adjunct professor of mechanics at the same

Michael Idvorsky Pupin, professor of electro-mechanics, Columbia University. Served as president of AIEE and of the American Association for the Advancement of Science; and as chairman of the Engineering Foundation from 1917 to 1919 .

institution, and in 1901 he was promoted to the position of professor of electromechanics. In 1903 he became the director of Columbia's Phoenix Research Laboratory, a position that he held until his retirement in 1929.

Throughout his career, Pupin's research resulted in important advancements in the field of long distance communications. In 1902 he patented innovations in electric tuning which would prove to be of great importance to the new radio industry. He was also the first researcher to accomplish the rectification of high frequency electrical waves. However, his most widely known research was the development of the principle of loading telephone and telegraph lines by lumped inductance through the use of his patented 'loading coils' which compensated by their inductance for the capacitance between the two wires of a circuit. To best utilize loading coils, he also developed a formula for their proper placement along a transmission line. As a result of his research, long distance telephone and telegraph transmission was greatly improved.

Pupin made also notable developments in the field of X rays. He was the first researcher to properly describe the nature of secondary radiation and he also greatly aided the development of X-ray technology as a medical diagnostic tool through the use of a fluorescent screen.

Pupin played an active role in the affairs of many professional engineering organizations. As the second chairman of the Engineering Foundation he was faced with the task of guiding the use of its resources in support of the U.S. war effort. During his 1917–19 term the Foundation provided critical financial support to the nascent National Research Council and it also funded war time projects designed to help detect enemy submarines. He served as the president of the AIEE and he enjoyed a long and beneficial association with the Institute of Radio Engineers. He served as vice president of this organization between 1892 and 1903, and as its president during 1925–26.

Pupin received numerous honors during the course of his distinguished career. In 1902 he received the Cresson Medal of Philadelphia's Franklin Institute and in 1916 he was awarded the Herbert Prize by the French Academy. During 1924 he was awarded the medal of the Institute of Radio Engineers, while in 1925 he received the Edison Medal of the AIEE. Perhaps the greatest honor that was given to Pupin was the John Fritz Medal, which was awarded to him in 1935.

Pupin died at New York City on March 2, 1935.

4. EDWARD DEAN ADAMS

Edward Dean Adams, a prominent entrepreneur in the field of electric power generation, served as the vice-chairman of the Engineering Foundation from 1915 to 1926 and as honorary board member until his tragic death in 1931.

Adams played a large role in the creation of an extensive hydroelectric generating system in the Niagara Falls area. His achievements were made possible by the development of the long-distance, high-voltage transmission line during the 1890s. This technological innovation made it possible to replace small local hydroelectric generators of limited capacity with huge plants which could harness the immense waterflows of large waterways. In this manner a small number of centrally located plants could meet the electrical power needs of an entire region.

During the early 1890s Adams brought together a consortium of financiers to form the Niagara Falls Power Company. This concern almost immediately secured the use of an alternating current generating system that had been developed by Nicola Tesla, and in 1892 it began the construction of a large hydroelectric generating station on a site that was located almost immediately upstream from Niagara Falls, New York. Named the Adams Station in honor of the corporation's founder, its machinery was housed in an elaborate structure that was designed by the noted New York architectural firm of McKim, Mead & White. The Adams station was placed in operation on November 15, 1896, and it soon supplied much of the power needs of Buffalo, New York. Within a few years it was joined by another large plant.

Realizing that the Canadian side of the Niagara River also offered prime sites for the construction of hydroelectric generating plants, Adams organized a Canadian subsidiary which was granted exclusive hydroelectric power generation rights by the Queen Victoria Niagara Park Commission in 1892. However, financial and technical difficulties prevented the construction of the plants which were proposed by the Canadian subsidiary. In 1898 this failure, combined with public pressure, caused the commission to revoke the monopoly and substitute for it a lease to Adams' Corporation for a fixed annual amount of waterflow from the Niagara River. Despite this setback, Adams and his partners did successfully construct a large hydroelectric generating plant on the Canadian side.

This plant was built upstream from the falls at Niagara Falls, Ontario, and it was placed in operation by 1905. During the next two decades, it was upgraded until by 1924 it housed eleven large-capacity generators which were individually capable of producing 12,000 volts. Almost all of the power that this plant generated was exported to Buffalo.

The profits that Adams derived from his organization and management of these hydroelectric enterprises contributed greatly to his personal fortune, a portion of which he contributed to the endowment of the Engineering Foundation (See Table 6).

REFERENCES

C.1 Mark Frame, "A Selective Guide to Industrial Archeology in the Niagara," Ontario Society for Industrial Archeology, Toronto, Ontario, 1984.

APPENDIX D

Popular Research Narratives

Contents:

A SERBIAN HERDSMAN'S CONTRIBUTION TO TELEPHONY

Conspicuous among hundreds of inventions which have brought America's telephone systems to their high development are those of Dr. Michael Idvorsky Pupin. They are highly scientific in character and based upon the wave transmission of sound and electricity. When the Edison Medal was presented to him in February, 1921, by the American Institute of Electrical Engineers, he told how he first became interested in sound transmission.

Although for many years an American citizen by adoption, Michael Pupin was born in a village near Belgrade, Serbia. At the age of twelve he began summer vacation service with other boys as assistant to the guardians of the villagers' herd of oxen, and at the same time his studying in Nature's own laboratory of the wave transmission of sound. Daytime duties were light; the hot sun and the hungry flies kept the wise ox in the shade. At night the cattle grazed. Moonless Serbian nights are so dark that the sky seems black even when the stars are blazing. Objects fifteen or twenty feet away cannot be seen. Only a few miles distant was the Rumanian border, and between lay extensive corn fields. When the wind blew from the corn fields to the grazing grounds, the pleasant fragrance tempted the cattle; but in the corn lurked many cattle thieves. The oxen must be kept out of the corn; on the dark nights, however, they could be followed only by sound.

Now, among the arts of the herdsmen in which the boys were trained, was the art of listening through the ground. A knife with a long wooden handle was stuck in the ground. One boy who was being trained would put his ear to the handle and listen, while another boy, thirty or forty yards away, would strike his knife similarly stuck in the ground. The first boy would have to tell the direction and guess the distance. This first lesson in wave transmission set young Pupin thinking. He soon observed, as herdsmen before him had, that sounds from the knife carry much farther through hard solid ground than through plowed ground.

The long nights of watching afforded much time for observing sounds and thinking about them. In the darkness the world seemed to have disappeared and the only signs of its existence were the messages of the low sounds from the grazing herd, the distant village clock, the rustling corn. Thoughts started in the lad's mind on those Serbian plains continued to evolve as he went from the village school to the academy at Prague; when he ran away from the unbearable confinement of the academy, after the freedom of the plains, and came to America, and as he made his way through many difficulties to a higher education in the sciences in the universities and laboratories of America and Europe.

Finally those germs of thought bore fruit in many scientific discoveries and inventions having to do with wave transmission, especially of sound and electricity. Among these inventions was the Pupin "loading coil," which greatly advanced the possibilities of successful long-distance telephony. A few years ago, when Chief Engineer Carty, of the Bell Telephone System, stretched his wires from the Atlantic seaboard to the Pacific coast of the United States, and President Theodore N. Vail, of "A. T. and T.," first made a human voice heard across a continent, there were Pupin coils at intervals of eight miles in that transcontinental line. In the whole world to-day there are more than three-quarter of a million Pupin coils in use in telephone lines, of which 600,000 are in the United States.

Based on information supplied by Dr. M. I. Pupin, Professor of Electro-Mechanics, Columbia University, New York.

SAFE EXPLOSIVES

Galen H. Clevenger

Liquid Oxygen Explosives are a fine example of the dependence of engineering and industry upon scientific research. They are a type of Sprengel explosive. The novel feature is that the oxidizing agent (liquid oxygen) and the combustible substance (carbonaceous matter alone or in combination with liquid hydrocarbons or even metallic powders, and, at times, inert absorbents) are brought together immediately before use. The components separately are non-explosive. Before L. O. X. could be seriously considered for commercial work, it was necessary to produce liquid oxygen economically in large quantity and to have satisfactory containers.

Liquefaction of so-called permanent gases, air, nitrogen, oxygen and others, long baffled those who attempted it, and yet required no unusual equipment once fundamental laws were known. Repeated endeavors to liquefy these gases by pressure alone, despite development of elaborate equipment for producing pressures up to

60,000 pounds per square inch, failed; one law had been overlooked, that all gases have not only a critical pressure, but also a critical temperature. Thomas Andrews, in 1869, was first to show that there is a temperature for every gas at and below which it can be liquefied and above which it cannot be liquefied by pressure; further, that when gases heretofore regarded as permanent were at their critical temperatures, they could be liquefied by comparatively moderate pressure.

Louis Cailletet, in 1877, produced the first liquid air by allowing the pressure on previously compressed air to fall 4500 pounds per square inch, thus lowering the temperature. In 1895, or somewhat earlier, Linde in Germany, Hampson in England and Tripler in America, demonstrated that liquid air could be produced upon a large scale. Later Claude found it more economical to cause the air to do work in expanding by passing it through an engine instead of expanding it through a nozzle.

Not until 1902 did Linde demonstrate that liquid air could be separated into its constituents by passing through a special still, similar in principle to that used for separating alcohol-water mixtures. This rendered possible the production of liquid oxygen on a large scale. To-day it is practicable to produce it upon any scale desired, employing air pressures of 900 to 3000 pounds per square inch. Liquid oxygen boils at—182.93 degrees Centigrade; to be used without excessive evaporation loss, satisfactory storage and transport containers were necessary. Dulong and Petit were probably first to discover that passage of heat through glass by conduction could be greatly reduced by a vacuum wall; d'Arsonval, in 1887, was first to make practical use of glass-walled vacuum containers, reducing evaporation loss to 1/10 that in plain glass. It was soon discovered that a vacuum did not prevent passage of radiant heat. Dewar, in 1892, found that silvering the inner walls of vacuum chambers of containers reduced evaporation loss to 1/200 of a plain glass container's loss. These containers were admirable for laboratory use, but fragile. There was needed a metallic container with low evaporation loss. Dewar discovered that properly treated charcoal, at the temperature of liquid oxygen, had the property of adsorbing large quantities of gases, including air, so perfectly as to produce a very high vacuum. He secured a British patent in 1904, and in 1906 pointed out that a high vacuum could be maintained between metallic walls by this means. Modern metallic containers for liquid oxygen are rugged, highly efficient devices.

These scientific developments and actual trial of L. O. X. had been made before the world war. When Germany was cut off and it became necessary to find a substitute for large quantities of explosives used for civilian purposes, incentive came for rapid development of these new explosives.

L. O. X. under certain conditions, have distinct advantages: substantial saving in cost, greater safety, freedom from noxious gases, no possibility of explosive in ore or waste rock, which may occasion trouble; elimination of danger from drilling into unexploded charges. In cities, where large quantities of explosives are used in excavating, there is ever present danger attendant upon transportation through the streets and risk of their falling into the hands of miscreants. These hazards can be eliminated through use of L. O. X. L. O. X. have certain disadvantages. They are not so convenient in inaccessible parts of a mine, nor have they been used successfully for shaft-sinking, or under water. Under no circumstances should they be used in gaseous or dusty coal mines. We may still look forward to important improvements through research in this interesting development of a valuable commercial use for one of the constituents of the air which surrounds us.

Based upon information supplied by Galen H. Clevenger, Consulting Metallurgist, U. S. Smelting, Refining & Mining Co., Boston, Mass. Details can be found in a paper by Michael H. Kuryla and Galen H. Clevenger, presented to February, 1923, meeting of American Institute of Mining and Metallurgical Engineers.

REFLECTIONS ON THE MOTIVE-POWER OF HEAT

C. H. Peabody

It is fitting that engineers of the United States should commemorate the centenary of the publication of Sadi Carnot's work on the Motive-Power of Heat.

Sadi's father was the organizer of victory for the French people at the time of the Revolution. Sadi was born June 1, 1796, in the smaller Luxembourg and named Nicolas-Leonard-Sadi. Prenatal conditions gave him a delicate constitution, which he so fortified by correct living and judicious exercise, that he showed no lack of energy, but on occasion exhibited courage, physical strength and dexterity. At sixteen he entered the Polytechnic School and the next year he left it, first in artillery.

A portrait at the age of 17 years shows a slender sub-lieutenant of engineers with a firm, refined face and a mathematical head. Being too young for the school at Metz, he continued another year at Paris, and took part in military exploits at Vincennes in the Polytechnic battalion. He was appointed lieutenant at 23 and managed to continue his studies at Paris for a time. He appears to have hidden extreme sensibility by a reserved manner, but to have had a just appreciation of his own ability unbiased by egotism.

In 1824, at the age of 28, he published his *Reflections*. This modest work by an unknown young officer, dealing with ideas far in advance of his times and couched in language that few engineers could appreciate, passed out of print and appeared to be lost, until Lord Kelvin, then a young and brilliant mathematician, discovered it, revealed its extra-ordinary merits and restated the propositions in the forms of modern thermodynamics.

In 1832, Carnot had an attack of scarlet fever, followed by cholera, which carried him off on August 24. Thus was lost to France and to the world one of the clearest and most brilliant intellects of all time.

Let us try to follow the working of his mind as shown in his *Reflections*, premising that the experiments of Gay-Lussac and Dalton were then young and that before Regnault (1811–1878) the properties of steam were but dimly known. He accepted the caloric theory that heat was a substance and though his investigations led him toward the modern energetic theory of heat, his data were insufficient to reach such a conclusion.

His attention is given first to the steam engine, then the only known heat engine. He questions whether there is a natural limit to improvements in economy and concludes that general principles must include all possible forms of heat engines. He determines at once that the work of heat is due to its fall in temperature similar to the fall of water on a water wheel. The idea itself leads in the right direction and he appears not to be misled by the unfortunate parallelism. The production of motive power is, in his mind, due to the transportation of caloric from a hot to a cold body, with steam as an agent.

Then he questions whether the action varies with the agent. To find an answer he invents his ideal heat engine and thereby so clarifies the problem that he comes wondrously near the modern theory of thermodynamics and its quantitative results. He describes the combination of his hot body A, his cold body B, his working cylinder of non-conducting material, and his working substance; he develops his cycle of operations with steam as the working substance, consisting of supply of heat from the boiler, adiabatic expansion, and rejection of heat to the condenser.

He then introduces the idea of the reversible cycle, perhaps the most important of all for thermodynamics. With the steam engine he gets into trouble with the feed water when he reverses the cycle, and so turns to the idea of the hot-air engine, for which the cycle is clearly expressed and the reversibility demonstrated. From study of the reversible cycle, he demonstrates that the efficiency of the ideal engine is independent of the working substance; otherwise a combination of a reversed engine with a more efficient direct engine would make power, an idea which he rejects as unthinkable.

He concludes his discussion of the ideal engine with the requirements that all heat absorbed must be taken at the temperature of the hot body A, and heat must be rejected only at the temperature of the cold body B. This clearly stated ideal cycle is perhaps the most practical invention relating to heat engineering.

Continuing his study of the properties of air, he just missed discovering the constant ratio of the specific heats at constant volume and at constant temperature. He makes a computation of what we now recognize as the mechanical equivalent of heat and comes within 15 per cent of the Joule function.

He concludes that the fall of caloric produces more motive power at inferior than at superior temperatures. In other words, it is better to get a more perfect vacuum than to raise steam pressure; a fact but recently realized by the modern development of steam turbines.

No wonder that Carnot, unknown to his contemporaries, is the head of the corner of the edifice of heat engineering.

Contributed by C. H. Peabody, Dr. Eng., Professor Emeritus of Naval Architecture and Marine Engineering, Massachusetts Institute of Technology, Cambridge.

OXYGEN, IRON AND STEEL

F. W. Davis

Oxygen, in the total by weight, exceeds all the other chemical elements put together in our earth and its atmosphere. It is the great supporter of combustion. In ordinary combustion processes, however, air is used as it occurs in nature, a mixture of a little less than 23 per cent of oxygen, 77 per cent of nitrogen and small quantities of moisture and several rare gases. But nitrogen is one of the most inert of all gases and so takes no active part in combustion.

Iron is obtained from ores by smelting, a process involving combustion and therefore requiring fuel and oxygen. With the advance of civilization the consumption of fuel and iron has increased greatly and has reached an enormous total. Nature's supplies of these minerals is not renewed. Man has used the best first, quite properly, but has not been so economical of high-grade raw materials as posterity in successive generations could wish. In more fields than one, the necessity is beginning to force more intelligent use of the best and development of methods for utilizing the inferior, while advancing standards for final products.

This Narrative differs from those which have been issued before in that it deals with a study in progress rather than the result of a finished research. Advantages to be gained by substituting oxygen for air, or enriching air with oxygen, for use in metallurgical furnaces are great. There are possibilities in other industries also, for example: ceramics and city gas supply. Likewise there have been great difficulties.

One difficulty was the high cost of oxygen. This has been surmounted. A committee of chemists and metallurgists co-operating with the United States Bureau of Mines has learned that the big factors in the price of commercial oxygen are compression into cylinders, storage and transportation. Large oxygen plants can be built to serve metallurgical purposes directly, capable of delivering oxygen through pipes at a cost not to exceed $3 per gross ton.

Iron was first smelted from oxide ores in forge furnaces by burning charcoal. This process was slow and wasteful of fuel, a large volume of gases passing off at high temperatures. In the first half of the 15th century, Germans attempted to use these hot gases for preheating the charge of ore, flux and fuel. This led to the first shaft, or blast, furnace. The shaft, or upper part, of the furnace was intended to reclaim part of the heat carried off, largely by the inert nitrogen. Although there have been revolutionary changes in mechanical equipment for smelting iron, the metallurgical features have remained almost the same for four centuries.

Application of oxygen will revolutionize smelting, probably changing the whole operation and equipment. Now, material changes in the heat at the hearth, the most active part of the furnace, can be accomplished only by charging fuel in the top, and will not be effective until the added fuel reaches the hearth 10 to 15 hours later. Were oxygen in use, the hearth heat would be continually and quickly controllable.

From an extended study of oxygenated air for blast furnaces the conclusions are: Decrease of production costs; increase of output per furnace; increase of flexibility of process—better control; increase in uniformity of product; utilization of lower grades of ore and cheaper fuel; reduction of sulphur, an objectionable impurity.

Bessemer steel making may also be helped. There are now two major processes, known as acid and basic from the nature of their slags. For best results the basic process requires pig iron containing from 2.5 to 3 per cent of phosphorus. In making basic steel, oxidation of the phosphorus contributes a large percentage of the total heat. The phosphorus, which is undesirable in the steel, is thus removed. In the acid process phosphorus is not removed. In the United States there are ores containing too little phosphorus for the basic process and too much to make acceptable steel by the acid process. With oxygen a good steel could be produced from these ores and a high-phosphorus slag obtained, valuable for fertilizer.

Although thermal efficiency of the Siemens-Martin, or open hearth, process of steel making is very low, improvement by aid of oxygen is impeded by serious difficulties. A furnace lining ("refractory") which will be infusible at much higher temperature is needed,—a "super-refractory." Given this, the major portion of the refining of the metal now occupying 3 to 5 hours could be done in less than half an hour; plant and materials cost could be substantially reduced, and steel of superior quality could be produced equal in every respect to that now made in electric furnaces.

Based on information supplied by F. W. Davis, metallurgist, U. S. Bureau of Mines, and printed with permission of Director H. Foster Bain.

A UNIQUE RESEARCH IN CIVIL ENGINEERING

Charles D. Marx

Water serves humanity in many ways, yet few persons know how engineers control it. Rain falls and snow melts irregularly; there are floods and droughts. In many places water must be stored, or its level raised, or its course changed, for development of hydroelectric power, for water supply of communities, for irrigation, for flood control, for navigation in rivers and canals.

Dams are means to these ends. Some dams are walls of concrete or other masonry, of a variety of shapes. As engineers have been emboldened by experience, dams have been built higher. Many now exceed 200 feet and a height of even 700 feet has been considered. Because stored water becomes vitally necessary to those dependent upon it and because the breaking of a dam would in many places destroy life and property, dams must be unquestionably safe. Yet they may not be too costly.

A few centuries ago an engineer conceived that curving a dam upstream would make it stronger. A curved dam would in some situations act as an arch stood on its edge; hence engineers took advantage of the arch strength as revealed in bridges and buildings. A single curved wall will act as an arch only in a narrow canyon, or gorge, with steep rock sides. Also engineers used for dams across wider valleys series of arches, between buttresses.

Arch dams developed on assumption and theory. With safety in mind many curved dams were made so thick they could have but little arch action. About thirty-five years ago, an engineer built a dam in Bear Valley, near Redlands, California, so thin that conservative dam builders gasped. It still stands. Other thin arch dams were built in several countries. None built by an engineer has failed.

There was divergence of engineering opinion. Experimental knowledge was much desired, but obstacles were forbidding. In 1922, Engineering Foundation undertook an investigation. A committee of engineers was organized. Observations on existing dams were undertaken, but this method was difficult and slow. Available instruments were not satisfactory. Consideration was given to models and to methods and instruments. A meeting of the committee was enthused by a proposal from a large power company to build an experimental dam fifty to a hundred feet high, and the offer of a substantial contribution.

Financing and planning went along together. Power companies, industries, bankers, governmental bureaus, universities, American and Foreign engineers joined Engineering Foundation in carrying on the investigation. Better instruments became available, some being devised especially by sub-committees and collaborators.

For this full-size experiment favorable conditions existed not likely to be duplicated in many years. In the Sierras, 60 miles east of Fresno, California, Stevenson Creek tumbles down the mountain through a granite gorge into the canyon of San Joaquin River. On this creek was an ideal spot. It was near a tunnel through the mountain conveying a great volume of water to a power plant.

A small pipe, controlled by valves, led from the tunnel to the creek, above the dam site, so that the reservoir could be filled as desired, like a tank in a laboratory. The reservoir was so small that the quantity of water needed was unimportant. Even if the dam should break while the reservoir was full, no damage would result. The power company was constructing extensive works in the region. Hence a works railroad, a highway, electric light and power, equipment, supplies, and skilled men were at hand.

When the excavation in the rock was almost ready for the dam, an untimely flood washed rocks and sand into the trench, causing delay and expense. June 4, 1926, the dam was completed to a height of 60 feet. It was shaped like a symmetrical, triangular piece of the side of a round can, stood on its point, curved upstream.

The dam was built of portland cement concrete without steel reinforcement. Thousands of parts of instruments were embedded for making tests. Its thickness was reduced gradually from 7½ feet at the bottom to only 2 feet 30 feet above and is 2 feet for the upper 30 feet. It is 140 feet long on top.

Thousands of measurements of strain, deflection and temperature were made with the water at various depths up to 60 feet, and with no water in the reservoir, according to a program, from June to October. To have practically no change of temperature and moisture during a test, measurements were made at night.

Late in November a flood overtopped the dam at least three feet, destroying the observation platforms and many instrument connections and partially filling the reservoir with sand and rock. This prevented cold weather tests planned for January. There was another flood in February. The dam still stands.

A celluloid model one-fortieth the dimensions of Stevenson Creek dam was tested with mercury instead of water. The results agreed remarkably. Models of larger sizes, of mortar and other substances, are being planned of dams of the same shape as the Stevenson

Creek dam and several other shapes. Stevenson Creek dam should be built higher and tested until it breaks. Useful information has been gotten and is being made into a book.

Contributed by Professor Charles D. Marx, Past-President, American Society of Civil Engineers, Chairman, Committee on Arch Dam Investigation, Stanford University, California.

Index of Names

Indices do not include appendices or prefatory material.

Subject Index